Fundamentals of Ecotoxicology

Fundamentals of Ecotoxicology

Michael C. Newman

ANN ARBOR PRESS

Cataloging-in-Publication Data:

Newman, Michael C.
    Fundamentals of ecotoxicology / Michael C. Newman.
       p. cm.
    Includes bibliographical references (p. 281) and index.
    ISBN 1-57504-013-1
    1. Pollution—Environmental aspects. 2. Pollution—Health aspects. I. Title
QH545.AIN49 1998   97-48518
571.9′5—dc21   CIP

Sleeping Bear/Ann Arbor Press
121 South Main Street
Chelsea, MI 48118
www.sleepingbearpress.com

Ann Arbor Press is an imprint of Sleeping Bear Press

Printed in the United States of America
10 9 8 7 6 5 4 3 2 1

Cover photo: The Reddish Egret (*Egretta rufescens*), an uncommon heron that feeds on crustaceans, small fish, and amphibians in saltwater marshes and coastal lagoons. It is shown here hunting in the Florida Everglades, an extensive and valued wetland system (see figure 12.3) that has been impacted by various human activities including water diversion and widespread agricultural activities. Photograph courtesy of David Scott, University of Georgia's Savannah River Ecology Laboratory.

Then proudly smiled that old man
　To see the eager lad
Rush madly for his pen and ink
　And for his blotting-pad—
But, when he thought of publishing,
　His face grew stern and sad.

Phantasmagoria (Carroll, 1865)

Contents

Section Three. TOXICANT EFFECTS

Section Five. RISK FROM POLLUTANTS

Section Six. SUMMARY

Preface

Tenor of This Book

> *Nothing is new under the sun. Even the thing of which we say, "See, this is new!" has already existed in the ages that preceded us.*
>
> Ecclesiastes 1: 10-11

In contrast to the conceptual doldrums described above by Coheleth, we live in exciting times, rich in new discoveries and novel applications of familiar concepts: tremendous opportunity exists for intellectual growth. This is so obvious in fields such as molecular genetics and computer sciences that no elaboration is required beyond this simple statement. Several immediately recountable examples affecting our understanding of the world range widely from discoveries in planetary biogeochemistry, mathematics, and geology. Lovelock's **Gaia hypothesis** (the earth's albedo, temperature, and surface chemistry are homeostatically regulated by the biota) (Lovelock, 1988; Margulis and Lovelock, 1989) has stretched our perspective beyond the ecosystem to consider contaminants in a global context. Recently, we learned that a **global distillation** has moved volatile and persistent pesticides from warmer areas of use to cooler areas of the globe where they had been banned for decades or seldom used (Simonich and Hites, 1995).

Only a few decades ago, nonlinear dynamics and chaos theory revealed the limitations of determinism. We now know that it is impossible to predict precisely the behavior of all but the simplest of systems. Such new concepts allowed me to understand that the bioaccumulation models presented in Chapter 3 do not necessarily predict pollutant accumulation until a single equilibrium concentration is reached. In sharp contrast to current thinking, oscillations in concentrations are expected under certain conditions (Newman and Jagoe, 1996).

Just a few decades ago, French scientists discovered that life's activities had modified the geochemistry in Oklo (Gabon, Africa) 1.8 billion years ago and brought together enough fissile uranium (^{235}U) to reach critical mass (Lovelock, 1988; 1991). Proterozoic life set the stage by influencing the hydrological cycle, creating an oxidizing atmosphere, and enhancing uranium accumulation through **biomineralization** (biologically mediated deposition of minerals) (Lovelock, 1988; Milodowski et al., 1990). These **natural Oklo reactors**, producing power for nearly a million years (Choppin and Rydberg, 1980), are clear evidence that the nuclear consequences of life's activites began long before humankind appeared on the earth! The present distribution of residue from these natural reactors also provides environmental scientists with clues about the long-term fate and migration of modern waste fission products.

My reasons for writing this book stem from the realization that new concepts and novel applications of existing ideas surround us every day. This book is my effort to expose you to new and useful concepts of ecotoxicology. Not only are these concepts and facts fascinating in themselves, they also provide us with the tools to avoid or solve environmental problems.

> ... *alert and healthy natures remember that the sun rose clear. It is never too late to give up our prejudices. No way of thinking or doing, however ancient, can be trusted without proof. What everyone echoes or in silence passes by as true to-day may turn out to be a falsehood to-morrow, mere smoke of opinion* ...
>
> THOREAU (1854)

Content and Format of This Book

This book is intended for use in an upper level undergraduate or graduate level course, and as a general reference. It is a treatment of ecotoxicology ranging from molecular to global perspectives. Reflecting our present imbalance of knowledge and effort, the book retains some bias toward lower levels of ecological organization, e.g., biochemical and organismal topics. Yet, it does extend discussion beyond the ecosystem to include landscape, regional, and biospheric topics. The intent of this extension is to impart a perspective as encompassing as the problems facing us today: many problems transcend the ecosystem context.

Human and ecological health issues are interwoven here. The reason for taking this unconventional and potentially distracting approach is simple: to do otherwise would be inconsistent with our present approach to assessing hazards and risks from pollutants. Also, many mechanisms of action, internal dynamics, and effects are similar for all animal species, including humans. Finally, although we prefer the delusion of transcendency in environmental issues,[1] our decisions regarding ecological effects of contaminants are made with motives as profoundly anthropocentric as those involving human health. Why not discuss them together?

Topics are divided among 15 chapters. The first two chapters provide general perspective. In Chapters 3 to 12, bioaccumulation and effects of contaminants are

---

[1]Decisions about ecological effects are often perceived as arising from selfless protection of the earth. In fact, our motives are based on the value we give to the heretofore free services provided by intact ecological systems. These services include generation of clean water and air, production of food and game species, provision of biological materials for medical and genetic uses, and provision of pleasing settings for living and recreation. Decisions are based on the perceived value of these services relative to those of technological services and goods. The delusion of selfless motivation in environmental stewardship and advocacy is sufficiently widespread as to be named, the **Lorax Incongruity**. (The Lorax is a character in the popular children's book by Dr. Seuss [Geisel and Geisel, 1971] who "speaks for the trees, for the trees have no tongues.") This well-intended but intransigence-inducing delusion is pervasive in Society today.

detailed at increasing levels of ecological organization. The framework of these 10 chapters is scientific, not regulatory. Chapters 13 and 14 address the technical issues of risk assessment. The final chapter summarizes the volume. There is a bias toward North American examples in all chapters. This reflects the author's background, not a quiet snub of non-North American interests. I recognize, but have no solution for, this shortcoming of the book relative to British, European, and other non-North American students.

To aid the reader, important terms are highlighted in the text and summarized in the glossary. Some important material has been included in the footnotes for purposes of continuity; consequently, the reader is cautioned not to unintentionally overlook these materials. Opinionative insights of the author, not generally shared by ecotoxicologists, are clearly identified as "incongruities" throughout the chapters, e.g., the Lorax incongruity discussed above. The student may decide to ignore most of them with no danger of becoming an uninformed ecotoxicologist. Suggested readings are provided at the end of each chapter.

About the Author

Michael C. Newman is a Professor in the Department of Environmental Sciences at the College of William and Mary's Virginia Institute of Marine Science. Before joining the faculty at the Virginia Institute of Marine Science, he was a Senior Research Scientist at the University of Georgia's Savannah River Ecology Laboratory and head of its Environmental Toxicology, Remediation, and Risk Assessment (ETRRA) Group.

After receiving B.A. (Biological Sciences) and M.S. (Zoology) degrees from the University of Connecticut, he earned M.S. and Ph.D. degrees in Environmental Sciences from Rutgers University. After postdoctoral fellowships at the University of Georgia and the University of California-San Diego, he joined the research faculty at the University of Georgia (1983). His research interests include toxicity and bioaccumulation models, toxicant effects on populations, factors modifying toxicity and bioaccumulation, quantitative methods for ecological risk assessment, statistical toxicology, and inorganic water chemistry. He has published more than 70 scientific articles on these topics. He authored the book, *Quantitative Methods in Aquatic Ecotoxicology* (1995) and was senior editor of *Metal Ecotoxicology: Concepts and Applications* (1991, with A.W. McIntosh), *Ecotoxicology: A Hierarchical Treatment* (1996, with C.H. Jagoe), and *Risk Assessment: Logic and Measurement* (1998, with C.L. Strojan). He also directed the development of UNCENSOR, a program that produces univariate statistics for data sets containing "below detection limit" observations.

Dr. Newman is active in professional societies and teaching. He founded and was first president of the Carolinas Chapter of the Society of Environmental Toxicology and Chemistry (SETAC). He chaired (1992–1994) and served on (1988–1996) the SETAC awards committee. He is an editor (aquatic toxicology) for the journal *Environmental Toxicology and Chemistry*. He also sits on editorial boards of the journal *Archives of Environmental Contamination and Toxicology*, and two books series, *Advances in Trace Substances Research* and *Current Topics in Ecotoxicology and Environmental Chemistry*. Dr. Newman has taught at the University of Connecticut, the University of California-San Diego, the University of South Carolina, the University of Georgia, and Royal Holloway University of London. He teaches a summer short course (Quantitative Methods in Ecotoxicology) annually.

Acknowledgments

I am grateful to Dr. T. Hinton for his intelligent and thorough contribution to this book. Drs. Thomas La Point, Peter Landrum, Alan McIntosh, Donald Mackay, Margaret Mulvey, Carl Strojan, and Glenn Suter II provided excellent reviews of the manuscript. My gratitude is extended to Robert (Skip) DeWall at Ann Arbor Press for his considerable efforts and encouragement.

Introduction

On the day of the patients' victory at court, someone wrote a headline:
"The Day that Tomoko Smiled." She couldn't possibly have known.
Tomoko Uemura, born in 1956, was attacked by mercury
in the womb of her outwardly healthy mother. No one knows
if she is aware of her surroundings or not.
SMITH AND SMITH (1975)

HISTORIC NEED FOR ECOTOXICOLOGY

As pressures mount for fiscal restraint, it is natural and responsible to reconsider the wisdom of our complex system of costly environmental regulations. At this time, it may be difficult to understand why significant amounts of these monies shouldn't be redirected to the national deficit, critical social problems, medical research, technological innovation, education, space exploration, or other worthwhile endeavors. But, just a few decades ago, it was not hard to understand the need for such expenditures: Tomoko Uemura's mother understood (Figure 1.1).

As World War II ended, the **dilution paradigm** (the solution to pollution is dilution) was slowly replaced by the **boomerang paradigm** (what you throw away can come back and hurt you). Two horrible epidemics of heavy metal poisoning from contaminated food had occurred in Japan. In the 1950s, organic mercury was transferred through the marine foodweb to poison hundreds of people. Nearly a thousand people fell victim to **Minamata Disease** before Chisso Corporation halted discharge of mercury into Minamata Bay. From 1940 to 1960, Japanese in the Toyama Prefecture were poisoned by cadmium in their rice. This **itai-itai disease** was linked to irrigation water contaminated from metal mine wastes. The name, itai-itai, reflects the extreme joint pain associated with the disease and literally means "ouch-ouch."

In 1945, open air testing of nuclear weapons began at Alamogordo, New Mexico, and nuclear bombs exploded over Hiroshima and Nagasaki later that year. Nine years later, the "Project Bravo" bomb exploded at Bikini Atoll, dropping fallout on thousands of square kilometers of ocean including several islands and the fishing vessel, *Lucky Dragon* (Woodwell, 1967). The islands of Ailingae, Rongelap, and Rongerik received radiation levels of 300 to 3000

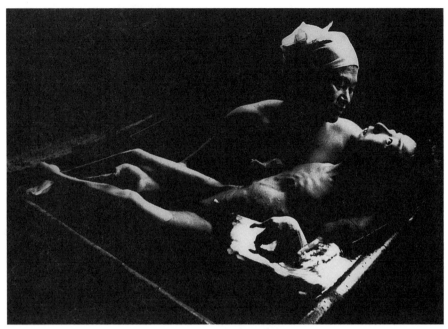

FIGURE 1.1. Tomoko Uemura, victim of *in utero* poisoning by organic mercury, being cared for by her mother (© Heirs of W. Eugene Smith, courtesy of Black Star).

rem[1] within four days of detonation (Choppin and Rydberg, 1980). The rapid hemispheric dispersal and unexpected accumulation of fission products in food-stuffs from these and subsequent detonations prompted much concern about possible long-term effects to humans. From 1960 to 1965, human body burdens of ^{137}Cesium increased rapidly worldwide and then slowly decreased as the U.S., former Soviet Union, France, and China ceased open air testing (Shukla et al., 1973).

Unreported discharge of radionuclides occurred prior to these overt releases. All were kept from the general public for reasons of national security. On the northwest coast of England, a fire in the Windscale plutonium processing unit released 20,000 curies of radioactive iodine (^{131}I) to the surrounding area (Dickson, 1988). Radioactive iodine is of particular concern because it concentrates in the thyroid, causing cancer. After atmospheric release, ^{131}I contaminates local vegeta-

---

[1]A **rem** or Roentgen equivalent man is a measure of radiation that takes into account the differences in potential biological effects of various types of radiation. It relates the dose received to potential damage. As such, it is a convenient unit for defining allowable radiation exposures, e.g., the average person receives approximately 0.360 rem (360 mrem) of radiation annually, and the average radiation worker cannot exceed exposures of 600 mrem/month. The rem has been replaced as the official unit by the Sievert (Sv). (1 rem = 0.01 Sv.) In contrast, the **curie** used later is simply a measure of radioactivity. One curie is 2.2×10^6 dpm (disintegrations per minute). Although still used widely as in this book, the curie has been replaced by the **becquerel** (Bq) as the official unit of radioactivity. One curie is 3.7×10^{10} Bq.

tion, may be taken up by dairy cattle, and accumulates in thyroids of humans after consumption of dairy products. At a secret Soviet military plant ("Chelyabinsk 40") in the Urals, plutonium processing had secretly discharged 120 million curies to a nearby lake and enough down the Techa River to induce radiation poisoning in citizens living downriver (Medvedev, 1995). In September of 1957, a storage tank explosion at Chelyabinsk 40 released 18 million curies of radioactive material and forced the evacuation of approximately 11,000 people from a 1000 km² area (Trabalka et al., 1980; Medvedev, 1995). From 1944 to 1966, releases from the U.S. Atomic Energy Commission's Hanford Site in Washington State were kept from the general public. Between 1944 and 1947, the complex released 440 thousand curies of radioactive iodine (^{131}I) into the atmosphere (Stenehjem, 1990). On May 12, 1963 at the Hanford K-East reactor, 20,000 curies were released to the Columbia River (Stenehjem, 1990).

Concern about pollutant effects on nonhuman species was also growing. Pesticides such as DDT[2] (dichlorodiphenyltrichloroethane or 2,2-bis-[p-chlorophenyl]-1,1,1-trichloroethane) accumulated in wildlife to alarming concentrations, resulting in direct toxicity and sublethal effects. From 1957 to 1960, Hunt and Bischoff (1960), and Dolphin (1959) documented deaths of Western grebes (*Aechmophorus occidentalis*) resulting from bioaccumulation of the pesticide, DDD (1,1-dichloro-2,2-bis[p-chlorophenyl] ethane) from a freshwater foodchain (Clear Lake, California). These pesticides accumulated in the brain until enough was present to cause axonic dysfunction and death. Dolphin (1959) described the 1949 administration of DDD to control a nonbiting gnat of Clear Lake as "involving introducing approximately 40,000 gallons of a 30% DDD formulation . . . from drum-laden barges"!

Silent Spring (1962), the extraordinary book by Rachel Carson, drew the attention of the public to these and less obvious consequences of pesticide accumulation in wildlife. Although relatively nontoxic to humans, DDT and DDE (dichlorodichloroethylene or 1,1-dichloro-2,2-bis-[p-chlorophenyl]-ethene) inhibit Ca-dependent ATPases in the shell gland of birds resulting in shell thinning and increased risk of egg damage after being laid (Cooke, 1973; 1979). Birds at higher trophic levels were extremely vulnerable because DDT and its degradation product DDE are relatively resistant to degradation and accumulate in lipids. These qualities result in an increase in concentration with each trophic exchange in a food web. Reproductive failure of raptors and fish-eating birds became a widespread phenomenon. The average number of offspring per pair of osprey (*Pandion haliaetus*) nesting on Long Island Sound dropped from 1.71 young/nest (1938–1942) to only 0.07 to 0.40 young/nest by the mid-1960s (Spitzer et al., 1978). Reproductive success of raptor populations decreased in Alaska (Cade et al., 1971) and other regions of the U.S. (Hickey and Anderson, 1968). Ratcliffe (1967, 1970) reported the

---

[2]DDT was an extremely important tool for disease and agricultural pest control throughout the world. Indeed, Paul Müeller was awarded the 1948 Nobel Prize in medicine for discovery of its value as an insecticide. Its importance in this context is often overshadowed by our present understanding of its adverse effects on nontarget species if used indiscriminately.

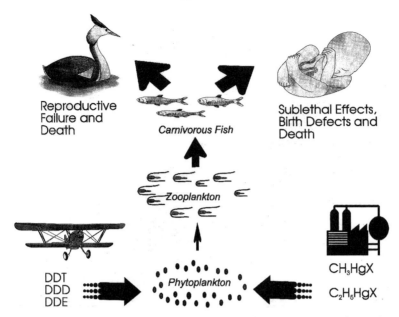

FIGURE 1.2. Two of the first contaminants to draw attention to the inadequacies of the dilution paradigm were DDT and methylmercury. They became watershed examples of the boomerang paradigm. Both were returned to humans and to valued wildlife species by transfer through foodwebs.

same downward trends for falcons (*Falco peregrinus*) and other raptors in the United Kingdom. Reproduction of brown pelicans (*Pelecanus occidentalis*) on the South Carolina coast from 1969 to 1972 fell below that needed to maintain the population (Hall, 1987).

The two watershed events that most captured the public's attention and resulted in a paradigm shift (dilution paradigm to boomerang paradigm) were Minamata disease and DDT accumulation in raptors and fish-eating birds (Figure 1.2). Together, they drew some attention away from giddy industrialization and the Green Revolution to the consequences of ignoring pollutants in ecological systems. They were among the first issues to give impetus to the science of ecotoxicology.

CURRENT NEED FOR ECOTOXICOLOGICAL EXPERTISE

Everyone would like to feel that the problems described above reflect mistakes made earlier in the techno-industrial revolution that will not be repeated. This is not the case. Environmental problems continue to occur despite our increased awareness and complex regulations. Indeed, problems seem to extend more and more frequently to transnational and global scales.

Nuclear materials still require our attention and expenditure of monies. The core of Three Mile Island Reactor Unit 2 (Harrisburg, PA) melted on March 28, 1979, releasing approximately 3 curies of radiation and incurring an estimated $965 million in cleanup costs (Booth, 1987). Nearly 30 years after the Chelyabinsk 40 explosion in the Urals, the largest radioactive release in history (301 million curies as estimated by Medvedev [1995]) occurred in the Ukraine on April 26, 1986 as the Chernobyl Reactor 4 core melted down. Fallout from Chernobyl spread rapidly across the Northern Hemisphere. High-level radioactive waste storage tanks like, but much larger than, that which exploded at Chelyabinsk 40 remain unresolved post-Cold War problems for the U.S. Department of Energy nuclear complex. Despite worldwide protest, French underground testing of nuclear devices resumed briefly in Micronesia in 1995.

Chemical wastes continue to require attention and funds. A myriad of Soviet environmental issues remain as part of the Cold War legacy (Tolmazin, 1983; Edwards, 1994). Tributyltin (TBT), a widely-applied antifouling agent in marine paints, has harmed estuarine molluscs throughout the world (Bryan and Gibbs, 1991). Mercury levels in fish and game remain a concern with new sources appearing such as mercury used for South American gold mining (e.g., de Lacerda et al., 1989; Branches et al., 1993; Reuther, 1994). Recently, subsurface agricultural drainage in the San Joaquin Valley of California brought selenium in the Kesterson Reservoir and Volta Wildlife Area to concentrations causing avian reproductive failure (Ohlendorf et al., 1986). Efforts to reduce lead in products such as gasoline (human poisoning, e.g., Ember, 1980; Miller and Cooney, 1982; Settle and Patterson, 1980) and lead shot (dabbling duck poisoning, e.g., Hawkes, 1977) have only been effective since the late 1970s. Even into the 1980s, debate continued about effects of lead and the need for federal regulation (Anderson, 1978; Anonymous, 1984a; Ember, 1984; Marshall, 1982; Putka, 1992).[3] At the same time, the controversy about the Hooker Chemicals and Plastics Corporation's dump sites at Hyde Park and Love Canal became hysterical as the public watched (Anonymous, 1981; 1982; Culliton, 1980; Smith, 1982). On December 2, 1984, a storage tank at a Union Carbide pesticide plant (Bhopal, India) exploded and released a cloud of methyl isocyanate killing 2,000 people and harming an estimated 200,000 more (Anonymous, 1984b; Heylin, 1985; Lepkowski, 1985). On March 16, 1978, the *Amoco Cadiz* supertanker ran aground at Portsall (France) and released 200,000 tons (roughly equivalent to 209,000 m^3) of crude oil (Ellis, 1989). On March 24, 1989, the *Exxon Valdez* spilled 41,340 m^3 of crude oil into Prince William Sound. The oil covered an estimated 30,000 km^2 of Alaskan shoreline and offshore waters (Piatt et al., 1990). From August 2, 1990 until February 26, 1991, the largest oil release to have ever occurred was deliberately spilled by Iraqi troops occupying Kuwait. Half a million tons (roughly equivalent to 522,000 m^3) of crude oil from

---

[3]Relative to our slow acceptance of lead's adverse effect, Tackett (1987) provides a revealing quote by Benjamin Franklin (July 31, 1786). "This my dear Friend is all I can at present recollect on the Subject. You will see by it, that the Opinion of this mischievous Effort from lead is at least above Sixty Years; and you will observe with Concern how long a useful Truth may be known and exist, before it is generally receiv'd and practis'd on." As recently as one decade ago—two hundred years after this quote was made, the value of reducing lead in gasoline was being actively questioned.

the Mina Al-Ahmadi oil terminal were pumped into the Arabian Gulf (Sorkhoh et al., 1992). Plumes of contaminating smoke from the intentional ignition of Kuwaiti oil wells by the Iraqi troops were visible from space (Figure 1.3).

Other more diffuse, but incrementally more damaging, events also require expertise in ecotoxicology. Beyond the intentional release described above, the Arabian Gulf receives 160,000 tons (roughly equivalent to 167,000 m^3) of oil annually from leaks and spills (Sorkhoh et al., 1992). The average number of oil spills and volume per spill in or around U.S. waters from 1970 to 1989 were 9,246 and 47,000 m^3, respectively, with no obvious downward trend in either through time (Table 8 in Gorman, 1993). At the time Rachel Carson was writing *Silent Spring* (*circa* 1960), annual production of synthetic organic chemicals was 43.9 billion kg. By 1970, it had reached 145.1 billion kg (Corn, 1982). By 1985, U.S. use of pesticides roughly doubled from the 227 million kg used in 1964 (Figure 7 in Gorman [1993]). Many persistent pesticides restricted in developed countries are still used in the Third World (Simonich and Hites, 1995).

Contaminants amenable to atmospheric transport have become especially disconcerting. Acid rain is now a transnational problem (Likens and Bormann, 1974; Likens, 1976; Cowling, 1982) damaging both aquatic (Baker et al., 1991; Glass et al., 1982) and terrestrial (Cowling and Linthurst, 1981; Ellis, 1989) ecological systems. Chlorofluorocarbons (CFCs) used as propellants and coolants have been linked to ozone depletion in the stratosphere (Kerr, 1992; Zurer, 1987; 1988),

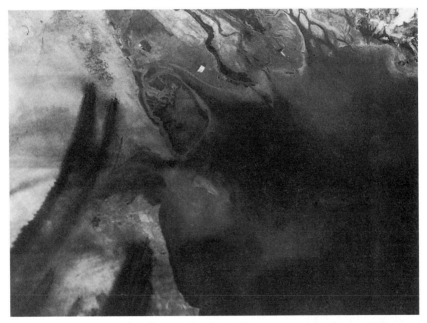

FIGURE 1.3. Kuwaiti oil wells set afire by Iraqi troops as seen from a U.S. space shuttle flight. Oil wells are seen burning north of the Bay of Kuwait and immediately south of Kuwait City (Courtesy of NASA).

and efforts are being made to greatly reduce their use (Crawford, 1987). Despite the 1987 Montreal Protocol (with signatures from 70 countries including the U.S.) that calls for complete elimination of CFC use by 2000, efforts by lawmakers were underway in the mid-1990s to delay, and perhaps avoid, any U.S. reduction of CFC emissions (Lee, 1995).

This long litany of problems does not imply that techno-industrial advancement is an Icarian endeavor incompatible with environmental and human health. Rather, it demonstrates two simple points. First, approximately 50 years ago, the dilution paradigm failed with clearly unacceptable consequences to human health and ecological systems. Second, expertise in ecotoxicology is now critical to our well-being. Major environmental problems remain and new challenges arise daily which are as significant as the historical problems just described. Therefore, expertise in ecotoxicology is essential for determining the costs and benefits of the innumerable technological and industrial decisions affecting our lives. Consideration of nonmarket goods and services, and natural capital (Odum, 1996) must be incorporated into these decisions. Such services provided *pro bono* by nature are estimated to be in the range of $33 trillion annually, twice the annual gross domestic product of the earth's 194 countries (Rousch, 1997). Complex and costly environmental regulations save human lives and allow responsible environmental stewardship.

MAJOR CLASSES OF CONTAMINANTS

What are the contaminants of most concern today? The most prominent classes of contaminants are detailed in Table 1.1. They are divided into organic and inorganic contaminants. These terms, organic and inorganic, were initially based on whether the chemical was obtained from living organisms (organic) or mineral sources (inorganic). However, this distinction is imprecise. For example, carbon dioxide is considered an inorganic gas yet it is produced by organisms. Organic compounds could generally be considered those containing carbon and involving at least one C-H covalent bond, i.e., methane (CH_4) but not carbon dioxide (CO_2). But graphite is an exception to this rule. One might then begin to make the distinction by including reduced carbon as part of the definition, but carbon tetrachloride (CCl_4) is considered an organic compound. Fortunately, the distinction becomes much clearer with compounds composed of carbon chains or rings. These are obviously organic compounds. (The interested reader is referred to Laron and Weber [1994, p. 2] for further discussion of this ambiguity.)

The organic contaminants include those used intentionally as poisons (e.g., insecticides and herbicides). They become a problem only if sufficiently high concentrations of these poisons come into contact with nontarget species. Other organic contaminants are not designed as poisons but still have an adverse effect when released into the environment, e.g., degreasers, solvents, industrial by-products, and by-products of other human activities. Similarly, inorganic contaminants are composed of intentional and unintentional poisons. Some are released during

TABLE 1.1. Summary of Contaminants of Current Concern

Category	Contaminant	General Details
Organic	Chlorofluorocarbons (CFCs)	CFCs are used in refrigeration and in the production of foams, e.g., Styrofoam. These gaseous contaminants contribute to ozone depletion in the stratosphere.
	Organochlorine Alkenes	These compounds are often used in large amounts as solvents and degreasers. Examples include tetrachloroethene, a dry cleaning solvent that sometimes contaminates drinking water, and trichloroethene, a degreaser that is denser than water, very insoluble in water, and frequently associated with groundwater contamination.
	Chlorinated Phenols	These contaminants are used as wood preservatives (e.g., trichlorophenol and pentachlorophenol) and fungicides (e.g., pentachlorophenol). They can be produced during Kraft pulp mill operations as chlorine in the bleaching step reacts with natural phenolic compounds.
	Chlorination Products	Chlorine compounds or gas used to disinfect drinking waters can produce chlorinated organic compounds with potential toxic or carcinogenic effects. For example, chlorination produces trihalomethanes, i.e., the carcinogen, chloroform ($CHCl_3$).
	Organochlorine Insecticides	These pesticides degrade slowly and tend to be very soluble in lipids such as those of organisms. This results in bioaccumulation and possible increase in concentrations through food webs. Examples include DDT, chlordane, aldrin, dieldrin, and heptachlor.
	Organophosphate Insecticides	These pesticides degrade faster than organochlorine pesticides and are less persistent in the environment. However, they tend to be more toxic to mammals than organochlorine pesticides. They also accumulate in fats and oils of organisms. The mode of action for these compounds is the inhibition of acetylcholine esterase activity. Examples include Parathion, Malathion, and Sevin.
	Carbamate Insecticides	Like organophosphate insecticides, carbamate insecticides degrade rapidly in the environment and cause neural dysfunction by inhibiting acetylcholine esterase, an enzyme essential to neuron function. Examples include carbofuran and aldicarb.
	Pyrethroid Insecticides	These neurotoxicants are similar to natural pesticides produced by plants. They degrade quickly and, consequently, most problems are associated with acute exposure. Examples are allethrin and cyclethrin.
	Herbicides	These biocides include products such as Paraquat and Diquat. They include triazines such as atrazine, that can enter waterbodies from agricultural fields. Phenoxy herbicides (e.g., 2,4-D) are also important in the control of dicots and function by disrupting plant growth regulation.

TABLE 1.1. Summary of Contaminants of Current Concern (Continued)

Category	Contaminant	General Details
	Dioxins	Dioxins enter the environment as contaminants in herbicides (e.g., Agent Orange) and wood preservatives. Polychlorinated dibenzo-p-dioxins (PCDDs) are contaminants in commercial products such as PCBs and chlorophenols. Dioxins are also formed as combustion products and during the bleaching process of Kraft pulp mills. Dioxins are extremely toxic.
	Polynuclear Aromatic Hydrocarbons (PAHs)	PAHs are produced by incomplete burning of wood, coal, or petroleum products such as during internal combustion in automobiles. They are associated with air, soil, and water pollution. PAHs can cause cancer. Examples include anthracene, phenanthrene, and benzo-[α]pyrene.
	Polychlorinated Biphenyls (PCBs)	PCBs are used as lubricants, heat conductors in electrical transformers, plasticizers, and in a variety of other applications. Their toxic effects to humans and wildlife are a major concern and some biota (e.g., mink consuming tainted fish) are very sensitive. They degrade very slowly and are soluble in fats and oils. Consequently, they can bioaccumulate to high concentrations. PCBs are produced commercially as mixtures of compounds such as Aroclor 1221.
	Polychlorinated dibenzofurans (PCDFs)	PCDFs are released as contaminants of other commerical products such as PCB mixtures and chlorophenols. They may also be produced during combustion and bleaching associated with Kraft pulp mills.
	Petroleum Related Compounds	These are complex mixtures including some of the compounds listed separately above. They have numerous routes of introduction to the environment including oil drilling activities, oil spills, refining, and the myriad of petroleum product uses.
	Oxygen Demanding Compounds	Putrescible materials possess a high biochemical oxygen demand and can, in aquatic systems, decrease oxygen concentrations to stressful levels. Sewage treatment plant effluents are one, but certainly not the only, important source of such materials.
Inorganic	Gaseous (CO_2, NO_x, SO_2)	Carbon dioxide (CO_2) from combustion is of concern because concentrations in the atmosphere are slowly increasing through time. This increase has been linked to global warming. Nitrogen oxides (NO_x) and sulfur dioxide (SO_2) are also produced by combustion at stationary (e.g., coal power plants) and mobile (e.g., automobiles) sources. Nitrogen oxides and sulfur dioxide react in the atmosphere to produce low pH precipitation or "acid rain." There is epidemiological evidence of adverse health effects of these gases such as linkage to various lung diseases. High levels of sulfur dioxide also can cause plant damage, e.g., leaf necrosis.

TABLE 1.1. Summary of Contaminants of Current Concern (Continued)

Category	Contaminant	General Details
	Nutrients (N, P)	An excess of one or more of nitrogen and phosphorus nutrients in aquatic systems, and some terrestrial systems, can change the structure and functioning of associated ecological communities. Cultural eutrophication is the classic example of this process.
	Nitrogen Species	Nitrogen species can contribute to the dysfunction described above for ecosystems receiving an excess of nutrients. They have other adverse effects if present at sufficiently high concentrations.
	1. Nitrate	1. Nitrate can enter water bodies from runoff or sewage discharges. High concentrations in drinking water can cause **methemoglobinemia** ("blue-baby syndrome" resulting from the reaction of nitrite to hemoglobin to convert it to methemoglobin which is incapable of normal transport of oxygen) in newborn infants. (The nitrite is produced from nitrate in the baby's stomach.) Nitrosamines, potent carcinogens, may form from nitrogen compounds in drinking waters.
	2. Nitrite	2. Nitrite is very toxic to aquatic biota and can also cause methemoglobinemia in babies.
	3. Ammonia	3. Ammonia can cause toxicity to aquatic biota near sources such as sewage discharges. Ammonia toxicity is very pH dependent.
	Metals and Metalloids	
	1. Aluminum	1. A naturally abundant metal in the environment that, under low pH conditions such as those resulting from acid precipitation or mine drainage, can increase to unusually high dissolved concentrations that can kill aquatic species.
	2. Arsenic	2. This metalloid and its compounds are used in numerous products including metal alloys, pesticides (e.g., $Pb_3[AsO_4]_2$), wood preservatives, plant desiccants, and herbicides (e.g., $Na_3As_3O_3$). It is associated with coal fly ash and is also released during gold and lead mining. It is often present as an oxyanion, e.g., $HAsO_4^{2-}$, AsO_4^{3-}, and $H_2AsO_4^{-}$. It is toxic and carcinogenic.
	3. Cadmium	3. This metal is used in alloy production, electroplating, and galvanizing. It is also used in pigments, batteries, and numerous other products. It is used as a plastic stabilizer. It is generated during zinc ore processing. Smokers are exposed to high levels of cadmium in cigarettes. It is toxic and carcinogenic.

TABLE 1.1. Summary of Contaminants of Current Concern (Continued)

Category	Contaminant	General Details
	4. Chromium	4. Chromium is used in alloys, catalysts, pigments, and wood preservatives. It also is used in tanning. It may be present as Cr(VI) or Cr(III). These are referred to as hexavalent and trivalent chromium, respectively. Cr(IV), but not Cr(III), is carcinogenic and is the more toxic of the two forms. Chromium is often present as an oxyanion, e.g., CrO_4^{2-} and CrO_7^{2-}.
	5. Copper	5. Copper is used extensively for wiring and electronics, and plumbing. It is also used to control growth of algae, bacteria, and fungi. It is toxic at high concentrations.
	6. Lead	6. This poisonous metal is ubiquitous due to its widespread use in gasoline, batteries, solders, pigments, piping, ammunition, paints, ceramics, caulking, and numerous other applications. It causes anemia and neurological dysfunction with chronic exposure. Poisoning of birds can occur by ingestion of lead shot.
	7. Mercury	7. Used in electronics, dental amalgams, chlorine-alkali production, gold mining, and paints. It is an excellent industrial catalyst and, because it is liquid at ambient temperatures, as a component in electrolysis. It was used extensively as a biocide, involving seed treatments to prevent fungal growth and growth inhibition in numerous industries such as pulp mills. This highly toxic metal can accumulate to high concentrations in biota.
	8. Nickel	8. This metal is used in alloys such as stainless steel and for nickel plating. It also has innumerable other uses including battery production (Ni-Cd batteries). At sufficiently high concentrations, nickel is both toxic and carcinogenic.
	9. Selenium	9. The metalloid, selenium, is used in the production of electronics, glass, pigments, alloys and other materials. It is a byproduct of gold, copper and nickel mining. It is also associated with coal fly ash.
	10. Zinc	10. This metal is used extensively in protective coatings and galvanizing to prevent corrosion. It is also used in alloys. It is less toxic than most metals listed here.
Organometals	1. Tin	1. Tributyltin (TBT) was and, in some situations, is still used as an antifouling paint on hulls of ships and boats. At very low concentrations, it can cause extensive damage to molluscan populations, resulting in shell abnormalities in oysters and modification of sexual characteristics of snails. Organotins such as trimethyltin (TMT) and triethyltin (TET) are neurotoxicants.

TABLE 1.1. Summary of Contaminants of Current Concern (Continued)

Category	Contaminant	General Details
	2. Lead	2. Organic compounds of lead such as tetraalkyllead have been used extensively as anti-knock additives to gasoline. At high concentrations, these compounds (e.g., trialkyllead produced via liver metabolism from tetraalkyllead) can cause neurological dysfunction and other problems.
	3. Mercury	3. Methylmercury and other organomercury compounds are used or produced as a consequence of several industrial processes. They are used as biocides as in seed coatings. Methylmercury is found in many species consumed by humans. These compounds cause neurological damage as detailed previously in this chapter.
	Radionuclides	Radioactive elements are generated in nuclear weapons production and testing, nuclear energy production, and numerous medical, research, and industrial uses. An important example is ^{137}Cesium from nuclear weapons production via nuclear fission.

a very specific use, e.g., TBT, but others are introduced by many human activities, e.g., lead. Contaminants such as nitrite in drinking waters become a problem only if our activities bring concentrations to abnormal levels. Indeed, several of the metals listed in Table 1.1 are essential to life but, above certain concentrations, produce adverse consequences. A mammalian toxicologist would not consider some of the chemicals listed in the table to be toxicants as they do not directly poison individuals. However, contaminants such as excessive amounts of P and N nutrients in a lake can have pronounced, adverse consequences to the ecological community of that lake. Similarly, global changes in atmospheric gases may have a pronounced influence on ecosystems of the earth but no direct toxicity to humans. Putrescible compounds also may not directly kill aquatic biota but can indirectly kill large numbers of organisms by removing dissolved oxygen from receiving waters. Consequently, these contaminants are listed, along with the more conventional toxicants, as toxicants in the context of ecotoxicology.

ECOTOXICOLOGY

The subject of this book could have been developed under either of the terms "environmental toxicology" or "ecotoxicology" because definitions of both are rapidly converging. Some definitions of ecotoxicology seem to exclude discussion of humans except as the source of contaminants, but the original definition given to ecotoxicology by Truhaut (1977) includes effects to humans (Table 1.2). Ecotoxi-

TABLE 1.2. Definitions of Ecotoxicology and Environmental Toxicology

Definition	Reference
Environmental Toxicology	
1. the study of the effects of toxic substances occurring in both natural and man-made environments	Duffus (1980)
2. the study of the impacts of pollutants upon the structure and function of ecological systems (from molecular to ecosystem)	Landis and Yu (1995)
Ecotoxicology	
1. the branch of toxicology concerned with the study of toxic effects, caused by natural and synthetic pollutants, to the constituents of ecosystems, animals (including human), vegetable and microbial, in an integrated context	Truhaut (1977)
2. the natural extension from toxicology, the science of poisons on individual organisms, to the ecological effects of pollutants	Moriarty (1983)
3. the science that seeks to predict the impacts of chemicals upon ecosystems	Levin et al. (1989)
4. the study of the fate and effect of toxic agents in ecosystems	Cairns and Mount (1990)
5. the science of toxic substances in the environment and their impact on living organisms	Jørgensen (1990)
6. the study of toxic effects on nonhuman organisms, populations, and communities	Suter (1993)
7. the study of the fate and effect of a toxic compound on an ecosystem	Shane (1994)
8. the field of study which integrates the ecological and toxicological effects of chemical pollutants on populations, communities, and ecosystems with the fate (transport, transformation, and breakdown) of such pollutants in the environment	Forbes and Forbes (1994)
9. the science of predicting effects of potentially toxic agents on natural ecosystems and nontarget species	Hoffman et al. (1995)

cology was selected here to include consideration of effects to humans as well as ecological systems. Although the term "ecotoxicology" was chosen, a strong and confining ecosystem emphasis is apparent in many of its definitions. Because the ecosystem framework is now insufficient to contain all germane subjects, ecotoxicological discussion in this book extends beyond the ecosystem to the biosphere. Therefore, as used here, **ecotoxicology** is the science of contaminants in the biosphere and their effects on constituents of the biosphere, including humans.

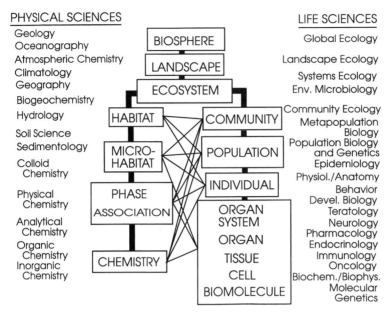

FIGURE 1.4. Hierarchical organization of topics addressed by ecotoxicology. Disciplines contributing to understanding abiotic interactions are listed on the left side of the diagram and those contributing to understanding biotic interactions are listed on the right. Important interactions, denoted by lines connecting components, occur between biotic and abiotic components.

ECOTOXICOLOGY: A SYNTHETIC SCIENCE

Ecotoxicology is a synthetic science drawing from many disciplines (Figure 1.4). Questions of effect are posed from the molecular (e.g., enzyme inactivation by a contaminant) to the population (e.g., local extinction) to the biosphere (e.g., global warming) levels of biological organization. Questions of fate and transport are addressed from the chemical (e.g., dissolved metal speciation) to the habitat (e.g., contaminant accumulation in depositional habitats) to the biosphere (e.g., global distillation of volatile pesticides) levels of physical scale. Sometimes, this can produce a confusing complex of scales and associated specialties. The key to maintaining conceptual coherency in this complex of interwoven and hierarchical topics was articulated by Caswell (1996), ". . . processes at one level take their mechanisms from the level below and find their consequences at the level above . . . Recognizing this principle makes it clear that there are no truly 'fundamental' explanations, and makes it possible to move smoothly up and down the levels of the hierarchical system without falling into the traps of naive reductionism or pseudo-scientific holism." All levels are equally important to effective environmental stewardship.

　　Although all levels are equally important, they contribute differently to our efforts and understanding (Figure 1.5). Questions dealing with lower levels of the

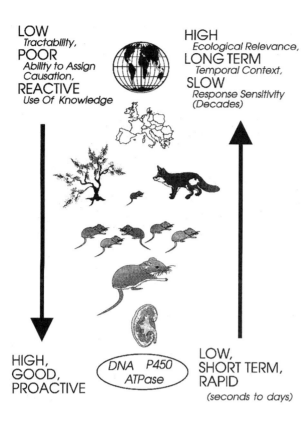

LOW
  Tractability,
POOR
  *Ability to Assign
  Causation,*
REACTIVE
  Use Of Knowledge

HIGH
  Ecological Relevance,
LONG TERM
  Temporal Context,
SLOW
  *Response Sensitivity
  (Decades)*

HIGH,
GOOD,
PROACTIVE

*DNA   P450
ATPase*

LOW,
SHORT TERM,
RAPID
  (seconds to days)

FIGURE 1.5. Hierarchical organization of topics in ecotoxicology relative to ecological relevance, general tractability, ability to assign causation, general use of knowledge, temporal context of consequence, and temporal sensitivity of response.

conceptual hierarchy, e.g., biochemical effects of toxicants, are more tractable and have more potential for linkage to a specific cause than effects at higher levels such as the biosphere. Depressed δ-aminolevulinic acid dehydratase (ALAD) activity in red blood cells can be assayed inexpensively and quickly linked to lead exposure. The observation of ozone depletion over Antarctica was much more challenging to measure and a causal linkage to chlorofluorocarbon release was extremely difficult to make with a high degree of certainty. As a consequence, effects at lower levels of the ecological hierarchy are used more readily in a proactive manner. They can indicate the potential for emergence of an adverse ecotoxicological effect, whereas effects at higher levels are useful in documenting or reacting to an existing problem. Although highly tractable and sensitive, the ecological relevance of effects at lower levels is much more ambiguous than effects at higher levels of organization. A 50% reduction in species richness is a clear indication of diminished health of an ecological community, but a 50% increase in metallothionein in adults of an indicator species provides an equivocal indication of the health of populations in the associated community. Relative to those at higher levels of biological organization, effects at lower levels tend to occur more rapidly after the stressor appears and disappear more quickly after it is removed. Considering all of these points, it's clear that information from all relevant levels of biological organization are best combined in the manner suggested here and by Caswell (1996).

Suggested Readings

Adriano, D.C. *Trace Elements in the Terrestrial Environment*. Springer-Verlag New York, Inc., New York, 1986, p. 533.

Carson, R. *Silent Spring*. Houghton-Mifflin Co., Boston, 1962, p. 368.

Marco, G.J., R.M. Hollingworth, and W. Durham. *Silent Spring Revisited*. American Chemical Society, Washington, DC, 1987, p. 214.

Smith, W.E. and A.M. Smith. *Minamata*. Holt, Rinehart and Winston, New York, 1975, p. 192.

Woodwell, G.M. Toxic substances and ecological cycles. *Sci. Am.* 216, pp. 24–31, 1967.

┌─────┐
│ 2 │
└─────┘

Ecotoxicology:
Science, Technology and Practice

Now the world's turned foul; my happiness rots
. . . [but] there's a text on repentance, I seem to recall.
But what? What is it? I've forgotten the words,
Don't have a book; and there's no one to guide
My footsteps here in this trackless wood.

IBSEN (1964)

OVERVIEW

There is a general lack of unanimity in ecotoxicology that leads to much confusion. Much like the quote above, there appears to be no single "text" or clear voice indicating the direction one should go. The main objective of this chapter is to describe how the outwardly inconsistent activities of ecotoxicologists come together to address three intermeshing and equally important goals.

Ecotoxicology has diverse goals (scientific, technological, and practical) in addition to diverse contributing disciplines (Slobodkin and Dykhuizen, 1991; Newman, 1995). The diversity of subjects prompts most ecotoxicologists to specialize in a particular area and only look peripherally at other information. Subsets of ecotoxicologists must also block out from consideration major portions of our collective knowledge structure in order to move effectively toward their respective and more focused goals. For example, distinct but overlapping subsets of information and methodologies are used by the scientist, analyst, and regulator. Whereas a scientist may rely heavily on the hypothetico-deductive method, this approach would be a hinderance to a regulator who, instead, might employ a weight-of-evidence approach to expeditiously assess the need for remediation at a specific site. The use of a standard method may hinder achievement of a scientific goal (e.g., determining the processes involved in cycling of a micropollutant) yet be necessary for achieving another (e.g., generation of a consistent water quality database for regulatory purposes).

Slobodkin and Dykhuizen (1991) assert that much confusion is generated if the distinct goals and approaches are not understood and respected by the professionals in any applied science. Difficulties arise if methods of one ecotoxicologist are judged unacceptable by another without recognition of their differing goals.

Because the goals of ecotoxicologists are tightly intermeshed and boundaries are not easily drawn, there is sometimes the appearance of inconsistency or confusion in the field. Again, the purpose of this chapter is to dispel some of this apparent inconsistency by delineating these three goals (scientific, technological, and practical) and associated means of achieving them. Although one extremely important aspect of practical ecotoxicology, environmental regulation and law, will not be discussed, key U.S. laws are summarized in Appendix 3.

SCIENTIFIC GOAL

The goal of any science is to organize knowledge based on explanatory principles (Nagel, 1961). It follows that the scientific goal of ecotoxicology is to organize knowledge, based on explanatory principles, about contaminants[1] in the biosphere and their effects. The approaches used to reach this scientific goal are well established. But they are worth reviewing because they are taught informally and, consequently, many unkept opinions exist regarding the conduct of science. The discussion below is condensed from Newman (1995, 1996) who synthesized the works of Sir Karl Popper, and extraordinary articles by Platt (1964) and Chamberlin (1897) in the context of ecotoxicology.

In the early history of science, untested or weakly tested theories were used to explain specific phenomena (Chamberlin, 1897). A question was presented to an acknowledged expert and explanation given based on some prevailing or "ruling" theory. This was all that was required to fit the phenomena or observation into the existing knowledge structure. Facts gradually accumulated around the ruling theory, fostering a sense of consistency. The cumulative effects of such uncritical acceptance of an explanation based on a ruling theory (**precipitate explanation**) is considered inappropriate in modern science yet it reappears occasionally in most disciplines. Consequently, it is important to recognize precipitate explanation in any field and avoid it in your own behavior.

Modern sciences have replaced the ruling theory with the working hypothesis. The **working hypothesis** is never accepted as true and only serves to enhance the development of facts and their relations by functioning as the focus of the falsification process (Chamberlin, 1897). Experiments and less-structured experi-

---

[1]Terms such as pollutant, contaminant, xenobiotic, and stressor have specific and distinct connotations. A **pollutant** is "a substance that occurs in the environment at least in part as a result of man's activities, and which has a deleterious effect on living organisms" (Moriarty, 1983). A **contaminant** is "a substance released by man's activities" (Moriarty, 1983). There is no implied adverse effect for a contaminant although one may exist. A **xenobiotic** is "a foreign chemical or material not produced in nature and not normally considered a constitutive component of a specified biological system. [It is] usually applied to manufactured chemicals" (Rand and Petrocelli, 1985). A **stressor** is that which produces a stress. **Stress** "at any level of ecological organization is a response to or effect of a recent, disorganizing or detrimental factor" (Newman, 1995). As will be discussed, the terms have slightly different legal definitions too.

ences are used to test the working hypothesis. The working hypothesis approach still has a proclivity toward precipitate explanation because a central theory or hypothesis tends to be given favored status during testing. Chamberlin (1897) suggested the **method of multiple working hypotheses** to reduce this tendency. The method of multiple working hypotheses reduces precipitate explanation and subjectivity by considering all plausible hypotheses simultaneously so that equal amounts of effort and attention are provided to each. In fields where multiple causes or interactions are common, it also reduces the tendency to stop after "the cause" has been discovered.

In any modern science, a hypothesis is never assumed to be true regardless of the approach used. But it can gain enhanced status after repeated survival of rigorous testing. Status is not legitimately enhanced unless tests also have high powers to falsify. Unfortunately, consistent application of weak testing can lead to progressive dominance of an idea by repetition alone. Weak testing is occasionally used to promote an idea or approach; consequently, members of any science such as ecotoxicology must be able to recognize false paradigms that emerge from weak testing and avoid weak testing in their own work. Further, tests involving imprecise or biased measurement should be avoided as they frequently generate false conclusions and foster confusion.

Gradually, observational and hypothesis testing methods produce a framework of explanatory principles or paradigms about which facts are organized. These **paradigms** (generally accepted concepts in a healthy science that withstood rigorous testing and, as a result, hold enhanced status as explanations) are learned by members of a discipline and define the major directions of inquiry in the field. They act as nuclei around which ancillary concepts are formulated and as a framework for further testing and enrichment of fact. Unlike ruling theories, these paradigms remain subject to future scrutiny, revision, rejection, or replacement. They are explanations that are currently believed to be the most accurate and useful reflections of truth, but they are not absolute truths. This is an important distinction to keep in mind. For example, the paradigm of matter conservation (matter cannot be created nor destroyed) was an adequate explanation of phenomena until Einstein demonstrated its conditional nature. It was then incorporated into a more inclusive paradigm (relativity theory) with the qualification that relativistic mass (mass + mass equivalent of energy) is constant in the universe, but mass may be converted to energy and vice versa ($\Delta E = \Delta m(c^2)$).

It follows that two general and interdependent types of behavior occur in any science: normal and innovative science (Kuhn, 1970). **Normal science** works within the framework of established paradigms, and increases the amount and accuracy of our knowledge within that framework. The contribution of normal science is the incremental enhancement of facts and articulation of ideas with which paradigms can be reaffirmed, revised, or replaced by new paradigms (Kuhn, 1970). Most scientific effort is normal science. In contrast, **innovative science** questions existing paradigms and formulates new paradigms. Innovative science can only occur after the incremental enrichment of knowledge brought about by normal science has uncovered inconsistencies between facts and an established paradigm.

Although normal science tends to be more important in a young field, an excessive preoccupation with details ("tyranny of the particular" [Medawar, 1967]) or measurement (*idola quantitatus* [Medawar, 1982]) can slow maturation and progress of a science. Conversely, insistence on rigorous hypothesis testing prior to the accumulation of sufficient facts and establishment of accurate measurement techniques can lead to premature rejection of a hypothesis that might otherwise be accepted. Both normal and innovative science must be balanced in a healthy science. In ecotoxicology, many areas still require more normal science before innovative science can be applied effectively. In many other areas of this maturing science, the tyranny of the particular and *idola quantitatus* exist at the expense of much-needed innovative science (Newman, 1996). A balance between normal and innovative science is essential to effectively achieve the scientific goal of ecotoxicology. The long-term benefit of such a healthy balance will be optimal efficiency and effectiveness in environmental stewardship.

TECHNOLOGICAL GOAL

The goal of ecotoxicology as a technology is to develop and apply tools and methods to acquire a better understanding of contaminant fate and effects in the biosphere. Often some activities in technology are indistinguishable from normal science; however, their goals are distinct. Relative to plainly scientific endeavors, the benefits to society of technology are more immediate but slightly less global. Although analytical instrumentation (e.g., advancements in gas chromatography or atomic absorption spectrophotometric techniques) is a key component of technology, other components include standard procedures and approaches, and computational methods. Many of these technologies can also become pertinent to the practical goals of ecotoxicology *when used to address specific problems*. Consequently, the distinction between technology and practice is also based on context.

The development of analytical instrumentation able to detect and quantify low concentrations of contaminants in complex environmental matrices has been essential to the growth of knowledge. Following the first description of atomic absorption spectrophotometry (AAS) capabilities in the 1950s, the number of commercial AAS units increased exponentially through the early 1960s. This made possible the rapid and widespread measurement of trace element contamination in diverse environmental materials (Price, 1972). Flameless AAS methods lowered detection limits for most elements and enhanced analytical capabilities further. Now, a wide range of atomic emission, atomic absorption, atomic fluorescence, and mass spectrometric techniques are available for the study of elemental contaminants at levels ranging from mg g^{-1} to pg kg^{-1} concentrations. Also, gas chromatography (GC) techniques allowed study of the more volatile organic contaminants. Techniques including GC coupled with a mass spectrometer (GC-MS), more effective columns for separation, and improved detectors have all enhanced our understanding of fate and effects of organic contaminants. For organic compounds less amenable to GC-related techniques, innovations such as advanced separation columns and high pressure pumps have quickly improved high pressure, liquid chromatographic (HPLC) methods. Overarching all of these advances have been computer-enhanced sample processing, analytical control, and signal processing. These and a myriad of instrumental techniques have appeared in the last few decades and allowed rapid advancement of this science.

Again, procedures and protocols are important components of environmental technology too. Pertinent procedures vary widely. They may include such activities as the mapping of **ecoregions** (relatively homogeneous regions in ecosystems or associations between biota and their environment) as a means for defining sensitivity to contaminants of U.S. waters and lands (Omernik, 1987; Hughes and Larsen, 1988). These naturally similar regions of the country are grouped for study or development of a common management strategy. Also, crucial procedures are developed in seminal papers such as the series of papers defining the generation and analysis of aquatic toxicity data (e.g., Sprague, 1969, 1970; Buikema et al., 1982; Cherry and Cairns, 1982; Herricks and Cairns, 1982). The recent establishment of a procedural paradigm for ecological risk assessment (e.g., EPA, 1991a) constitutes a technological advance as well as a contribution to ecotoxicology's practical goals. General methods for **biomonitoring** (use of organisms to monitor contamination and to imply possible effects to biota or sources of toxicants to humans) (e.g., Phillips, 1977; Goldberg, 1986) and applying **biomarkers** (cellular, tissue, body fluid, physiological, or biochemical changes in extant individuals that are used quantitatively during biomonitoring to imply presence of significant pollutants or as early warning systems for imminent effects) (e.g., McCarthy and Shugart, 1990) are also important technologies developed in the last several decades. Most biomonitoring programs are only possible now because of the advances in analytical instrumentation described above.

Experimental design, statistical methods, and computer technologies are also important here. Valuable descriptions of experimental designs and statistical methods are provided by professional organizations (e.g., APHA, 1981) and government agencies (e.g., EPA 1985a, 1988a, 1989a,b) to enhance our understanding of contaminant fate and effects. These often have easily-implemented computer programs associated with them (e.g., EPA, 1985, 1988, 1989b). Other computer programs have been developed by the EPA to enhance scientific progress. An example is the MINTEQA2 program (EPA, 1991b) which predicts speciation and phase association of inorganic toxicants such as transition metals. Numerous programs for statistical analysis of toxic effects data are available from the EPA (e.g., EPA, 1985, 1988, 1989b) and commercial sources. As discussed in Chapter 12, geographic information system (GIS) technologies have been developed to study nonpoint source contamination over large areas such as watersheds (e.g., Adamus and Bergman, 1995).

Some technology-related approaches are difficult to understand if an inappropriate context is forced upon them. For example, standard or operational definitions (e.g., acute versus chronic effect, sublethal versus lethal exposure) may have dubious scientific value relative to predicting impact in an ecological context. Yet they are invaluable in applying our technology to various scientific subjects, e.g., to determining if the free metal ion is the most toxic species of a metal by using standard acute toxicity endpoints.

Qualities valued in technological works are effectiveness (including cost-effectiveness), precision, accuracy, an appropriate level of sensitivity, consistency, clarity of results, and ease of application. As discussed below, several of these qualities are also important in practical ecotoxicology.

PRACTICAL GOAL

The practice of ecotoxicology has as its goal the application of available knowledge, tools, and procedures to solving or documenting specific problems. Many of the technological contributions are relevant here; however, the goal of their application is to solve or document a particular environmental situation. Techniques appropriate for the practical ecotoxicologist may be general, such as methods for determining contaminant leaching from wastes (e.g., Anonymous, 1990). Predictive software such as the QUAL2E program (EPA, 1987) which estimates stream water quality under specific discharge scenarios may also be important tools in achieving practical goals. Other tools include specific steps to take during implementation of a technology, e.g., biomarker-based biomonitoring on U.S. Department of Energy sites (McCarthy et al., 1991). They may involve guidelines for the practice of risk assessment on hazardous waste sites (EPA, 1989c) or for waste basin closure. In each of these instances, the goal is not to understand the phenomena more completely but to address and resolve specific problems. Indeed, attempts to conduct scientific work in such efforts may delay progress toward the practical goal—solving an immediate problem and removing potentially harmful pollutants.

Practical tools may also include criteria and standards for regulation of specific discharges or water bodies. **Criteria** are estimated concentrations of toxicants based on current scientific information that, if not exceeded, are considered protective for organisms or a defined use of a water body (or some other environmental media). Criteria are developed for individual contaminants, e.g., aluminum (EPA, 1988b), cadmium (EPA, 1985b), copper (EPA, 1985c), lead (EPA, 1985d), and zinc (EPA, 1987) using a standard approach (i.e., EPA, 1985e). Based on scientific knowledge, they are used to recommend toxicant concentrations not to be exceeded by discharges into waters. Discussed here for water, criteria are also defined for air (see Clean Air Act in Appendix 3) and solid media such as sediments (Shea, 1988; Di Toro et al., 1991).

Based on criteria and the specified use of a water body, water quality standards may be set for contaminants. **Standards** are legal limits permitted by each state for a specific water body and thought to be sufficient to protect that water body. Both criteria and standards are designed with the intent to avoid specific problems, but they are based partially on existing scientific knowledge. Consequently, a healthy growth of scientific knowledge in ecotoxicology is essential to improving our progress toward this practical goal of ecotoxicology. Indeed, criteria and standards (EPA, 1983) are revised periodically to accommodate new knowledge.

Effectiveness, precision, accuracy, sensitivity, consistency, clarity, and ease of application are valued in practical ecotoxicology as well as in technical ecotoxicology as discussed above. Also important to practical ecotoxicology are the following: unambiguous results, safety, and clear documentation of progress during application.

SUMMARY

Ecotoxicologists have overlapping yet distinct scientific, technological, and practical goals that must be understood and respected. Our current knowledge available for achieving these goals (Figure 2.1, upper panel) requires further expansion and more integration. Although the knowledge applied to these goals overlaps, there remain numerous instances of inappropriate or inadequate integration. For example, present regulations are biased toward single species tests done in the laboratory yet our scientific knowledge indicates that results from multiple species tests

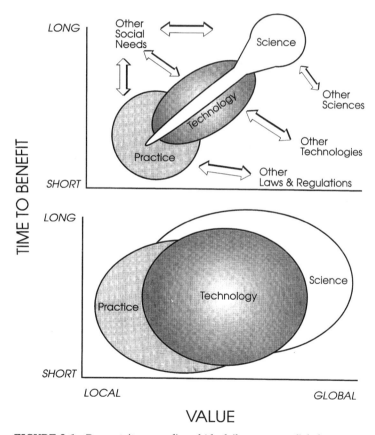

FIGURE 2.1. Present (top panel) and ideal (bottom panel) balance among scientific, technological, and practical components of ecotoxicology. The relative amount of effort in each is reflected by area on the plots of scale of value (local to global) and time to realize (short- to long-term) benefit from these components of the field.

are at least as valuable to understanding risk. Recognizing the continual need for reintegration of knowledge, lawmakers have incorporated periodic review and revision into major legislation and associated regulations. Further, new or improved technologies are continually being drawn into our scientific efforts, e.g., new molecular technologies applied to assay genetic damage by pollutants. The scientific foundations of the field must also expand and come into balance with technology and practice. This is done most effectively using the methods described at the beginning of this chapter. Passé and flawed behaviors such as precipitate explanation, overdominance of normal science, the tyranny of the particular, and *idola quantitatus* should be avoided as impediments to scientific progress.

Ecotoxicology is rapidly becoming a mature discipline and, hopefully, will soon achieve an effective balance of knowledge to best address its scientific, technological, and practical goals (Figure 2.1, lower panel). Scientific understanding must expand in all directions, especially toward more global and long-term phenomena. Technology and practice must do the same. The sound laws presently implemented in the U.S. should and do accommodate this evolution of technology and science.

Suggested Readings

Crane, M. and M.C. Newman. Scientific method in environmental toxicology. *Environ. Rev.* 4, pp. 112–122, 1996.

Chamberlin, T.C. The method of multiple working hypotheses. *J. Geol.* 5, pp. 837–848, 1897.

Mackenthun, K.M. and J.I. Bregman. *Environmental Regulations Handbook*. Lewis Publishers, Boca Raton, FL, 1992, p. 297.

McGregor, G.I. *Environmental Law and Enforcement*. Lewis Publishers, Boca Raton, FL, 1994, p. 239.

Newman, M.C., *Quantitative Methods in Aquatic Ecotoxicology*. Lewis Publishers, Boca Raton, FL, 1995, p. 426.

Newman, M.C., Ecotoxicology as a Science, in *Ecotoxicology: A Hierarchical Treatment*, Newman, M.C. and C.H. Jagoe, Eds., CRC/Lewis Publishers, Boca Raton, FL, 1996.

Platt, J.R. Strong inference. *Science (WASH DC)*. 146, pp. 347–353, 1964.

Uptake, Biotransformation, Detoxification, Elimination, and Accumulation

Models are, for the most part, caricatures of reality, but if they are good, then, like good caricatures, they portray, though perhaps in distorted manner, some of the features of the real world.

Kac (1969)

INTRODUCTION

It is important to understand and be able to predict the accumulation of contaminants in biota because effects are a consequence of concentrations in target organs or tissues. Also, because human exposure often occurs through the consumption of tainted food, prediction of accumulation in these potential vectors of exposure is key to avoiding human poisonings. The basic processes resulting in bioaccumulation (uptake, biotransformation, and elimination) will be discussed in this chapter. Chapters 4 and 5 will contribute detail and depth to this framework.

Bioaccumulation is the net accumulation of a contaminant in and on an organism from all sources including water, air, and solid phases in the environment. Solid phases include food, soil, sediment, and fine particulates suspended in the air or water. Bioaccumulation is not the same as the more restricted term, **bioconcentration** that has come to mean the net accumulation in and on an organism of a contaminant from water only. These two terms and their distinctions arise from an earlier, somewhat-dated debate in aquatic toxicology regarding the relative importance of water and food sources of contaminants to aquatic organisms.

The treatment of bioaccumulation has always relied heavily on mathematical models. This has greatly enhanced progress; understanding of these tools has become essential to prediction of bioaccumulation and consequent effects. However, as expressed in the quote above, it is important to understand also that these models are only useful caricatures and understanding of the models does not constitute a full and accurate understanding of the associated phenomena.

Figure 3.1 depicts the conceptualization of bioaccumulation as a simple

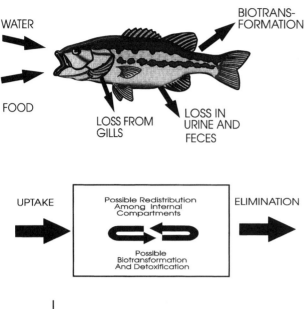

WATER

BIOTRANS-
FORMATION

FOOD

LOSS FROM
GILLS

LOSS IN
URINE AND
FECES

UPTAKE

Possible Redistribution
Among Internal
Compartments

ELIMINATION

Possible
Biotransformation
And Detoxification

CONCENTRATION

U   E

U   E

U   E

DURATION OF EXPOSURE

FIGURE 3.1. A simplified conceptualization of bioaccumulation. At the top of this figure, the fish (largemouth bass) is thought to potentially take in contaminants from its food and water, and lose contaminants via the gills, urine, and feces. There may be internal redistribution or biotransformation of the contaminant. This process is rendered to a simple box and arrow diagram. Here, only uptake from water is assumed to be significant, and all elimination processes are described by one elimination process. The most common mathematical description of this model predicts a gradual increase in contaminant in the fish until a steady state concentration is obtained as depicted in the graph at the bottom of this figure.

mathematical model. At the top of this figure, a fish is exposed to a contaminant through the water passing over its gills and the food passing through its gut. Dermal exposure may also be important depending on the contaminant qualities (Landrum et al., 1996) and the surface to volume ratio of the fish (Barron, 1995; Landrum, 1996). In this example of a largemouth bass, only one source of constant concentration is assumed for the sake of numerical expediency. All other sources are considered insignificant or adequately incorporated into the one uptake coefficient. The contaminant may be lost from the gills or pass over the gills without being taken up. It could also be excreted in the urine or eliminated in the feces after passage through the gut or entering the gut through biliary excretion. The contaminant may be transformed and redistributed to various compartments within the fish, e.g., from the gill to the plasma to the kidney. In this example, only one elimination constant is quantified. Multiple processes are mathematically combined in this one constant or assumed trivial relative to a single, dominating elimination process. Also, no biotransformation or effects to the organism are occurring. Kinetics and rate constants are assumed to remain constant over the time of accumulation. These details may be rendered to a box and arrow model (center of figure) with associated simplifying assumptions regarding the mathematics of uptake, elimination, and internal transformations and exchanges. In this example, uptake and elimination take place for one compartment in which the contaminant is mixed homogeneously and instantly. These assumptions and simplifications are subject to testing during model development. Mathematical expression of this simplest of models predicts a time course for bioaccumulation such as that shown at the bottom of Figure 3.1. This curve of a slow and monotonic increase to a maximum concentration is a result of the change in relative influences of uptake and elimination processes on change in concentration during the course of exposure. At the beginning of the exposure, little contaminant is contained in the fish and available for elimination: uptake (U) dominates relative to elimination (E) in the initial dynamics, and concentrations increase. But, as more and more contaminant accumulates in the fish, more contaminant becomes available for elimination. Elimination becomes increasingly important and the rate of increase in concentration begins to decline. Eventually, a balance between uptake and elimination results in a steady state[1] concentration in the fish that will be maintained as long as conditions remain constant.

Although described here in terms of concentration (amount of contaminant per amount of organism such as 25 µg of Pb per g of tissue), such models are also developed in terms of **body burden**, the mass or amount of contaminant in (and on) the individual (e.g., 2,500 µg of Pb per individual). The details and mathematical expression of this general model, and models derived from it, will be developed primarily in terms of concentration in this chapter.

UPTAKE

Introduction

Uptake (the movement of a contaminant into or onto an organism) can occur by several mechanisms and involve the dermis, gills, pulmonary surfaces, or gut. Simkiss (1996) categorized uptake by a cell into three general routes: the lipid, aqueous, and endocytotic routes. The lipid route encompasses the passage of lipophilic contaminants through the bilayer of membrane lipids. Small, uncharged polar molecules such as CO_2, glycerol, and H_2O also diffuse readily through the lipid bilayer (Alberts et al., 1983). The aqueous route employs two general types of

---

[1]Landrum and Lydy (personal communication, 1991; Landrum et al., 1992) comment that the terms steady state and equilibrium are often used incorrectly. Steady state refers to a constant concentration in an organism resulting from processes (e.g., uptake, elimination, and internal exchange among compartments) including those requiring energy. However, equilibrium concentrations resulting from chemical equilibrium processes do not require energy to be maintained. Steady state concentrations resulting from bioaccumulation can be considerably higher than those predicted for chemical equilibrium.

membrane transport proteins that either form channels (**channel proteins**) or act as **carrier proteins** in the membrane which transfer hydrophilic contaminants into cells. Some channels (**porins**) are nonspecific; others are specific relative to the substances passing through them. The functioning of some channels is influenced by the presence of other chemical substances including other contaminants (Simkiss, 1996).

Possible mechanisms for uptake include adsorption, passive diffusion, active transport, facilitated diffusion or transport, exchange diffusion, endocytosis, and solvent drag (Newman, 1995). Several of these processes may be important for any particular combination of contaminant, exposure scenario, and species.

Movement onto the organism, including the initial stages leading to movement into the organism, may involve adsorption. **Adsorption** is the accumulation of a substance at the common boundary of two phases (solution onto a solid surface), e.g., adsorption by ion exchange of metal dissolved in water to the integument of an insect. The more general term **sorption** may be used instead of adsorption if the specific mechanism by which a compound in solution becomes associated with a solid surface is unknown or undefined.

Two equations are used to define adsorption, the **Freundlich** and **Langmiur isotherm equations**. The Freundlich equation (Equation 3.1) is an empirical relationship and the Langmuir equation (Equation 3.2) is a theoretically-derived relationship.

$$\frac{X}{M} = KC^{\frac{1}{n}} \tag{3.1}$$

where X = amount adsorbed, M = the mass of adsorbent, K = a derived constant, C = the concentration of solute in the solution after adsorption is complete, and n = a derived constant.

$$\frac{X}{M} = \frac{abC}{1+bC} \tag{3.2}$$

where X, M, and C are as defined above, a = the adsorption maximum (amount), b = affinity parameter reflecting bond strength. A plot of C (abscissa or x-axis) versus X/M (ordinate or y-axis) will result in a curve shaped as the one at the bottom of Figure 3.1. Such relationships can be linearized with equations such as Equation 3.3 (Freundlich isotherm) or Equation 3.4 (Langmuir isotherm) to facilitate data fitting by linear regression.

$$\log \frac{X}{M} = \log K + \frac{\log C}{n} \tag{3.3}$$

$$\frac{C}{X/M} = \frac{1}{ab} + \frac{1}{a}C \tag{3.4}$$

Adsorption theory and these equations have been used successfully to define toxicant movement onto diverse biological surfaces such as those of unicellular

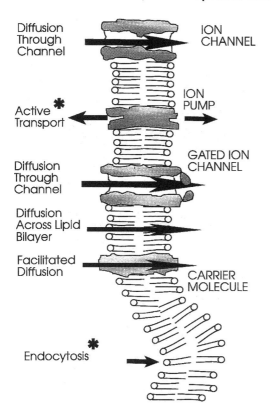

Diffusion Through Channel — ION CHANNEL

Active Transport * — ION PUMP

Diffusion Through Channel — GATED ION CHANNEL

Diffusion Across Lipid Bilayer

Facilitated Diffusion — CARRIER MOLECULE

Endocytosis *

FIGURE 3.2. Mechanisms of uptake of contaminants into cells. Simple diffusion can occur across the lipid bilayer or through an ion channel formed by a channel protein. Channels may be gated and their functioning influenced by chemical and electrical conditions. Facilitated diffusion occurs via a carrier protein. Active transport passes the solute up an electrochemical gradient. Here the Na^+, K^+ ATPase pump ion pump is illustrated. Potassium is pumped in as sodium is pumped out of the cell. The last mechanism for cellular uptake is endocytosis. As indicated by an *, endocytosis and active transport require energy.

algae (Crist et al., 1988), fish gills (Pagenkopf, 1983; Janes and Playle, 1995), periphyton (Newman and McIntosh, 1989), and zooplankton (Ellgehausen et al., 1980). Crist et al. (1988) found that uptake of H^+ by algae involved rapid adsorption followed by a slow diffusion into the cell. Langmuir equation constants (Equation 3.2) were used by Crist and coworkers to define relative metal interactions with algal cell surfaces.

Diffusion is the movement of a contaminant down an electrochemical gradient.[2] It may be simple diffusion of a charged ion via a channel protein or passage of a lipophilic molecule through the lipid route (Figure 3.2). Simple diffusion does not require expenditure of energy. If diffusion involves a protein channel, passage through a channel is influenced by ion charge and size including size of the hydration sphere about the ion. Channels may be gated and respond to various chemical or electrical conditions. Diffusion could also be facilitated by a carrier protein. **Facilitated diffusion** occurs down an electrochemical gradient, requires a carrier protein, does not require energy, and contaminant movement is faster than predicted for simple diffusion. Because a carrier protein is involved, facilitated diffusion may be subject to saturation kinetics and competitive inhibition. Some facilitated diffusion involves the exchange of ions across a membrane (**exchange diffusion**).

---

[2]The term electrochemical gradient means a concentration, activity, or electrical gradient.

Diffusion accurately describes uptake of many nonpolar organic compounds or uncharged inorganic molecules such as ammonia (NH_3) (Fromm and Gillette, 1968; Thurston et al., 1981) or $HgCl_2^0$ (Simkiss, 1983) by the lipid route (Spacie and Hamelink, 1985; Barber et al., 1988; Erickson and McKim, 1990), and many charged moieties such as dissolved metals and protons (H^+) via the aqueous route (Spacie and Hamelink, 1985). Facilitated diffusion is expected for some metals and organic contaminants (Landrum and Lydy, personal communication, 1991). Diffusion back and forth across a membrane may be complicated if the chemical is changed after crossing that membrane. For example, pentachlorophenol may pass though a membrane into the blood stream but then be converted to the charged pentachlorophenate. This compound is much less capable of passing back across the membrane and out of the organism.

Diffusion is described most readily by Fick's Law (Equation 3.5).

$$\frac{dS}{dt} = -DA\frac{dC}{dX} \tag{3.5}$$

where dS/dt = rate of contaminant movement across the surface, D = a diffusion coefficient, A = the area of surface through which diffusion is occurring, and dC/dX = the concentration (or some other type of) gradient across the boundary of interest, e.g., the difference in concentration between the two sides of a cell membrane. This equation is incorporated into many models of bioaccumulation.

Active transport requires energy to move the contaminant up an electrochemical gradient. Because it involves a carrier molecule, active transport may be subject to saturation kinetics and competitive inhibition. The best example of active transport is cation transport by membrane-bound ATPase. ATPase, using energy from the hydrolysis of ATP, acts as a coupled ion pump to simultaneously remove some ions from the cell (e.g., sodium) while moving other ions (e.g., potassium) into the cell. Radiocesium, a chemical analog of potassium, can be taken up by this mechanism (Newman, 1995). Some metals (e.g., cadmium taken up as an analog of calcium) and large, hydrophilic compounds can also be subject to active transport (Cockerham and Shane, 1994).

Endocytosis (pinocytosis and phagocytosis) may be important in uptake, especially for contaminants entering an organism in food. Simkiss (1996) details an excellent example of contaminant metals being taken up by the transferrin route for iron assimilation. Iron, and perhaps other metals, are bond to the membrane-associated transferrin protein. The metal-transferrin complex moves to a specific region of the cell surface where it becomes engulfed and incorporated into a vesicle. In the cell, the vesicle fuses with a lysosome and the associated metal is released.

Under certain circumstances, contaminants can move into an organism by **solvent drag**, the movement of a solute along with the physical movement of the bulk solution. Movement into organisms is possible by passage between adjacent cell junctions. For example, H^+ can compete under low pH conditions with Ca^{2+} which normally binds to cell surface ligands and maintains tight cell junctures of fish gill epithelia. This could lead to increased gill permeability to metals at low pH conditions (Newman and Jagoe, 1994).

Reaction Order

The kinetics of uptake are often defined in terms of reaction order. In the context of a reaction involving only one reactant, reaction order refers to the exponent (n) to which the reactant concentration (C) is raised in the equation describing the reaction rate, $dC/dt = kC^n$. In the case of a zero order reaction, n is 0 and $dC/dt = kC^0$. This reduces to $dC/dt = k$ and k has units of C/h. The concentration will increase independent of reactant concentration with zero order kinetics. For first order kinetics ($dC/dt = kC$), the rate will change with reactant concentration and units for k are h^{-1}. First-order reaction kinetics are the most commonly observed and applied kinetics for bioaccumulation modeling. Higher order reaction kinetics are occasionally warranted. Also commonly used are saturation kinetics such as Michaelis-Menten kinetics. Saturation kinetics are relevant with enzymatically-mediated processes such as detoxification. Saturation kinetics may also occur for active or facilitated transport. Above a certain concentration of reactant, a system is saturated and cannot proceed any faster than a maximum velocity (V_{max}), i.e., zero order kinetics. However, the kinetics shift to first order as reactant concentrations drop below saturation conditions. For example, Mayer (1976, as reported in Spacie and Hamelink, 1985) saw evidence of saturation for the elimination of di-2-ethylhexylphthalate from fathead minnows (*Pimephales promelas*). Equation 3.6 describes the change in reactant concentration by Michaelis-Menten kinetics.

$$-\frac{dC}{dt} = \frac{V_{max} C}{k_m + C}$$

(3.6)

where V_{max} = the maximum rate of reactant change (C/h), and K_m = half-saturation constant. A plot of the velocity of concentration change (V, y-axis) against concentration (C, x-axis) would look very much like the plot at the bottom of Figure 3.1. The velocity of concentration change increases with reactant concentration only to a certain point. There would be a maximum velocity (V_{max}) which could not be exceeded regardless of any further increase in concentration. The concentration at which V was equal to $V_{max}/2$ would be the k_m in Equation 3.6. Again, although zero, first, or saturation kinetics are applied to uptake, first order uptake is the most common.

BIOTRANSFORMATION AND DETOXIFICATION

General

Once a contaminant enters the organism, it becomes available for **biotransformation**, the biologically-mediated transformation of one chemical compound to another. Biotransformation can involve enzymatic catalysis and, as a consequence, can be subject to saturation kinetics and competitive inhibition. Biotransformation may lead to enhanced elimination, detoxification, sequestration, redistribution, or activation. Biotransformation may enhance the rate of loss from the or-

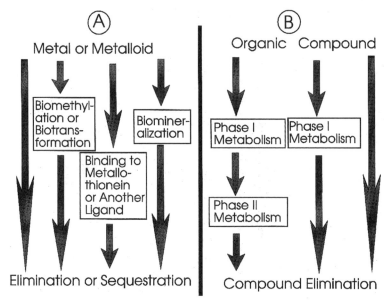

FIGURE 3.3. General mechanisms of biotransformation and detoxification of inorganic (A) and organic (B) contaminants. See text for explanation.

ganism as is often the case if a lipophilic xenobiotic is converted to a more hydrophilic compound, e.g., naphthalene oxidation to naphthalene diol. The contaminant may be rendered to a nontoxic form. Some contaminants may be transformed to a form that is retained within the organism but is sequestered away from any site of possible adverse effect. With **activation**, the adverse effect of a contaminant is made worse by biotransformation or an inactive compound is converted to one with high, adverse bioactivity. For example, the organophosphorus pesticide, parathion, undergoes oxidative desulfuration to form the very potent paraoxon (Hoffman et al., 1995). General processes associated with biotransformation (Figure 3.3) are described below.

Metals and Metalloids

Metals and metalloids are subject to biotransformation, resulting in elimination from or sequestration within the individual (Figure 3.3A). However, an ion may quickly bind to a plasma-associated ligand and become available for removal from the organism without any transformation. Lyon et al. (1984) demonstrated the role of ligand binding to metal ions in the blood of crayfish in determining the relative rates of such elimination for a series of metal ions.

Microbes genetically adapted to metal contaminated environments may have enhanced ability to add methyl or ethyl groups to metal ions (e.g., ionic mercury converted to methyl mercury) (Wood and Wang, 1983). Trivalent (As^{3+}) and pentavalent (As^{5+}) arsenic entering plants or animals may be rendered less toxic by methylation to monomethylarsonic and dimethylarsinic acids (Nissen and Benson,

1982; Peoples, 1983). Arsenic may also be converted to arseno-sugar (e.g., trimethylarsonium lactate), arsenobetaine, or phospholipid (O-phosphatidyl trimethylarsoniumlactate) (Cooney and Benson, 1980; Edmonds and Francesconi, 1981). Selenium entering a plant as selenite (SeO_3^{-2}) can be reduced and converted to selenocysteine. If incorporated into proteins, such selenoamino acids result in protein dysfunction, providing one mechanism for selenium toxicity (Brown and Shrift, 1982). In selenium-tolerant plants, large amounts of the non-protein amino acids (Se-methylselenocysteine and selenocystathionine) are produced, implying that high rates of synthesis of these amino acids serve as a means of detoxification (Brown and Shrift, 1982).

Metals may be bound and sequestered from target sites by metallothioneins and similarly functioning molecules. **Metallothioneins** are a class of relatively small (circa 7,000 daltons) proteins with approximately 25 to 30% of their amino acids being sulfur-rich cysteine and possessing the capacity to bind six to seven metal atoms per molecule (Hamilton and Mehrle, 1986; Shugart, 1996). They are commonly induced by metals, including cadmium, copper, mercury, and zinc.[3] Silver, platinum, and lead also have been reported as inducers of metallothioneins (Garvey, 1990). In addition to their role in essential metal homeostasis, they also can be induced by toxic metals, bind these metals, and reduce the amount available to cause a toxic effect. Metallothioneins and metallothionein-like proteins are found in many vertebrates and invertebrates, and also have been reported in higher plants (Grill et al., 1985). A class of metal-binding polypeptides, **phytochelatins**, are similarly induced in plants by metals (Grill et al., 1985). Grill et al. (1985) found that phytochelatins were induced by the metal ions, Cd^{2+}, Cu^{2+}, Hg^{2+}, Pb^{2+} and Zn^{2+}. They may also function in regulation and detoxification of metals.

Finally, metal and other cations may be sequestered or eliminated through biomineralization. Metals such as lead (e.g., Bercovitz and Laufer, 1992; Newman et al., 1994) and radionuclides such as radiostrontium or radium (Grosch, 1965) can be incorporated into relatively inert shell, calcareous exoskeleton, or bone. Some metals are incorporated into exoskeleton (including the gut lining) of soil invertebrates and lost upon molting (Beeby, 1991). Other metals such as lead can be incorporated into molluscan shell (Beeby, 1991; Newman et al., 1994) and rendered unavailable for interaction at target sites. Radiostrontium from atmospheric fallout accumulates in bone along with calcium. Children whose bodies are actively building and reworking bone during development are particularly vulnerable. The consequence of such sequestration in bone may be extensive damage, such as in the case of another bone-seeking radionuclide, radium. In 1925, an outbreak of radium poisoning was reported among young women employed at brush-

---

[3]As pointed out by Simkiss and Taylor (1981), metallothioneins are involved primarily with the detoxification of Class B metals. **Class B metal cations** (Nieboer and Richardson, 1980) have filled d orbitals of 10 to 12 electrons and low electronegativity. They are "soft" spheres readily deformed by adjacent ions. They easily form covalent bonds with donor atoms such as sulfur. Class B metals include Cu^+, Ag^+, Au^{+1}, Au^{+3}, Cd^{2+}, and Hg^{2+}. **Class A metal cations** (e.g., Li^+, Na^+, K^+, Mg^{2+}, Ca^{2+}, and Sr^{2+}) have inert gas electron configurations, high electronegativity, and hard spheres. Intermediate between Class A and B metals are **borderline metal cations** such as Fe^{2+}, Co^{2+}, Ni^{2+}, Cu^{2+}, Mn^{2+}, and Fe^{3+} (Stumm and Morgan, 1970; Brezonik et al., 1991).

ing radium-laced paint onto watch faces.[4] They ingested radium while licking the paint brush bristles to a fine point as they worked. The gamma-emitting radium accumulated in bone, causing extensive bone lesions or fatal anemia by damaging the bone marrow (Grosch, 1965).

Metals may be sequestered by incorporation into a variety of granules or concretions in addition to sequestration in structural tissues (Mason and Nott, 1981; Simkiss and Taylor, 1981; Pynnönen et al., 1987). Such granules are usually associated with the midgut, digestive gland, hepatopancreas, Malpighian tubules, and kidneys of invertebrates (Roesijadi and Robinson, 1994). They are also found in other specialized cells of invertebrates, and connective tissues of vertebrates and invertebrates (Roesijadi and Robinson, 1994).

Hopkin (1986) described four general categories of granules in invertebrates. Type A granules are intracellular granules 0.2 to 3 µm in diameter built as concentric layers of calcium (and magnesium) pyrophosphate and an organic matrix of lipofuscin. Metal precipitation with phosphate is mediated by pyrophosphatase (Howard et al., 1981). These Type A granules are found in most invertebrate phyla (Roesijadi and Robinson, 1994). Type A granules accumulate Class A and intermediate metals such as manganese and zinc. Type B granules are also intracellular granules of roughly the same size as Type A granules and have high concentrations of sulfur. Consequently, they have high concentrations of Class B and intermediate metals (copper, cadmium, mercury, silver, and zinc) that avidly bind to sulfur-containing groups. Type C granules are intracellular granules rich in the iron-containing products of ferritin. The last granule (Type D) is extracellular, can be 20 µm in diameter, is composed of calcium carbonate, and functions to buffer the hemolymph (blood) of molluscs.

The various types of granules differ in their locations in the organism. In the marine snail, *Littorina littorea*, Type A granules are found in basophil cells of the digestive diverticulum and Type D granules are located in calcium cells of connective tissues of the foot and other tissues. Type A, B, and C granules may be storage sites of metals. They can be instrumental in the elimination of metals by their discharge from cells into the gut (Hopkin, 1986; Beeby, 1991). Elimination may involve **exocytosis** (fusion of intracellular vesicles with the cell membrane and emptying of vacuole content to the cell exterior) or cell lysis with release of associated granules.

Organic Compounds

Depending on their qualities, organic contaminants can be eliminated rapidly or be subjected to metabolism[5] with subsequent excretion of metabolites. During me-

---

[4]Our collective ignorance of the effects of radionuclides such as radium was staggering at this time. Even more shocking from today's perspective is the intentional ingestion of radium as the patent medicine, Radithor. Its consumption was responsible for numerous painful ailments and unnecessary deaths. Its use was banned after the fatal poisoning of a prominent New York socialite (Macklis, 1993).

[5]Lech and Vodicnik (1985) object to using the term "metabolism" in this context and suggest "biotransformation" as the preferred word for biochemically-mediated conversion of xenobiotics. They feel that the term "metabolism" should be restricted to biochemical reactions of "carbohydrates, proteins, fats and other normal body constituents."

tabolism, lipophilic compounds are often, but not always, made more amenable to excretion by conversion to more hydrophilic products. Metabolism of organic contaminants can be broken down into Phase I and II reactions (Figure 3.3B). Generally, reactive groups such as —COOH, —OH, —NH$_2$, or —SH are added or made available by **Phase I reactions**, increasing hydrophilicity. Although predominantly oxidation reactions, hydrolysis and reduction reactions are also important in Phase I (George, 1994). One of the most common Phase I reactions involves addition of an oxygen to the xenobiotic by a **monooxygenase** (mixed function oxidases or MFOs) (Hansen and Shane, 1994). After formation, the products of Phase I reactions may be eliminated or enter into **Phase II reactions** (Figure 3.3B). Conjugates are formed by Phase II reactions which inactivate and foster elimination of the compound. Compounds conjugated with xenobiotics include acetate, cysteine, glucuronic acid, sulfate, glycine, glutamine, and glutathione (Hansen and Shane, 1994; Landis and Yu, 1995).

These Phase I and II reactions can be illustrated with the metabolism of naphthalene (Figure 3.4). Phase I oxidation (naphthalene → naphthalene epoxide) and hydrolysis (naphthalene epoxide → naphthalene 1,2-diol) are shown in this example to produce more water soluble metabolites of naphthalene. Subsequent Phase II conjugation with glucuronic acid is shown for naphthalene 1,2,-diol. Many more reactions than shown here are involved in the metabolism of xenobiotics as discussed in more detail in Chapter 6. Many involve inducible enzymatic systems subject to the complex regulation typical of biochemical pathways. The modeling of bioaccumulation and elimination of the original xenobiotic may be complicated by the dynamics of associated metabolites, especially if a radiotracer is used in the experimental design. But, in the case of an activated compound, the dynamics of a metabolite may be more important than that of the original compound. Regardless, with the potential for metabolism, it is important if using radiotracers to measure all biotransformation products, not only total radioactivity.

ELIMINATION

Elimination Mechanisms

Elimination is the excretion or metabolism of a contaminant resulting in a decrease in the amount of contaminant within an organism. Although often used synonymously, depuration and clearance do not have precisely the same meaning as elimination (Barron et al., 1990). **Depuration** is a term associated with a particular experimental design in which the organism is placed into a clean environment and allowed to lose contaminant through time. **Clearance**, as will be discussed shortly, is a term used when modeling bioaccumulation kinetics and reflects the rate of substance movement between compartments normalized to concentration. Clearance has units of volume time^{-1}, e.g., mL h^{-1}. If contaminant concentration is monitored in individuals over time, another phenomenon resulting in a decrease in concentration must be defined, i.e., **growth dilution**. The con-

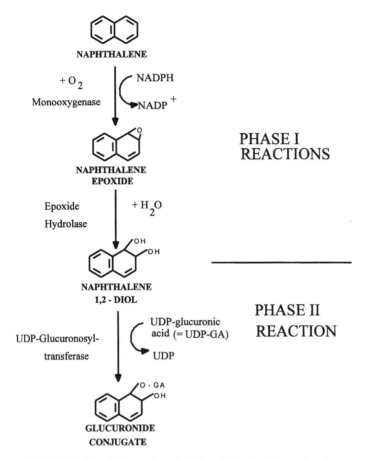

NAPHTHALENE

$+ O_2$

Monooxygenase

NADPH

NADP$^+$

PHASE I
REACTIONS

NAPHTHALENE
EPOXIDE

Epoxide
Hydrolase

$+ H_2O$

NAPHTHALENE
1,2 - DIOL

UDP-Glucuronosyl-
transferase

UDP-glucuronic
acid (= UDP-GA)

UDP

PHASE II
REACTION

GLUCURONIDE
CONJUGATE

FIGURE 3.4. Metabolism of naphthalene including Phase I and Phase II reactions. The first Phase I reaction (naphthalene → napthalene epoxide) is an oxidation reaction and the second Phase I reaction (naphthalene epoxide → naphthalene 1,2-diol) is a hydrolysis reaction. UDP-glucuronic acid is formed by condensation of UTP (high energy nucleotide, uridine triphosphate) with glucose-6-phosphate. UDP is uridine diphosphate and GA is glucuronic acid. (Composite figure of Figures 1, 5, and 14 of Lech and Vodicnik [1985].)

centration of contaminant may decrease in a growing organism because the amount of tissue in which the contaminant is distributed has increased. Growth dilution is not a component of elimination because the total amount of contaminant (body burden) has not changed as a result of growth.

The relative importance of specific elimination mechanisms varies among plants, vertebrates, and invertebrates as well as among contaminants. Plants may lose contaminant by leaching, evaporation from surfaces, leaf fall, exudation from roots, or herbivore grazing (Duffus, 1980; Newman, 1995). Animals may eliminate contaminants by transport across the gills, exhalation, secretion of bile from the gall bladder, secretion from the hepatopancreas, secretion from the intestinal mucosa, shedding of granules, molting, excretion via the kidney or an analogous structure, egg deposition, or loss in hair, feathers, and skin. Aluminum bound to gill mucus is rapidly lost by mucus sloughing (Wilkinson and Campbell, 1993). In higher animals, elimination may involve loss in sweat, saliva, and genital secretions (Duffus, 1980). The liver bile, gills, and kidney are often the primary elimination routes for animals.

Elimination from the gills can be rapid for xenobiotics with low lipophilicity, i.e., log K_{ow}[6] values of 1 to 3 (Barron, 1995). Nonpolar organic compounds resistant to metabolism tend to diffuse slowly across the gill (Spacie and Hamelink, 1985). Spacie and Hamelink (1985) note that DDT, di-2-ethylhexylphthalate (DEHP), phenol, and pentachlorophenol can be eliminated in significant amounts this way.

Large, nonpolar molecules and associated metabolites can be eliminated from the liver, into the bile, and lost in the feces (Duffus, 1980; Spacie and Hamelink, 1985; Barron, 1995). In humans, compounds with molecular weights exceeding approximately 300 daltons are eliminated in significant amounts in the bile (Gibaldi, 1991). Many metals (e.g., aluminum, cadmium, cobalt, mercury, and lead) and metalloids (e.g., arsenic and tellurium) complexed with proteins or other biochemical compounds in the plasma are incorporated into bile (Camner et al., 1979).

After passage into the liver, a contaminant enters the hepatic sinusoids where it may be absorbed by parenchymal cells and metabolized. The compound or its metabolites either return to the sinusoids or become incorporated into bile. Those incorporated in bile may be eliminated in the feces; however, some compounds entering the small intestine in bile may be subject to reabsorption and repeated passage through the liver. This **enterohepatic circulation** increases persistence of some compounds in the body and may increase damage to the liver (Duffus, 1980; Gibaldi, 1991). Although a minor complication of most elimination processes, compounds incorporated into saliva can also establish a similar cycle (Wagner, 1975). A Phase II reaction metabolite may be reabsorbed after being deconjugated, e.g., hydrolyzed, in the intestine (Wagner, 1975) and lead to cycling. Metals may exhibit enterohepatic circulation depending on the metal complex size. Trivalent arsenic and methyl mercury exhibit significant enterohepatic circulation (Camner et al., 1979). For arsenic, this involves active transport. Competition between compounds during incorporation into bile, and factors affecting bile formation and enterohepatic circulation will dictate the effectiveness of bile elimination.

Kidney excretion tends to be important for compounds with molecular weights lower than approximately 300 daltons. This is the primary route of metal (e.g., cadmium, cobalt, chromium, magnesium, nickel, tin, and zinc) excretion by mammals (Roesijadi and Robinson, 1994). However, there are exceptions in which

---

[6]K_{ow} is the partition coefficient for a compound between n-octanol and water. It is used to reflect the lipophilicity of a compound and to imply relative partitioning of a xenobiotic between aqueous phases of the environment and lipids in the organism. It is sometimes designated by the capital letter, P.

gastrointestinal excretion dominates (Camner et al., 1979). Metals such as cadmium and mercury can be excreted directly through the intestinal mucosa by active or passive processes (Camner et al., 1979). Gastrointestinal excretion may also involve loss by normal cell sloughing of the intestine wall.

Renal elimination involves three processes: glomerular filtration, active tubular secretion, and passive reabsorption. Filtration through capillary pores allows passage of most xenobiotics except those bound to plasma proteins. Elimination of some metals and lipophilic compounds by this passive filtration mechanism is inhibited by such binding. Weak organic acids and bases can be actively secreted to the urine or reabsorbed depending on urine pH (Spacie and Hamelink, 1985). Beryllium elimination involves tubular secretion (Camner et al., 1979). In contrast to filtration, renal secretion is an energy-requiring movement up a concentration gradient that depends on the contaminant concentration, concentrations of competing compounds, pH of the urine, delivery rate to carrier proteins in the proximal tubule, and the relative affinity of the compound for the carrier proteins of the tubule and plasma proteins (Gibaldi, 1991). Because secretion involves carriers, it can be subject to competitive inhibition and saturation kinetics (Gibaldi, 1991).

After a high concentration gradient is established, compounds can be passively reabsorbed into the blood: this process is more likely for lipid soluble compounds than for ionized or water soluble compounds (Gibaldi, 1991). Cadmium bound to metallothionein is reabsorbed in the renal tubules (Camner et al., 1979). Some compounds can be actively reabsorbed (Gibaldi, 1991). Obviously, pH of the urine in the distal tubule will strongly influence reabsorption of weak acids and bases (Spacie and Hamelink, 1985), and metals such as lead and uranium (Camner et al., 1979).

Other routes of elimination may be important depending on the organism and contaminant. Volatile organic compounds can be lost in expired breath, and lipophilic compounds can be lost from species such as fish by deposition in lipid-rich eggs. Arthropods have the capacity to eliminate contaminants such as metals by molting (Lindqvist and Block, 1994), peritrophic membrane sloughing from the midgut and hindgut cuticle loss during molting (Hopkin, 1989), or by discharge of granules. Birds may incorporate metals into feathers, e.g., lead (Amiard-Triquet et al., 1992; Burger et al., 1994) and mercury (Becker et al., 1994). Mammals can eliminate lipophilic compounds (e.g., DDT) or calcium analogs (e.g., strontium) in milk. Fin whales transfer some of their body burden of PCB and DDT to their young by normal placental processes prior to giving birth and, after birth, transfer more to the calf in milk (Aguilar and Borrell, 1994). Some metals and metalloids are incorporated into skin and hair (e.g., Roberts et al., 1974).[7]

---

[7]Incorporation of elements into hair is used to monitor exposure as typified by two recent studies of historical significance. Analysis of hair from Napoleon Bonaparte (Maugh, 1974) indicates that, sometime during his exile on Elba or in Saint Helena, he was slowly being poisoned with arsenic, a common political tool of the time. Hair from Sir Isaac Newton had extremely high levels of mercury, a common element used in his experiments. Newton habitually tasted his chemicals during experimentation. Poisoning is now identified as the most probable reason for his "year of lunacy" (1693), a year filled with abnormal behaviors characteristic of mercury's neurological effects (Broad, 1981).

Modeling Elimination

Given the variety of elimination mechanisms, it is surprising that first order kinetics are used in most models of elimination. Zero order and saturation kinetics are used less frequently and tend to be used more in complex model formulations.

Compartment models of elimination similar to that shown in Figure 3.1 can be formulated as rate constant, clearance volume, or fugacity models. All three approaches are equivalent for the simple (single source) models shown here and associated constants may be interconverted (Newman, 1995). Statistical moments methods for quantifying elimination without formulation of a specific compartment model are also common in pharmacology (Yamaoka et al., 1978) and have application to ecotoxicology (Barron et al., 1990; Newman, 1995).

Rate constant-based models employ constants, such as the first order rate constant described earlier, to quantify the rate of change of concentration or amount of toxicant in one or more compartments. Frequently, compartments are not physical (e.g., liver excretion) but are mathematical compartments (e.g., the "fast" elimination component). But some interpretations of mathematical compartments imply associated physical compartments; consequently, it is important to determine exactly the type of compartment model being described when assessing this type of research.

A simple, first-order rate constant model can be used to describe elimination of contaminant from an organism after it has been moved to a clean environment and allowed to depurate (top of Figure 3.5). The model can be expressed in terms of concentration (C) or body burden (amount in the individual or X).

$$\frac{dC}{dt} = -kC \tag{3.7}$$

where C = concentration in the compartment, k = rate constant for the concentration-based formulation (h^{-1}), and t = time or duration of depuration.

$$\frac{dX}{dt} = -kX \tag{3.8}$$

where X = amount in the compartment, k = rate constant for the amount-based formulation (h^{-1}), and t = time or duration of depuration. Equations 3.7 and 3.8 are integrated to Equations 3.9 and 3.10 to allow prediction of elimination from an initial concentration (C_0) or amount (X_0) to any time (t) during depuration.

$$C_t = C_0 \, e^{-kt} \tag{3.9}$$

$$X_t = X_0 \, e^{-kt} \tag{3.10}$$

Figure 3.5 (center) shows the exponential elimination of a contaminant described by these equations. To fit depuration data to these equations, the ln of concentration or body burden is plotted against time (bottom of Figure 3.5). The resulting line has a y-intercept equal to the ln of the initial concentration or body

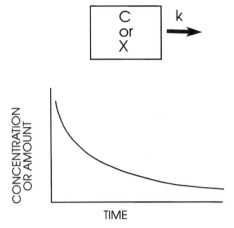

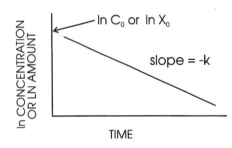

FIGURE 3.5. Elimination of a contaminant under a depuration scenario including linearization to extract k and C_0 or X_0.

burden. The slope is an estimate of -k. If linear regression of the logarithm of concentration or logarithm of body burden versus time is used to estimate C_0 or X_0, a commonly ignored bias exists in these estimates that should be corrected (Newman, 1995).

The time required for the amount or concentration to decrease by 50% ($t_{1/2}$, **biological half-life**) is (ln 2)/k. The **mean residence time** of a particle of compound (τ) in the compartment is $1.44t_{1/2}$ or, more directly, k = 1/τ.

If two or more elimination mechanisms are responsible for elimination from the compartment, they are easily included in the models.

$$C_t = C_0\ e^{-\Sigma k_i t} \tag{3.11}$$

where k_i = the individual elimination rate constants.

$$X_t = X_0\ e^{-\Sigma k_i t} \tag{3.12}$$

For models as described by Equations 3.11 and 3.12, the **effective half-life** (k_{eff}) is (ln 2)/Σk_i. Equations 3.11 and 3.12 are useful if a radiotracer is used to quantity elimination because the (first order) radioactive decay rate constant (λ) can be in-

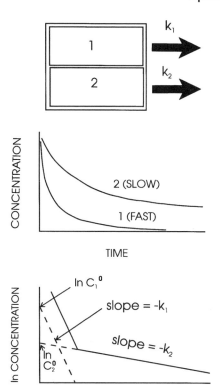

FIGURE 3.6. Elimination involving two compartments and two elimination constants including backstripping to calculate k_1, k_2, C_1, and C_2.

cluded in the formulation, i.e., $\Sigma k_i = k + \lambda$ for a one component elimination model with radioactive decay of the tracer.

Rate constant-based models can be developed for biexponential or multiexponential elimination. One such conceptual model (top and middle of Figure 3.6) has elimination from two different compartments within an organism displaying distinct ("fast" and "slow") elimination kinetics. The integrated forms of this model are Equations 3.13 and 3.14.

$$C_t = C_1^0 e^{-k_1 t} + C_2^0 e^{-k_2 t} \tag{3.13}$$

where C_t = total concentration assuming equal volumes or masses for compartments 1 and 2, C_1^0 = initial concentration in compartment 1, C_2^0 = initial concentration in compartment 2, k_1 = elimination rate constant for compartment 1, and k_2 = elimination rate constant for compartment 2. Obviously, the assumption of equal sizes for the two compartments is an inconvenient assumption of this model. This constraint will be resolved later using clearance-volume models. However, the assumption of equal compartment sizes is not a difficulty in the case of the model based on amount in the various compartments,

$$X_t = X_1^0 e^{-k_1 t} + X_2^0 e^{-k_2 t} \tag{3.14}$$

where X_t = total amount of contaminant in the organism, $X_1{}^0$ = initial amount in compartment 1, $X_2{}^0$ = initial amount in compartment 2, k_1 = elimination rate constant of compartment 1, and k_2 = elimination rate constant for compartment 2.

With the linearizing method just described for monoexponential elimination, biexponential elimination produces a line with a distinct break in slope (bottom of Figure 3.6, solid line). A **backstripping or backprojection procedure** can be used to extract the model parameters from a multiexponential elimination curve. First, the region of the ln concentration vs time curve where predominantly slow elimination occurs (the straight part of the line after the break in slope in Figure 3.6, bottom) is used to fit a line for component 2. The slope of this line is $-k_2$ and the y-intercept estimates ln $C_2{}^0$. The linear equation just derived for this second component (ln C_2 = ln $C_2{}^0 - k_2t$) is then used to predict concentrations of contaminant in compartment 2 for all sampling times during depuration. These predicted concentrations in compartment 2 are subtracted from the original data noted for all times prior to the break in the line to estimate the concentrations present in compartment 1 during elimination; these "stripped away" data for component 1 produce a straight line (Figure 3.6., bottom panel, dashed line) with a slope of $-k_1$ and y-intercept of ln $C_1{}^0$. Wagner (1975) and Newman (1995) detail this and more accurate means of backstripping exponential curves of elimination. Computer programs are available to implement these methods and assess the results statistically.

Other multiple compartment models can be fit with this rate constant-based approach. One of the more common, two compartment elimination models (Figure 3.7, top) can be fit in a similar manner. In this model, a contaminant is introduced as a bolus into a central compartment (e.g., a single injection of the dose into the blood) and the compound is then subject to passage into and out of another peripheral or storage compartment. The compound can only be eliminated from the central compartment. The resulting elimination curves for these two compartments are illustrated at the bottom of Figure 3.7. Initially, the concentration in the central compartment (1) is high and there is no compound in the peripheral compartment (2). The concentration in the central compartment decreases and the concentration in the peripheral compartment begins to increase. The concentration in compartment 1 eventually describes a biexponential curve and the curve for compartment 2 eventually runs parallel to the end component of compartment 1 because elimination is occurring from compartment 1 only. If the two dashed lines shown in the bottom panel of Figure 3.7 are used to obtain intercepts and slopes for the apparent biexponential elimination dynamics from compartment 1, the microconstants (k_{10}, k_{21}, and k_{12}) can be estimated with the following equations (Equations 3.15–3.17). For example, one could monitor blood concentrations through time and fit the results to Equation 3.13 with backstripping methods. The resulting model is the same as Equation 3.13 except A, B, α, and β are substituted into the equation. Obviously, these constants do not retain the same physical interpretation as C_1, C_2, k_1, and k_2.

$$k_{21} = \frac{A\beta + B\alpha}{A + B} \qquad (3.15)$$

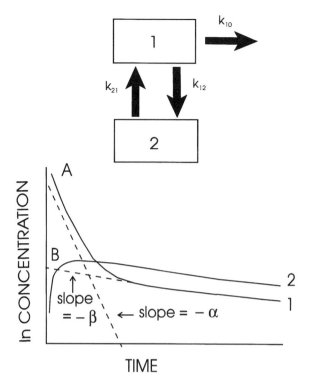

FIGURE 3.7. Elimination from a two-compartment model after bolus introduction of the contaminant into compartment 1. A, B, α, and β are used to estimate k_{12}, k_{10}, and k_{21} as described in the text.

$$k_{10} = \frac{\alpha\beta}{k_{21}} \tag{3.16}$$

$$k_{12} = \alpha + \beta - k_{21} - k_{10} \tag{3.17}$$

where A and B = antilogs of the y-intercepts for components 1 and 2, respectively, and α and β = slopes of components 1 and 2, respectively. The biological half-life for a contaminant using this model is $(\ln 2)/\beta$ (Barron et al., 1990).

Many similar multiple compartment models can be developed under the rate constant-based modeling scheme. However, modeling can be made easier as more compartments are added if one uses the **clearance volume-based** formulation. In this approach, the substance is distributed in and cleared from compartments of different volumes. Clearance (Cl) is expressed as flows, the volume of compartment completely cleared per unit time (volume time^{-1}).

The apparent volume of distribution concept is used in clearance volume-based models. It can be envisioned with the example of the introduction of a dose of compound into a compartment. After introduction, the dose distributes itself within the compartment and a compartment concentration is realized. The compartment volume is V = dose/concentration. For example, a dose is injected into the blood and allowed to distribute in the circulatory system. The **apparent volume of distribution** (V_d) for blood would be dose/concentration. Estimation becomes more involved as the compound becomes distributed in many compartments, but methods developed in pharmacology for estimation of apparent volumes and clearance rates are available (Newman, 1995). If other compartments are involved, their apparent volumes of distribution are derived mathematically and expressed in units of volume of the reference compartment, often the blood or plasma compartment. Consequently, these volumes are not physical volumes per se, but mathematical volumes. The total dose (D_T) in the organism is distributed among compartments as defined by the concentrations in the compartments (C) and the compartment V_d's (V). For example, if blood is the reference compartment and there are n additional compartments,

$$D_T = C_b V_b + \sum_{i=1}^{n} C_i V_i \qquad (3.18)$$

where subscripts of b and i denote blood and nonblood compartments, respectively. The k_i values of the rate constant formulation are equal to Cl_i/V_i for the clearance volume-based formulation.

The clearance volume-based formulation allows development of more complex models and parameterization of these models. They have been used in pharmacology to describe the internal kinetics of drugs (**pharmacokinetics**) and poisons (**toxicokinetics**) for many years. Consequently, a wealth of data, techniques, software, and expertise have been developed around this approach. Also, clearance volume-based models allow direct incorporation of physiological parameters into models, and thus, enhance predictive capabilities of models under different conditions and for different species. (Pharmacokinetic models that include physiological and anatomical features in describing internal kinetics are called **physiologically-based pharmacokinetics or PBPK models**).

A final compartment model that uses different units but is equivalent to the rate constant-based and clearance volume-based formulations is the **fugacity model**. It is based on the escaping tendency of a compound in a compartment or its fugacity. Fugacity (f) is expressed as a pressure (Pa) and is related to concentration in a phase, $C = fZ$ where Z is the fugacity capacity (mol $[m^{-3} Pa^{-1}]$ of the phase. The rate of transport between two compartments $(1 \rightarrow 2)$ is N (mol h^{-1}) = $D(f_1-f_2)$ where D is a transport constant (mol $[h^{-1} Pa^{-1}]$). The k_i's for the rate constant-based model are equivalent to the $D_i/V_i Z_i$ of fugacity models where V is the compartment volume. The major advantage of the fugacity model is that units are the same regardless of the phases being considered. Consequently, wide differences in concentrations are easily accommodated for complex models such as those including water, sediment, food, and biological compartments (Mackay and Paterson, 1982; Gobas and Mackay, 1987; Mackay, 1991).

Finally, there exists in the pharmacology literature a body of methods for calculating elimination qualities such as mean residence time for substances in individuals without assuming a specific model. Yamaoka et al. (1978) introduced a statistical moments approach to pharmacokinetics that uses only the area under the curve (AUC) of a plot of concentration (y-axis) versus time (x-axis). The mean residence time, its associated variance, and other parameters can be estimated with the AUC. Although underutilized in ecotoxicology (Barron et al., 1990; Newman, 1995), these methods could be very useful if an exact model is judged to be unnecessary, impractical, or impossible to define. Also, because of their general utility, these methods may be used if the exact model is known.

ACCUMULATION

Bioaccumulation is the net consequence of uptake, biotransformation, and elimination processes within an individual. The simplest, rate constant-based model includes first order uptake from one source into one compartment and first-order elimination from that compartment.

$$\frac{dC}{dt} = k_u C_1 - k_e C \tag{3.19}$$

where C_1 = concentration in the source (e.g., 1= water), C = concentration in the compartment (e.g., fish), k_u = uptake clearance (mL $[g^{-1} h^{-1}])^8$, and k_e = elimination rate constant (h^{-1}). This equation integrates to Equation 3.20.

$$C_t = C_1 \left(\frac{k_u}{k_e} \right) \left(1 - e^{-k_e t} \right) \tag{3.20}$$

The concentration in organisms can be predicted at any time (t) based on Equation 3.20. The resulting bioaccumulation curve is shown in Figure 3.1 (bottom panel) and in the accumulation phase of Figure 3.8. The clearance volume-based and fugacity equivalents of Equation 3.20 are the following,

$$C_t = V_d \, C_1 (1 - e^{-\frac{Cl}{Vd} t}) \tag{3.21}$$

where V_d and Cl = V_d and C_1 for the organism, respectively, and

$$C_t = C_1 \left(\frac{Z}{Z_1} \right) (1 - e^{-\frac{D\,t}{V\,Z}}) \tag{3.22}$$

where Z and Z_1 = fugacity capacities for the organism and source, respectively; D = transport constant for the organism, and V = the volume of the organism.

As the concentration in the organism approaches equilibrium (the bracketed term, $[1-e^x]$ in Equations 3.20 to 3.22 approaches 1), the relationship between the

---

[8]The units of k_u are often expressed as h^{-1} under the assumption of equal density of the source (e.g., water) and compartment (e.g., tissue). A mL of source volume and g of tissue then cancel each other out and units become h^{-1}. This makes k_u appear as a rate constant. However, k_u reflects a clearance from the source (Barron et al., 1990; Landrum et al., 1992), and consequently, should retain units of flow normalized to mass. Details supplied by P. Landrum are provided in Appendix 3 for the student interested in the derivation of these units for k_u.

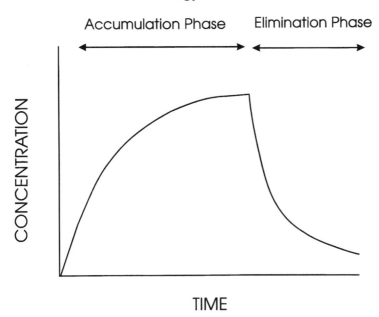

FIGURE 3.8. Simple bioaccumulation through time as described by Equation 3.20 followed by elimination as described by Equation 3.9. Data of this type are generated with an accumulation-elimination experiment. In the accumulation phase, kinetics are dictated by both uptake and elimination. The organism is then allowed to depurate and kinetics in this elimination phase are solely a consequence of elimination.

concentration in the organism (C at steady state or C_{ss}) and the source (C_1) is described as the following,

$$\frac{C_{ss}}{C_1} = \frac{k_u}{k_e} = V_d = \frac{Z}{Z_1} \qquad (3.23)$$

These terms define the steady state concentration in the organism relative to the source. If the environmental source is water, the k_u/k_e, V_d, or Z/Z_1 estimate the **bioconcentration factor** (BCF). This ratio is defined by many workers as the **bioaccumulation factor** (BAF or BSAF)[9] if the source of contaminant is sediment. Normalization may be done for some nonpolar contaminants in these models. If the

___

[9]As discussed in Chapter 5, the term, bioaccumulation factor is also used in a more general context to mean accumulation from other sources as well. It is often used if the exact source remains undefined or poorly defined as in a field survey. Some workers will use the more specific term, **biota-sediment accumulation factor (BSAF)** to make clear that the factor relates accumulated contaminant to that in sediments (Spacie et al., 1995).

concentrations are normalized to amount of lipid in the organism and to the amount of organic carbon in sediments, the ratio is referred to as the **accumulation factor** (AF).

Often bioaccumulation models are fit to the curve of concentration in the organism versus time of exposure using nonlinear regression methods. For the rate constant-based formulation, this method estimates k_u and k_e simultaneously. Although often adequate, this can lead to some difficulty in computations (poor convergence on a solution, lowered estimate precision, and a degree of covariance between estimates of k_u and k_e). A combined accumulation-elimination design (Figure 3.8) is used to avoid these difficulties as in the studies of Yamada et al. (1994), and Murphy and Gooch (1995). In this design, organisms are exposed to a contaminant and allowed to accumulate until internal concentrations approach practical steady state concentration (i.e., 95% of C_{ss}). Then, the organisms are removed from the contaminated arena and allowed to depurate in a clean setting. The drop in internal concentration is measured with time during this elimination phase. With these elimination phase data, the k_e may then be estimated independent of k_u. This estimate of k_e can then be used in the nonlinear fitting for the accumulation phase data. Now, the nonlinear regression must estimate only k_u from the accumulation data.

Other formulations of these bioaccumulation models allow inclusion of more detail. If there had been an initial concentration of contaminant in the organism prior to the trial exposure, the concentration of contaminant can be predicted through time by combining Equations 3.20 (bioaccumulation from source 1) and 3.9 (elimination of initial concentration, C_0).

$$C_t = \frac{k_u}{k_e} C_1 (1 - e^{-k_e t}) + C_0 e^{-k_e t} \tag{3.24}$$

Multiple elimination components can be included (Equation 3.25).

$$C_t = \frac{k_u}{k_{e1} + k_{e2}} C_1 (1 - e^{-(k_{e1} + k_{e2}) t}) \tag{3.25}$$

Under the expedient, but often excessively optimistic, assumption that growth dilution can be described as e^{-gt} (g = a rate constant akin to the elimination rate constant), growth dilution can be included as a component in a model such as Equation 3.25.

Uptake from food and water can be incorporated into these models using an estimated assimilation efficiency (α, amount absorbed per amount ingested in food) and specific ration (R, amount of food consumed per amount of organism). Assuming concentrations in water (C_1) and food (C_2) are constant,

$$C_t = \frac{k_u C_1 + \alpha R C_2}{k_e} (1 - e^{-k_e t}) \tag{3.26}$$

A general model incorporating multiple sources (k_{ui}), many elimination components (k_{ei}) and an initial concentration (C_0) can be defined.

$$C_t = \frac{\sum\limits_{j=1}^{m} C_j k_{uj}}{\sum\limits_{i=1}^{n} k_{ei}} (1 - e^{-(\sum\limits_{i=1}^{n} k_{ei})t}) + C_0\, e^{-(\sum\limits_{i=1}^{n} k_{ei})t} \tag{3.27}$$

As models become more complex, the data requirements become increasing difficult to meet. Consequently, most pharmacokinetic models are compromises between realism and expediency. This is important to understand in order to extract the fullest understanding from modeled systems without making the unintentional transition to naive overinterpretation.

SUMMARY

In this chapter, the methods and mathematics associated with the uptake, biotransformation, and elimination of contaminants were described. The general mechanisms of uptake were described: adsorption, passive diffusion, active transport, facilitated diffusion, exchange diffusion, endocytosis, and solvent drag. Biotransformation for metals, metalloids, and organic compounds was described briefly. Binding proteins and peptides, and biotransformation of and sequestration of metals and metalloids were described in detail. Phase I and Phase II reactions in the transformation of organic compounds were discussed. Elimination by a variety of mechanisms was discussed for plants and animals. Details of elimination via liver, kidney, and gill were highlighted.

Models were developed based on three different formulations: rate constant-based, clearance volume-based, and fugacity-based. All are equivalent but each formulation has its own advantages and disadvantages. Also, statistical moments methods exist that do not require a specified model. Although rate constant-based models have the longest history in ecotoxicology, clearance volume-based models have much promise in allowing linkage to the pharmacokinetics literature and techniques. They also are most amenable to generation of PBPK models. Fugacity models have an advantage for extremely different phases/compartments because they express contaminant levels in identical units for all compartments. Statistical moments methods require the least amount of information, and no model is required.

Suggested Readings

Barron, M.G., G.R. Stehly, and W.L. Hayton. Pharmacokinetic modeling in aquatic animals. I. Models and concepts. *Aquat. Toxicol. (AMST)* 18, pp. 61–86, 1990.

Gibaldi, M. *Biopharmaceutics and Clinical Pharmacokinetics*. Lea and Febiger, Philadelphia, PA, 1991, p. 406.

Hansen, L.G. and B.S. Shane. Xenobiotic Metabolism, in *Basic Environmental Toxicology*, Cockerham, L.G. and B.S. Shane, Eds., CRC Press, Inc., Boca Raton, FL, 1994.

Himmelstein, K.J. and R.J. Lutz. A review of the applications of physiologically based pharmacokinetics modeling. *J. Pharmacokinet. and Biopharm.* 7, pp. 127–145, 1979.

Landrum, P.F., H. Lee II, and M.J. Lydy. Toxicokinetics in aquatic systems: Model comparisons and use in hazard assessment. *Environ. Toxicol. Chem.* 11, pp. 1709–1725, 1992.

Lech, J.J. and M.J. Vodicnik. Biotransformation, in *Fundamentals of Aquatic Toxicology*, Rand, G.M. and S.R. Petrocelli, Eds., Hemisphere Publishing Corp., Washington, DC, 1985.

Mackay, D. *Multimedia Environmental Models. The Fugacity Approach*. Lewis Publishers, Inc., Chelsea, MI, 1991, p. 257.

Newman, M.C., *Quantitative Methods in Aquatic Ecotoxicology*. Lewis Publishers, Boca Raton, FL, p. 426, 1995.

Spacie, A. and J.L. Hamelink, Bioaccumulation, in *Fundamentals of Aquatic Toxicology*, Rand, G.M. and S.R. Petrocelli, Eds., Hemisphere Publishing Corp., Washington, DC, 1985.

4

Factors Influencing Bioaccumulation

Without question!
The chemical company called Chisso poisoned the fishing waters of
Minamata, poisoned the aquatic food chain, and eventually poisoned
a great number of inhabitants. Chisso poured industrial poisons through
waste pipes until Minamata Bay was a sludge dump, the heritage of
centuries destroyed.
SMITH AND SMITH (1975)

INTRODUCTION

General

The degree to which a contaminant accumulates in biota is a function of the qualities of the contaminant, the organism, and the environmental conditions under which the organism and contaminant are interacting. The qualities of the contaminant determine the chemical forms in which it is present in the environment and the degree to which it is available to be taken up, biotransformed, and eliminated. Physiological, biochemical, and genetic qualities determine an organism's ability to minimize uptake, biotransform, and eliminate the contaminant. Developmental or sex-related changes can influence bioaccumulation, e.g., age- and sex-correlated lipid content will influence accumulation of a lipophilic contaminant. Ecological and behavioral qualities of an organism determine routes of exposure and efficiency of uptake from each potential source: a predator is exposed through its prey but a pelagic species is not exposed directly to sediment-associated contaminant. The *milieu* in which interaction between a contaminant and organism takes place can affect speciation and phase association of the contaminant and, consequently, availability for accumulation. Environmental conditions may also directly modify the functioning of an organism. For example, temperature has clear effects on rates of pertinent physiological and biochemical processes taking place within the organism. Other factors such as salinity and pH strongly modify ion regulation and osmoregulation and, in so doing, influence the uptake of many contaminants. Qualities of the microenvironment at the site of interaction may be as important, or more important, than those of the general environment. The water chemistry at the surface microlayer of the gills (e.g., Janes and Playle, 1995) or that of interstitial waters surrounding an infaunal species (e.g., Campbell et al., 1988) strongly influence uptake.

Basic qualities of chemicals, organisms and the environment that have the strongest influence on bioaccumulation are outlined in this chapter. They are discussed first for inorganic contaminants and then for organic contaminants. Finally, general biological processes that transcend this dichotomy are discussed.

Bioavailability

Bioavailability is the extent to which a contaminant in a source is free for uptake (Newman and Jagoe, 1994). In many definitions, especially those associated with pharmacology or mammalian toxicology, bioavailability of a contaminant implies the degree to which the contaminant is free to be taken up *and to cause an effect at the site of action*.[1] This is a reasonable qualification in the context of these sciences but the more general definition given here seems warranted in ecotoxicology. In ecotoxicology, availability for bioaccumulation in a food or prey species in which no effect is seen could be as much a concern as availability to an organism that is directly affected. Even with this broader definition, the term may still be used in the context of the degree to which a toxicant is available to have an effect, e.g., availability of metal in sediments to kill benthic species (Carlson et al., 1991).

Bioavailability is measured or implied in a variety of ways. Availability of a contaminant from two types of food may be implied from differences in measured bioavailability between individuals exposed to similar concentrations of contaminant in the different foods. This qualitative approach has been used for contrasting different foods (Newman and McIntosh, 1983; Reinfelder and Fisher, 1991), sediments (Luoma and Bryan, 1978; Langston and Burt, 1991), and water chemistries (Wright and Zamuda, 1987; Driscoll et al., 1995). Landrum and Robbins (1990) suggested that bioavailability in different sediments can be compared by measuring the amount of a contaminant in sediments and that in organisms inhabiting the sediments, or comparing uptake clearance rates of organisms in the various sediment types. Using this approach, Hickey et al. (1995) compared the influence of feeding modes on bioaccumulation of polychlorinated biphenyls (PCB) and polynuclear aromatic hydrocarbons (PAH) in the deposit feeder, *Macomona liliana* and filter feeder, *Austrovenus stutchburyi*.

In Chapter 3, assimilation efficiency (Equation 3.26) was used to quantify bioavailability from food. Assimilation efficiency or percentage retention (efficiency expressed as a percentage) may be determined by feeding known amounts of contaminant, most often using a radiotracer, and measuring the increase in contaminant within the fed individual. For example, bioaccumulation of organotin antifouling agents (tributyltin and triphenyltin) in Red Sea bream (Yamada et al.,

---

[1]Bioavailability and concentration of a source dictate the exposure that an organism will experience. This exposure then dictates the probability and intensity of effects to the organism.

1994), and *trans* and *cis* isomers of chlordane in channel catfish (Murphy and Gooch, 1995) from food sources were determined in this manner. The mass incorporated into tissue divided by the mass fed to the individual is calculated assuming that all of the contaminant measured in the organism has become part of its tissues. Normally, much effort is made in designing this type of experiment to minimize the amount of unassimilated contaminant in the gut, hepatopancreas, or other similar sites. Without this precaution, the estimated assimilation efficiency would be biased upward from the true assimilation efficiency.

Bioavailability is measured another way in studies of drug pharmacokinetics. Such studies are concerned with determining the **effective dose** of a drug, the amount entering the blood and available to have an effect. To do this, a dose may be administered orally and the amount appearing in the blood compared to that of the same dose injected intravenously. Bioavailability measured in this manner involves both the amount of, and rate at which, the drug or toxicant enters the organism (Gibaldi, 1991): quantification of bioavailability must take both of these variables into account. This can be done by comparing the areas under the curves (AUCs) of concentration versus time plots obtained for the different routes of administration (Figure 4.1). The **absolute bioavailability** is estimated from the AUC

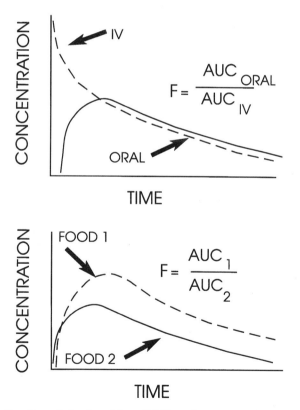

FIGURE 4.1. Estimation of absolute bioavailability of an ingested dose by comparison of AUCs for ingestion and intravenous injection (top panel), and relative bioavailability by AUC comparison for the same dose administered in two different foods (bottom panel).

for any route or form of the compound divided by the AUC of the compound after direct injection into the bloodstream. For example, Equation 4.1 might be used to estimate the bioavailability of a dose of ingested compound.

$$F = \frac{AUC_{oral}}{AUC_{iv}} \qquad (4.1)$$

where AUC_{oral} = AUC for dose D administered orally and AUC_{iv} = AUC for dose D administered intravenously. Figure 4.1 (top panel) shows two curves from which AUCs would be generated with methods such as trapezoidal estimation (Gibaldi, 1991; Newman, 1995). Similarly, **relative bioavailability** can be estimated for two different sources with one source serving as a reference. Figure 4.1(bottom panel) shows the blood concentration-time curves used to estimate relative bioavailability of a dose (D) administered in food 1 and then in food 2.

   With some modification, bioavailability can also be estimated if different doses are used for the various routes of introduction or administered forms (Newman, 1995). Assuming that AUC is a linear function of dose within the range of doses used,

$$F = \frac{AUC_{oral}\ D_{iv}}{AUC_{iv}\ D_{oral}} \qquad (4.2)$$

where D_{iv} and D_{oral} = the doses administered by intravenous injection and orally, respectively (Gibaldi, 1991). However, if half-lives are different for the two routes of administration, the associated difference in clearance is usually corrected by using the following equation instead of Equation 4.2 (Gibaldi and Perrier, 1982).

$$F = \frac{AUC_{oral}\ D_{iv}\ (t_{1/2})_{iv}}{AUC_{iv}\ D_{oral}\ (t_{1/2})_{oral}} \qquad (4.3)$$

   Other AUCs may be applied to estimate bioavailability under various conditions. For example, AUCs for concentration – time curves of urine or bile could be used effectively for some compounds.

   With statistical moments analysis of the AUCs or parameter estimation for the models described in Chapter 3, mean residence times (MRT, the mean time of drug residence in a compartment similar to τ described in Chapter 3) can be calculated (Yamaoka et al., 1978). The rate of drug or contaminant absorption is estimated with MRTs for various routes of introduction as the difference between MRTs. The **mean absorption time** (MAT) for a drug or contaminant in food is $MRT_{food} - MRT_{iv}$ (Gibaldi, 1991). Assuming first order kinetics, an **absorption rate constant** (k_a) can be calculated to be MAT = k_a^{-1} (Gibaldi, 1991). Such parameters are meaningful for estimating how quickly an ingested, inhaled, imbibed, or otherwise introduced toxicant becomes available at a target site within the organism.

CHEMICAL QUALITIES INFLUENCING BIOAVAILABILITY

Inorganic Contaminants

Water

Water chemistry affects bioavailability by modifying the chemical species present and the functioning of uptake sites. For example, pH has an obvious effect on the equilibrium, $NH_3 + H^+ \rightleftarrows NH_4^+$. The neutral NH_3 passes much more readily through the cell membrane than the charged NH_4^+ (Lloyd and Herbert, 1960), and consequently, is the more bioavailable form of ammonia. Similarly, the bioavailability of cyanide (HCN), a weak acid but extremely potent inhibitor of cytochrome oxidase, is influenced by the effect of pH on the equilibrium, $H^+ + CN^- \rightleftarrows HCN$ (Broderius et al., 1977). (Bioavailablity of cyanide may also be modified by iron (Fe[II]) which combines with cyanide to form a less toxic ferrocyanide, $Fe(CH)_6^{-4}$ [Manahan, 1993].) However, the situation is more complicated for both cyanide and dissolved sulfide (H_2S, HS^-, and S^{2-}) because more than one species contributes simultaneously to toxicity (Broderius et al., 1977).

The bioavailabilities of dissolved metals and metalloids are also affected by chemical speciation. Metal cations compete with other cations for dissolved **ligands**[2], anions or molecules that form coordination compounds and complexes with metals (Newman and Jagoe, 1994). Ligands combining with metals include dissolved organic compounds and inorganic species. Natural organic ligands have a wide range of functional groups, the most important in complexation are carboxylic and phenolic groups. The major inorganic species involved in metal complexation include $B(OH)$, $B(OH)_4^-$, Cl^-, CO_3^{2-}, HCO_3^-, F^-, $H_2PO_4^-$, HPO_4^{2-}, NH_3, OH^-, $Si(OH)_4$, and SO_4^{2-} (Öhman and Sjöberg, 1988; Newman and Jagoe, 1994). The ligands, NH_3, HS^- and S^{2-} are important in anoxic waters. Of course, H_2O is an important ligand that forms a hydration sphere around cations and, in so doing, influences bioavailability. As discussed in Chapter 3, size and charge of a hydrated cation influences its passage through membrane protein channels. At thermodynamic equilibrium, the distribution of a particular dissolved cation among its various species can be estimated as a function of competing cation concentrations, pH, ligand concentrations, temperature, and ionic strength. These predictions, or directly measured concentrations of free (aquated) ion, can be used to normalize metal concentrations under a variety of conditions to better estimate bioavailable metal.

As a general rule, bioavailability or toxicity is correlated with the free metal concentration (Allen et al., 1980; Andrew et al., 1977; Dodge and Theis, 1979; Borgmann, 1983). This common observation has led to the reasonable suggestion that the free ion is often the most bioavailable form of a dissolved metal. In fact, this concept is sufficiently prevalent as to be given a name, the **free ion activity**

---

[2]Ligands share electron pairs with metals. The ligand is a **monodentate ligand** if only one pair is shared and multidentate if more than one pair is shared. **Mulitdentate ligands** are also called **chelates**. Multidentate ligands are prefixed, e.g., bi- or tridentate to indicate the number of electron pairs involved.

model[3] or FIAM. Campbell and Tessier (1996) define the FIAM as "the universal importance of free metal ion activities in determining the uptake, nutrition and toxicity of all cationic trace metals." It must be kept in mind that the concentrations of other species are also correlated with that of the free ion concentration (or activity), and their bioavailability is often difficult to define independently as a consequence. Further, it is not always the case that the free ion is the most, or only, available form of a dissolved metal. Simkiss (1983) noted that neutral complexes of some Class B metals (e.g., $Hg(Cl)_2{}^0$) are extremely lipophilic relative to charged species (e.g., Hg^{2+}), and this extreme lipophilicity combined with the dominance of neutral chloro complexes in marine systems can greatly enhance their bioavailability.

Competition at uptake sites is also modified by water chemistry. Crist et al. (1988) described the competition of H^+ on the initial adsorption of metals to sites (carboxylic groups of pectin) on algal cells. They also showed the interaction of dissolved metals with Ca^{2+}, Mg^{2+} and Na^+ at algal cell binding sites. The influences of pH and major cations on silver (Janes and Playle, 1995) and aluminum (Wilkinson and Campbell, 1993) uptake on gills have been quantified based on competition for binding to gill surface sites (surface-associated ligand groups). Campbell and Tessier (1996) discussed several studies of competition of metals with hardness cations (Ca^{2+} and Mg^{2+}) at various biological surfaces. Although not the complete explanation, competition between hardness cations and metals is suggested as a reason for the decrease in bioavailability or toxicity of metals often measured with an increase in freshwater hardness.

The water chemistry of the microlayer at a biological surface can be extremely important. Excretion of $NH_4{}^+$, NH_3, $HCO_3{}^-$, and CO_2 from the gills can rapidly modify the chemistry of water as it passes over gill surfaces (Newman and Jagoe, 1994). Depending on the bulk water chemistry, the shift in water chemistry at surfaces can be sufficient to modify bioavailability of metals such as aluminum (Exeley et al., 1991; Neville and Campbell, 1988; Playle and Wood, 1989; 1990).

Solid Phases

The bioavailability of inorganic contaminants in aerosols, food, sediments, and other solid phases of the environment is difficult to predict precisely. However, some general themes do emerge from the literature. The direct availability from solid phases is only one part of the story; for example, availability from a phase such as sediments can be determined by the capacity of the metal to partition into the interstitial waters. These general phenomena will be discussed with examples here.

The bioavailability of metals or metalloids in aerosols and larger particulates

---

[3]The activity is used here instead of concentration to encompass situations where there is significant nonideal behavior of ion concentrations due to interionic interactions. In very dilute solutions, the distinction is not as important as in more concentrated solutions such as seawater where the activity coefficients are necessary to relate concentration and activity. By convention, the activity coefficient (activity/concentration) is 1 with infinite dilution.

suspended in air is determined not only by their chemical forms in the solid but also by the size of the particulates and the distribution of the element within the particulates. For example, arsenic tends to condense onto outer layers of smaller coal fly ash particles as they move up the smoke stacks of coal burning power plants and, because of this surficial deposition, the arsenic is more available than if it were uniformly distributed throughout small to large ash particles (Hulett et al., 1980; Wangen, 1981). Lead halides in automobile exhaust are more readily available for dissolution in the lung after inhalation than lead in road dust that has weathered to compounds such as lead sulfate (Laxen and Harrison, 1977). Also, lead is present at highest concentrations in small particles of road dust that gain deeper access to the lungs than large particles (Biggins and Harrison, 1980). In humans, larger particles are removed by nasal hair, and associated contaminants are unavailable as a consequence. Particles with diameters of 5 to 10 µm gain entry only to the region of the pharynx. However, those with diameters of 1 µm or less go much deeper into the terminal bronchioles and aveoli (Cordasco et al., 1995). Clearly, the depth of passage and consequent bioavailability of contaminants in inhaled particles are related to particle size.

Bioavailability of contaminants in food is a function of many factors. As with inhaled particulates, the size of a food particle can determine bioavailability as well as the chemical form of the affiliated contaminants. Particle size of materials passing though the human gut can modify bioavailability of some contaminants and drugs (Gibaldi, 1991). Bivalve molluscs have complex sorting mechanisms on the gills and palps, in the gut, and in the digestive diverticula that are strongly affected by particle size. The size of the food particle will determine the degree to which it participates in these different digestive processes. Some small particles such as iron oxide and iron saccharate can even be taken into phagocytes while still on the gills of oysters (Galtsoff, 1964). This process of sorting and digestion is further complicated by environmental factors such as temperature and tidal rhythm, which modify feeding behavior and digestive processes of bivalves (Morton, 1970).

Literature describing the bioavailability of elements ingested by humans provides several telling examples of additional factors modifying assimilation. Diet can have a strong impact on bioavailability; we have all been instructed about the do's and don'ts of eating while taking various medications. Chronic zinc deficiency and consequent dwarfism found in regions of the Middle East are linked to the poor zinc availability in the predominantly cereal diet of these peoples and the custom of eating clay (Sanstead, 1988). Zinc is more available in meats relative to cereals and clay in the diet sequesters zinc during its passage through the intestine. Protein in the diet increases zinc availability, likely due to enhanced absorption of zinc that is chelated by histidine and cysteine (Sandstead, 1988). In contrast, a protein-rich diet reduces calcium bioavailability.

Bioavailability of ingested lead to humans and nonhuman species is a topic of much deserved attention. The lead in paint chips is all too bioavailable to small children ingesting them. Many avian poisonings also involve lead. The bird species of concern are those feeding in wetlands or fields spattered with shotgun pellets from sporting activities. Lead sinkers from the bottom of fishing ponds may also be a source of lead to large waterfowl such as swans or geese (Pain, 1995). Raptors feeding on birds containing lead shot are exposed too (Pain, 1995). Indeed, 338 of

4,300 dead bald eagles that were examined in a survey by the U.S. Department of Interior were found to have been poisoned by lead in shot or bullet fragments from their prey (Franson et al., 1995). Shot is retained in the gizzard of birds as gizzard stones and lead is slowly released under the associated grinding and acidic conditions (Amiard-Triquet et al., 1992).

Bioavailability of metals and metalloids in sediments is not easily estimated (Luoma, 1989); however, it is thought to be determined by concentrations in interstitial water and concentrations in various solid phases. The solid phase concentrations influence bioavailability by dictating the concentrations of metal in the interstitial waters surrounding the biota and by direct ingestion of solids by benthic species.[4] Because total sediment-associated metal concentration can be a poor indicator of available metal (Tessier et al., 1984), most estimates of bioavailable metals depend on partial extractions such as a 1 N HCl extract (Krantzberg, 1994) or a series of sequential extractions of sediments thought to grossly separate particular metal-binding fractions of the solid sediments (e.g., Tessier et al., 1979; Babukutty and Chacko, 1995). As an example of a single extractant, sediment-bound metals extracted into an EDTA solution were correlated with bioavailability to several marine invertebrates (Ray et al., 1981). Using sequential extractions, Tessier et al. (1984) found that bioaccumulation of sediment-associated metals in a freshwater mussel was best correlated with metal concentrations in the more easily-extracted fractions in the extraction series shown in Figure 4.2.

For oxic sediments, several general trends can be identified regarding metal bioavailability. First, easily extracted (1 N HCl) iron, notionally reflecting iron hydrous oxides, tends to inhibit metal bioavailability (Newman, 1995). Presumably this reflects the avid binding of metals to oxides in oxic sediments. Consequently, metal concentrations in a 1 N HCl sediment extract may be normalized to the simultaneously extracted ("easily extracted") iron to account for this effect (e.g., Luoma and Bryan, 1978). Although less consistent than this effect of iron, an increase in sediment organic carbon can diminish bioavailability for some metals (Crecelius et al., 1982) and metal concentrations can be normalized to sediment organic carbon content. Finally, more easily-extracted fractions in sequential extractions tend to be more bioavailable than more tightly bound metals (Tessier et al., 1984; Rule and Alden, 1990; Young and Harvey, 1991; Newman, 1995).

For anoxic sediments, bioavailabilities of some metals (e.g., cadmium, chromium, lead, mercury, and nickel) are correlated with sulfide concentrations as reflected by the **acid volatile sulfides** (AVS, sulfides extracted with cold HCl believed to be predominately iron and manganese sulfides). The presence of sufficient amounts of AVS sequesters the sediment-associated metals as highly insoluble metal sulfides. An equilibrium between the extremely insoluble metal sulfides and the large amounts of iron and manganese sulfides is established, favoring

---

[4]Based on this premise, Campbell et al. (1988) classified sediment-associated organisms as Type A (those in contact with sediments but unable to ingest particles) and Type B (those capable also of ingesting particles). Examples of **Type A organisms** may be benthic algae or rooted macrophytes. Examples of **Type B organisms** are detritivores or many suspension feeders. With this distinction, bioavailability is discussed for the two classes relative to interstitial water concentrations only (Type A) or solid phase plus interstitial water (Type B).

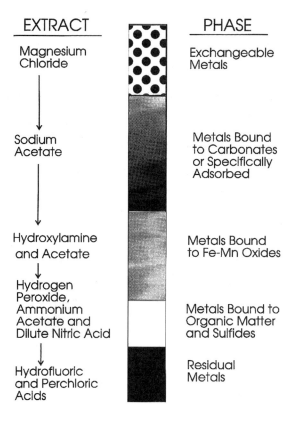

EXTRACT

Magnesium
Chloride

Sodium
Acetate

Hydroxylamine
and Acetate

Hydrogen
Peroxide,
Ammonium
Acetate and
Dilute Nitric Acid

Hydrofluoric
and Perchloric
Acids

PHASE

Exchangeable
Metals

Metals Bound
to Carbonates
or Specifically
Adsorbed

Metals Bound
to Fe-Mn Oxides

Metals Bound to
Organic Matter
and Sulfides

Residual
Metals

FIGURE 4.2. Fractionation of sediment solid phase metals by sequential extraction. In the scheme shown here (Tessier et al., 1984), an aliquot of sediment is sequentially extracted with magnesium chloride, sodium acetate, hydroxylamine and acetate, hydrogen peroxide, ammonium acetate and nitric acid, and, finally, strong acid to produce five different extracts. These extracts are thought to grossly reflect the amount of metal in the forms noted at the right side of the figure. These are operational definitions and the phase descriptions often do not accurately reflect the true phase association of the extracted metals.

metal precipitation (right side of the equation): $Cd^{2+} + FeS_{(S)} \rightleftarrows CdS_{(S)} + Fe^{2+}$. This maintains very low interstitial water concentrations of toxic metals and, according to Di Toro et al. (1990), renders them unavailable, especially those metals which bind avidly to sulfur. For anoxic sediments, a cold HCL extract may be analyzed for AVS and simultaneously extracted metals (SEM). The metal concentrations (SEM) may be normalized to AVS, e.g., Ankley et al. (1991), or Carlson et al. (1991). Di Toro and coworkers argue that, for values of SEM/AVS less than 1, the metal is precipitated as sulfide and relatively unavailable to have a toxic effect on associated benthic species.

Organic Contaminants

Water

Bioavailability of organic compounds from water and other sources has been described with **structure-activity relationships** (SARs) that use molecular qualities of the organic compound to predict activity, i.e., bioavailability. Often such qualitative relationships predict changes in activity of a drug or toxicant with changes such as the addition of a chloride atom or removal of a methyl group from a par-

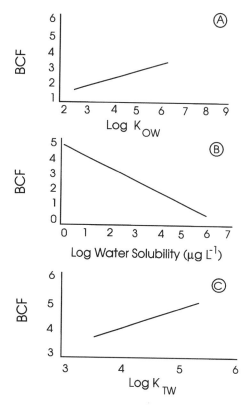

FIGURE 4.3. Relationships between bioconcentration factors (BCFs) for a variety of organic compounds and log K_{OW} (trout muscle), log water solubility (marine mussel), and K_{TW} or log of triolein-water partition coefficient (log K_{TW}, pooled data for rainbow trout and guppies). Panels A, B, and C are modified from Figure 1 of Neely et al. (1974), Figure 1 of Geyer et al. (1982), and Figure 3 of Chiou (1985), respectively.

ent molecular structure. If expressed quantitatively, SARs become **quantitative structure-activity relationships** or QSARs. A QSAR is a quantitative, very often statistical, relationship between molecular qualities and activity, i.e., bioavailability or toxicity. Molecular qualities include measures of lipophilicity, steric conformation, molecular volume, reactivity, etc. But, in ecotoxicology, the most commonly used are those based on measures of lipophilicity such as K_{OW}, water solubility, or K_{TW} of organic compounds (Figure 4.3). Historically, partitioning between n-octanol and water has been the most common phase partitioning procedure thought to reflect partitioning between water and lipids of organisms, e.g., Mackay (1982). Chiou (1985) used triolein (glyceryl trioleate)-water partitioning (i.e., K_{TW}) to better reflect partitioning between water and triglycerides in organisms (Figure 4.3C). N-octanol-water partitioning was found to accurately reflect partitioning between triolein-water to a log K_{OW} of 6. Above 6, the K_{OW} dropped slightly below expectations for perfect linear concordance between K_{TW} and K_{OW}.

In the simple K_{OW} approach, the organism is envisioned as a membrane-enveloped pool of emulsified lipids. In this conceptual model, uptake and elimination are controlled by permeation of the membrane and/or permeation through aqueous phases (Connell, 1990) with the predominant process being dictated by the qualities of the specific compound in question. Small, hydrophilic molecules are controlled by membrane permeation but large, hydrophobic compounds are

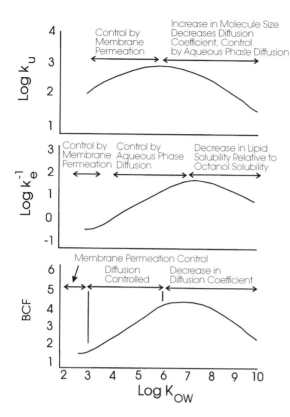

FIGURE 4.4. A summary of the changes in uptake, elimination, and bioaccumulation (bioconcentration factor or BCF) as related to K_{OW}. (Modified from Figures 3, 4, and 5 of Connell and Hawker [1988]).

controlled by permeation through aqueous phases. Very large molecules (e.g., molecular size ≥ 9.5 Å)(Landrum, personal communication) will not pass through the lipid membrane. Uptake across the membrane is dictated by Fick's law (Equation 3.5). The consequence of these behaviors of organic compounds varying in size and lipophilicity can be illustrated with the work of Connell and Hawker (1988) (Figure 4.4). In the top panel of this figure, log of the uptake constant is plotted against log K_{OW} for a series of chlorinated hydrocarbons accumulating in three species of fish. Among log K_{OW} values of approximately 3 to 6, uptake increases linearly with K_{OW} and is thought to be controlled by membrane permeation. Above log K_{OW} of 6, the rate of increase slows and eventually uptake begins to decrease with increasing K_{OW} as the large molecular size of the most lipophilic compounds begins to impede diffusion in aqueous phases of the fish. Elimination (log k_e^{-1}) is controlled by membrane diffusion at low K_{OW} values and then becomes linear (diffusion controlled, see Equation 3.5) between log K_{OW} values of approximately 2.5 to 6.5. Above this point, the lipid solubility begins to deviate from a perfect correlation with n-octanol-based predictions: K_{OW} becomes an increasingly poorer surrogate of lipid-water partitioning as compounds become more and more lipophilic. Chiou (1985) explained this deviation quantitatively with the **Flory-Huggins Theory** relating solubility to solvent molecular size. The discrepancy between molecular size of n-octanol and lipids of organisms becomes increasingly important for larger, more lipophilic compounds. The net result of these factors on uptake and elimination is shown in the bottom panel of Figure 4.4. Below approximately log K_{OW} of 3, bioaccumulation of the most water soluble compounds is controlled by permeation of the membrane. Bioaccumulation is determined primarily by diffusional processes between log K_{OW} values of 3 to 6. Above this range, the relationship is strongly influenced by effects of molecular size on diffusion and BCF drops with increasing log K_{OW}.

Other factors contribute to bioavailability of organic compounds from water. The above model would not apply for compounds undergoing extensive biotransformation. Bioavailability of ionizable organic compounds would be influenced by pH, as already described, and soon to be described in more detail for ingested, ionizable contaminants. QSAR models can be developed which include lipophilicity and ionization (e.g., Lipnick, 1985). Other factors may also be included in QSARs. Classic models based on the Hansch equation equate bioactivity to hydrophobicity, electronic, and steric qualities of molecules (Lipnick, 1995). **Linear solvation energy relationships** (LSERs) are based on molecular volume, ability to form hydrogen bonds, and polarity or ability to become polarized (Blum and Speece, 1990).

Solid Phases

For ionizable contaminants, gastric pH can influence availability with the direction of effect being determined by the pK_a of the contaminant.[5] (The pK_a is $-\log_{10}$ of the ionization constant (K_a) for a weak, Brønsted acid where $K_a = ([H^+][X^-])/[HX]$.) Weakly acidic, organic compounds with pK_a values greater than 8 are unionized in the human gastrointestinal tract but bioavailability of those with pK_a values between 2.5 and 7.5 is pH sensitive (Gilbaldi, 1991). According to the **pH-partition hypothesis** (Shore et al., 1957; Wagner, 1975), bioavailability is determined by diffusion of the unionized form from the gastrointestinal lumen across the "lipid barrier" of the gut lining and into the tissues as determined by pK_a and pH. The proportion of the compound remaining unionized can be estimated with the **Henderson-Hasselbalch relationship** for monobasic acids (Equation 4.4) and monacidic bases (Equation 4.5) (Wagner, 1975).

$$f_u = \frac{1}{1+10^{pH-pK_a}} \tag{4.4}$$

----

[5]Of course, the bioavailability as influenced by gut pH is more complicated than this because the pH conditions of various regions of the gut are quite different. Weakly basic drugs will be more rapidly absorbed in the small intestine than in the acidic stomach. However, many acidic compounds can also be absorbed more effectively in the small intestine than the stomach because of the large amount of surface area of the small intestine relative to the stomach. Consequently, **gastric emptying rates** (the rate at which the contents of the stomach are emptied into the small intestine) will also influence bioavailability for many substances, e.g., paracetamol (Heading et al., 1973) by modifying the time that a compound remains under different pH conditions.

$$f_u = \frac{1}{1+10^{pK_a-pH}} \tag{4.5}$$

However, the ionized form may also contribute to bioavailability (Wagner, 1975; Gibaldi, 1991). In such cases, the diffusion rates and amounts of both unionized and ionized forms of the contaminant contribute to estimation of bioavailability.

The K_{OW} influences bioavailability of lipophilic compounds in food. Spacie et al. (1995) noted a maximum availability of contaminant in food at log K_{OW} values of 6 with uptake being lower for very hydrophobic and large compounds. They discuss several studies indicating low bioavailability to fish of several lipophilic organic contaminants in food. Donnelly et al. (1994) indicated that organic compounds in soils with log K_{OW} values of 4 to 7 such as PCBs are quickly absorbed to soils and, consequently, are not readily available to terrestrial plants. In contrast, compounds such as many pesticides with log K_{OW} values of approximately 1 to 2 are taken up more easily by plants.

In sediments, bioavailability to benthic species usually decreases with increasing log K_{OW} (Landrum and Robbins, 1990). This is likely due to the enhanced partitioning of nonpolar organic compounds to the sediment solid phases with consequent low concentrations in the interstitial waters. Any increase in sediment organic carbon content can diminish the bioavailability of nonpolar organic compounds much as AVS decreases bioavailability of metals in anoxic sediments.[6] A maximum bioavailability has been noted at a log K_{OW} of approximately 6 for some series of chlorinated hydrocarbons (Landrum and Robbins, 1990).

BIOLOGICAL QUALITIES INFLUENCING BIOACCUMULATION

Temperature-Influenced Processes

Temperature is perhaps the most widely studied and important factor affecting the general physiology of individuals. This being the case, it should be no surprise that temperature has also been found to influence biochemical and physiological processes associated with bioaccumulation. Indeed, the strong positive relationship noted in the 1960s between metal (zinc or radiocesium) excretion rate and temperature-dictated metabolic rates of poikilotherms (Mishima and Odum, 1963; Williamson, 1975) and homeotherms (Pulliam et al., 1967; Baker and Dunaway, 1969) led researchers to explore ^{65}Zn elimination as a means of measuring metabolic rates of free-ranging individuals. Unfortunately, such use was confounded because free-ranging animals and laboratory-maintained animals often differed in

---

[6]This general partitioning of contaminants between solid and dissolved phases of sediments with consequent effects on bioavailability and toxicity (bioactivity) is the foundation for the equilibrium partitioning approach to sediment criteria development. See Shea (1988) and Di Toro et al. (1991) for more details.

other ways that significantly influenced elimination rates (e.g., Pulliam et al., 1967). These early studies identified much inexplicable variability and bias in results. It should be remembered that temperature also determines important rates such as those for feeding, growth, and egestion in addition to important cellular qualities such as membrane fluidity and lipid composition.

Generally, increases in temperature within normal physiological ranges have been shown to increase bioaccumulation, e.g., mercury in mayfly nymphs (Odin et al., 1994), cadmium and mercury in molluscs (Tessier et al., 1994), cadmium in Asiatic clams (Graney et al., 1984), and DDT in rainbow trout (Reinert et al., 1974). Cesium (^{134}Cs) uptake was highest at temperatures optimal for food consumption and growth of rainbow trout (Gallegos and Whicker, 1971). The biological half-life ($t_{1/2}$) for elimination of ^{134}Cs from rainbow trout increased with increased water temperatures according to an exponential relationship,[7] $t_{1/2} = (\text{Constant})e^{-0.106t}$ where t = temperature in °C (Ugedal et al., 1992). For the rainbow trout, retention of methylmercury was approximately 1.5 times longer at 0.5 to 4.0°C than at water temperatures of 16 to 19°C (Ruohtula and Miettinen, 1975).

Some studies report no effect of temperature changes on bioaccumulation kinetics. For example, methylmercury uptake and elimination by freshwater clams were not significantly affected by temperature (Smith and Green, 1975). Effects of temperature are sometimes complex and inconsistent with the general trends noted here.

Watkins and Simkiss (1988) gave a fascinating example of such a complication. They, like others, found enhanced accumulation of a contaminant (zinc in the marine mussel, *Mytilus edulis*) with an increase in water temperature (10°C increase to 25°C). But they also examined the effect of fluctuating temperatures between 15 and 25°C, and found that bioaccumulation of zinc was even higher than the constant 25°C treatment! They hypothesized that this result was linked to the shifts in zinc among various pools within the mussels. Both free and ligand-associated zinc are present in the mussels: $Zn^{2+} + L^{2-} \rightleftarrows ZnL$. When water temperatures increase, the equilibrium for this complexation shifts to favor more zinc existing free of the biochemical ligands. This zinc then becomes more available for incorporation into granules. Upon cooling, zinc entering from outside the animal establishes an equilibrium with ligands again, thus replenishing the zinc lost to granules. As temperatures increase and decrease, the process is repeated with enhanced accumulation of zinc in granules. The cyclic association with ligands, shift toward dissociation from ligands, and sequestering in granules ratchets zinc concentrations higher than if the mussel were left at a higher temperature of 25°C.

---

[7]Both power and exponential relationships will be discussed in this chapter. A **power relationship** is a mathematical relationship in which the Y variable is related to the X variable raised to some power. For example, $Y = aX^b$. A power relationship can be linearized and conveniently fit by linear regression by taking the log X and log Y in the regression ($\log Y = b \log X + \log a$). In contrast, an **exponential relationship** is a mathematical relationship in which the Y variable is related to some base raised to the X variable, i.e., $Y = a10^{bX}$. An exponential relationship is transformed to the form $\log Y = bX + \log a$ as done earlier to fit exponential (first order) elimination kinetics. Newman (1993) provides details on fitting these models using linear regression.

Allometry

Allometry, the study of size and its consequences (Huxley, 1950), can also be important to consider for bioaccumulation. Metabolic rate and a myriad of other anatomical, physiological, and biochemical qualities of organisms change with size (see Adolph [1949] and Heusner [1987] for more detail) and, in so doing, uptake, transformation, and elimination rates are modified. The commonly observed consequence is size-dependent bioaccumulation. Unfortunately, because age and size are correlated in most species, allometric effects are often confused with age or exposure duration effects in surveys of bioaccumulation. Regardless, many studies detail allometric effects on contaminant uptake (Newman and Mitz, 1988; Schultz and Hayton, 1994), biotransformation (Walker, 1987; Figure 4.5), elimination (Mishima and Odum, 1963; Reichle, 1968; Gallegos and Whicker, 1971; Ugedal et al., 1992), and general bioaccumulation (Boyden, 1974; 1977; Landrum and Lydy, 1991; Warnau et al., 1995). These have been reviewed for metals bioaccumulation by Newman and Heagler (1991), for linkage to metabolic rate by Fagerström (1977), and for scaling of pharmacokinetic parameters by Hayton (1989). Most resort to the classic power model for **scaling**, the manipulation of allometric data (size versus some physiological, morphological, or biochemical quality) to produce a quantitative relationship.

The general power function used in allometric scaling is the following,

$$Y = aX^b \tag{4.6}$$

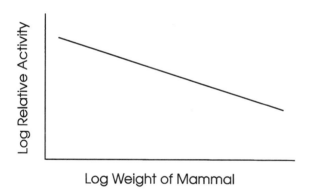

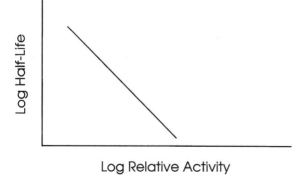

FIGURE 4.5. The scaling of monooxygenase activity (top panel) and the relationship between relative monooxygenase activity and biological half-life of xenobiotic metabolism (bottom panel). (Composite and modified Figures 2 and 3 from Walker [1978].)

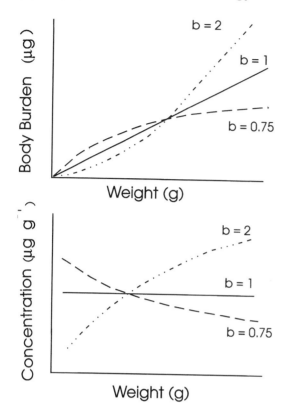

FIGURE 4.6. The relationships between organism size (weight) and body burden (amount per individual)(top panel) and concentration (amount per g of tissue) of metal. Modified from Figure 13 in Newman (1995) with slight exaggeration of curvature of the concentration versus size plots to make clear the nonlinear nature of two of these curves (b = 2 and b = 0.75). The point of intersection for both plots occurs where X = 1. (Reprinted with permission from *Quantitative Methods in Aquatic Ecotoxicology.* Copyright Lewis Publishers, an imprint of CRC Press, Boca Raton, Florida.)

where Y = some quality being scaled to size, X = some measure of individual (or species) size, and a and b are constants normally derived by regression analysis. This equation may be used to link morphological (gill surface area), physiological (blood flow rate or gill ventilation rate), or biochemical (monooxygenase activity) qualities to size. It has also been used to model contaminant body burden to animal size. To express the relationship for bioaccumulation in concentration units, Equation 4.6 can be easily converted by dividing both sides by mass to generate Equation 4.7.

$$Y = aX^{1-b} \qquad (4.7)$$

where X, a, and b = same as in Equation 4.6 but Y = concentration, not body burden. Much has been made of the values estimated for b for scaling bioaccumulation and associated parameters, as this parameter is a major theme in the general physiological literature.

In the mid-1970s, Boyden (1974, 1977) compiled metal body burden data for molluscs and established the power model for scaling contaminant accumulation. He identified three classes of models based on their associated b-values (Figure 4.6). One class had b-values of 1 and reflected a simple proportionality. There was a constant number of binding sites in the tissues regardless of size: body burden increased linearly with weight of the organism, and concentration was independent of weight. Another relationship with b-values in the general range of 0.77 had body burdens and concentrations that changed in a nonlinear fashion with weight. The concentration was higher for small individuals than large individuals. The b-value of 0.77 suggested a linkage to metabolic rate which, in general, has a b-value of approximately 0.75 during scaling. However, Fagerström (1977) quickly pointed out that one cannot compare "b-values" for states (body burden) and fluxes (metabolic rate) in this manner. He demonstrated mathematically that one would expect a b-value of 1 for contaminant burdens driven by metabolic rate. The elegant hypothesis of Boyden was an inadequate explanation. The final class noted by Boyden had a b-value of 2 or greater (burden related to the square of body weight) and reflected a gradual increase in concentration with an increase in size. He suggested that elements such as cadmium which are subject to rapid removal from circulation and very avid binding in some tissues will conform to this class of relationships. He suggested incorrectly that these relationships (b-values) would be constant for species-element pairs. Soon after Boyden proposed this constancy of b-values, Cossa et al. (1980) and Strong and Luoma (1981) demonstrated considerable spatial and temporal variation in b-values. Further, Newman and Heagler (1991) added to and reanalyzed Boyden's original data set and found no clear evidence for distinct classes of relationships based on b-values. They found a generally skewed distribution of b-values with medians in the range of 0.80 to 0.83. They also identified several sources of bias in Boyden's approach. Regardless, Boyden made a major contribution by clearly establishing the power model for scaling body burdens, and hypothesizing plausible and testable underlying mechanisms. His approach is used often to normalize body burden/concentration data taken during surveys of organisms of differing sizes. In such cases, the empirically-derived constants are adequate for normalization of data. Regardless of its value, it should not be forgotten that the foundation for such scaling is empirical.

Another equally important use of scaling is computer modeling to predict general behaviors of bioaccumulation dynamics and to isolate important factors controlling bioaccumulation kinetics. For example, a power model was used to model ^{137}Cs half-life in humans of different sizes (Eberhardt, 1967). In some such cases, estimation of the exact values for parameters may not be critical and the range of expected values may be sufficient. Today, more complicated allometric explanations and relationships than Boyden's, usually embedded in physiologically-based pharmacokinetics (PBPK) models, are used to predict size effects on bioaccumulation (Barber et al., 1988; Hayton, 1989; Schultz and Hayton, 1994). Examples of such complicated models are given in Figure 4.7. Temperature, scaling and other important factors can be included in these PBPK models. However, as models become more complicated, they require increasingly more and more estimated parameters for the system being modeled.

Other Factors

Many other factors can complicate prediction of bioaccumulation. Bioaccumulation may differ between sexes. The female fin whales studied by Aguilar and Borrell (1994) eliminate organochlorine compounds via transfer to young before birth and during nursing; this is obviously not an avenue of elimination for males. Conse-

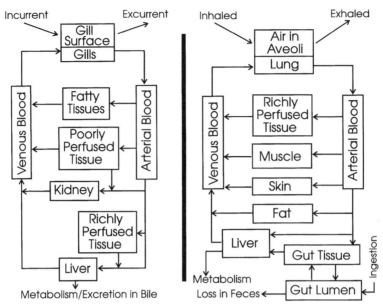

FIGURE 4.7. Two PBPK models. The model to the left involves the fish uptake of a xenobiotic across the gills, distribution among 5 compartments and blood, and loss via the liver (Nichols et al., 1990). The model on the right involves styrene inhalation and ingestion by a mammal, distribution into a series of compartments, and loss from liver metabolism and defecation (Paterson and Mackay, 1987). In the original publications, the fish model was formulated as a clearance volume-based model and the mammalian model was formulated as a fugacity-based model.

quently, older female whales have lower concentrations than older male whales. Diet, as alluded to above, can also influence bioavailability. Shifts in diet associated with ecological or developmental qualities of an organism have consequences relative to bioaccumulation. Acclimation of euryhaline species can change their ability to cope with a contaminant; killifish uptake of pentachlorophenol and excretion of a pentachlorophenol glucuronide conjugate decreased with the lowering of salinity (Tachikawa and Sawamura, 1994). Lipid content of an individual may also influence bioaccumulation. Rainbow trout with high total body lipid content (21%) accumulated more pentachlorophenol and had lower elimination rate constants than those with lower (13%) lipid content (van den Heuvel, 1991).

Inorganic toxicants can be influenced by normal regulatory processes associated with essential chemicals. For example, potassium body burden of humans has a strong influence on the accumulation of its analog, cesium (Leggett, 1986). Other essential elements[8] such as zinc are carefully regulated within the body. For example, a **cysteine-rich intestinal protein (CRIP)** exists to enhance the uptake of

---

[8]Mertz (1981) lists the **essential elements** as H, Na, K, Mg, Ca, V, Cr, Mo, Mn, Fe, Co, Ni, Cu, Zn, Cd(?), C, B, Si, Sn(?), N, P, As, O, S, Se, F, Cl, and I.

zinc by cells in the intestine wall of mammals (Roesijadi and Robinson, 1994). Regulation of essential elements influences availability or effects of nonessential analogs. An excess of an essential element analog such as cadmium can cause symptoms of an apparent (zinc) deficiency (Neathery and Miller, 1975): an excess of zinc may lessen toxic effects of cadmium (Leland and Kuwabara, 1985). Similarly, copper and zinc interact relative to effects on algal cell function (Rueter and Morel, 1981). Many essential elements and some of their analogs are defined as **biologically determinant**; their concentrations in the organisms remain relatively constant over a wide range of environmental concentrations (Reichle and Van Hook, 1970). Other elements are **biologically indeterminant** and their concentrations in organisms are directly proportional to environmental concentrations.

Perhaps more convoluted, but as telling, is the role of phosphate biochemistry on the accumulation of arsenate in marine species. In the phosphate deficient waters of the Great Barrier Reef, arsenate enters into the normal biochemical pathways involved in phosphate assimilation and is incorporated into zooxanthellae in the mantel of the giant clams, *Hippopus hippopus, Tridacna maima,* and *T. derasa* (Benson and Summons, 1981). Arsenic then accumulates to extraordinary levels, gaining entrance into the clam via the biochemistry of the symbiotic algae.

SUMMARY

In this brief chapter, factors influencing bioaccumulation and bioavailability of contaminants were discussed along with several methods of measuring bioavailability. Solid and dissolved sources of organic and inorganic contaminants were detailed relative to bioavailability. Finally, the importance of temperature- and size-dictated changes in biological functions and structures were discussed. Other important factors influencing bioaccumulation were identified including, sex, diet, lipid content, and elemental essentiality. Many details associated with accumulation from trophic exchange were omitted as they are discussed thoroughly in the next chapter.

Selected Readings

Campbell, P.G.C., A.G. Lewis, P.M. Chapman, A.A. Crowder, W.K. Fletcher, B. Imber, S.N. Luoma, P.M. Stokes, and M. Winfrey. *Biologically Available Metals in Sediments.* NRCC No. 27694, NRCC/CNRC Publications, Ottawa, Canada, p. 298, 1988.

Hamelink, J.L., P.F. Landrum, H.L. Bergman, and W.H. Benson, Eds., *Bioavailability. Physical, Chemical and Biological Interactions*, CRC Press, Inc., Boca Raton, FL, 1994, p. 256.

Hansen, L.G. and B.S. Shane. Xenobiotic Metabolism, in *Basic Environmental Toxicology*, Cockerham, L.G. and B.S. Shane, Eds., CRC Press, Boca Raton, FL, 1994.

Luoma, S.N. Can we determine the biological availability of sediment-bound trace elements? *Hydrobiolia.* 176/177, pp. 379–396, 1989.

Newman, M.C. and M.G. Heagler. Allometry of Metal Bioaccumulation and Toxicity, in *Metal*

Ecotoxicology, Concepts and Applications, Newman, M.C. and A.W. McIntosh, Eds., Lewis Publishers, Chelsea, MI, 1991.

Spacie, A., L.S. McCarty, and G.M. Rand. Bioaccumulation and Bioavailability in Multiphase Systems, in *Fundamentals of Aquatic Toxicology. Effects, Environmental Fate, and Risk Assessment*, 2nd Ed., Rand, G.M., Ed., Taylor & Francis, Washington, DC, 1995.

Bioaccumulation from Food and Trophic Transfer

Far too often food chains have been envisioned as mechanisms operating solely to concentrate pollutants as they move from prey to predator. Less often are they objectively recognized as ecological processes, with the net effect of concentration or dilution of materials during their transport along the food chain being dependent upon a complex of biological variables.
REICHLE AND VAN HOOK (1970)

INTRODUCTION

As evidenced by Minamata disease and DDT poisoning of birds, the transfer of contaminants through trophic webs[1] can have undesirable consequences to top predators. Some contaminants such as mercury and DDT display **biomagnification,**[2] an increase in contaminant concentration from one trophic level (e.g., prey) to the next (e.g., predator) due to accumulation from food. The possibility of biomagnification must be considered in any thorough assessment of ecological or human risk. However, because biomagnification played a pivotal role in our awakening to environmental issues, it is sometimes invoked as a ruling theory when equally plausible, alternate explanations are present (Moriarty, 1983; Beyer, 1986; Laskowski, 1991). For example, predators tend to live longer than prey species and have more time to accumulate contaminants than do prey. The result may be higher concentrations in predators than prey (Moriarty, 1983). Predators are often larger than prey and allometric effects on bioaccumulation can result in higher

---

[1]The term foodweb is preferable to food chain in discussions of trophic transfer. The concept of a web of interactions is often more accurate than that of an orderly transfer from one level to the next highest level only. Many species feed on an array of prey that, in turn, have equally complex feeding strategies.

[2]The terms, **bioamplification** and **trophic enrichment** are infrequently used instead of biomagnification. The general term, "bioaccumulation" is occasionally used incorrectly as a synonym for biomagnification. Sometimes "biomagnification" is used to describe field observations of increasing concentrations with increasing trophic level regardless of the ambiguity about the magnitude of uptake from food relative to water. It is not used in that context here as the historical literature clearly associates trophic transfer of contaminant with biomagnification. Results of biomagnification studies using this term otherwise can be needlessly confusing and should be scrutinized carefully to determine the inferential strength of associated conclusions.

concentrations of some contaminants in predators relative to prey (Moriarty, 1983). Also, for lipophilic contaminants, higher lipid content in predators than prey can also result in increases in contaminant concentrations with trophic level. Lower foodweb organisms tend to grow faster than those higher in the foodweb; growth dilution may be more pronounced at lower levels than at higher levels (Huckabee et al., 1979). Difficulties in defining trophic status, especially if species change their feeding habits with age (Huckabee et al., 1979), can confound and render subjective conclusions of biomagnification. Also most field studies of bioaccumulation do not distinguish between water and food sources, pressing many researchers toward unjustified speculation regarding biomagnification. Finally, some communities show such wide variation in concentrations among species of a particular trophic level that trends noted among trophic levels are often questionable (Beyer, 1986; Laskowski, 1991). Such calculations and their interpretations are frequently biased toward biomagnification. Reports of biomagnification must be read carefully in order to dissect away unintentional biases of investigators. Nevertheless, because of the important consequences of biomagnification, the concept is well worth investigating.

Biomagnification is not required in order to have adverse effects as a consequence of trophic transfer. If contaminant concentrations are very high in a food item, species farther up the trophic web may still be exposed to concentrations sufficient to produce an adverse effect. As discussed in the first chapter, this was the case with humans afflicted with Itai-Itai disease. Another example involves metals and metalloids that accumulate to extremely high concentrations in materials covering submerged surfaces in lentic and lotic systems (Newman et al., 1983, 1985). These materials, including the associated *aufwuchs* (periphyton), are ingested by an important grazer/scrapper guild of freshwater organisms and have the potential to cause adverse effects even in the absence of biomagnification (Newman and Jagoe, 1994). Also, the flux into an organism, not only the net concentration in that organism, may contribute to determining the adverse outcome of toxicant exposure.

Biomagnification is only one of three possible outcomes for trophic transfer of contaminants. Concentrations may be similar in both predator and prey: there may be no statistically significant upward or downward trend in concentrations. Alternatively, as is frequently the case, the contaminant concentration may decrease as trophic level increases. During each transfer, the required balance among ingestion rate, uptake from food, internal transformations, and elimination does not exist for the conservative transfer of a contaminant. In such situations, concentrations decrease with each trophic exchange. Diminution with increasing trophic level is called **trophic dilution** (Reichle and Van Hook, 1970) or **biominification** (Campbell et al., 1988). **Bioreduction** (Nott and Nicolaidou, 1993) has also been used in describing trophic dilution but the meaning shifts slightly to focus on rendering the contaminant less bioavailable in the prey biomass with a consequent ineffective assimilation in the predator.

This chapter was developed to specifically address bioaccumulation from food because of the importance of this phenomenon, especially in terrestrial systems where food can be the predominant source of contaminants to biota. The theme brought out by the above quote that bioaccumulation from food is a complex process that has been viewed too simplistically in the past, will be enriched in this chapter. Unlike the last chapter, the focus will be on the sequential transfer of contaminants among species rather than bioaccumulation in individuals. Details about differential transfer of radioisotopes and organic isomers will be provided. Means of measuring trophic status will be discussed too because of its crucial role in bioaccumulation studies.

QUANTIFYING BIOACCUMULATION FROM FOOD

Assimilation from Food

Assimilation of a contaminant in food by individuals is one estimate of contaminant transfer between members of different trophic levels. It can be quantified as already described for bioavailability from food and used to estimate the magnitude of trophic transfer between the levels represented by the food item (e.g., prey or primary producer) and consumer (e.g., predator or grazer). The amount of contaminant in the organism after a defined time, usually the time to reach a practical steady state concentration, is divided by the total amount of contaminant fed to the organism to estimate assimilation efficiency (see Chapter 3 for details). Similarly, the amount ingested and amount egested over a period of time can be used to estimate assimilation efficiencies. Less frequently, the area under the curve (AUC) method is used.

Assimilation can be measured by a **twin tracer technique**, a technique that introduces a radiotracer of the substance to be assimilated simultaneously with an inert radiotracer to which assimilation is compared (Weeks and Rainbow, 1990). The inert tracer is not assimilated to any appreciable amount and the retention of the assimilated tracer is quantified relative to it. For example, the assimilation of a ^{14}C-labelled organic compound may be compared to the amount of notionally inert ^{51}Cr fed simultaneously to a zooplankter (Bricelj et al., 1984). This technique is based on the assumption that the two radiotracers pass through the gut at approximately the same rate and the inert tracer is not absorbed to any significant extent. Both are incorporated into the same food items together to foster identical movement through the animal's gut.[3] (As with all such uses of radiotracers, the **specific activity concept** is central to accurate implementation: the radionuclide used to reflect movement of the stable nuclide [e.g., ^{14}C for C] behaves identically in chemical and biological processes as its nonradioactive nuclide [e.g., stable C].) In the absence of such an effective pairing of isotopes for exposures, a radioisotope may be fed to an organism and the difference between the amount of the single isotope ingested and egested over a time course used to estimate assimilation effi-

---

[3]With sediments, it may be difficult to incorporate dual tracers identically. The radiotracers may be distributed differently among sediment fractions. It then becomes important to account for feeding selectivity as some fractions (e.g., fine, organic carbon-rich particles) may be processed differently from others (e.g., large particles).

ciency. The time to pass through the gut is estimated and the amount remaining after gut evacuation is assumed to be assimilated. The advantage of the pairing with an inert tracer is that it clearly indicates when sampling of egested materials may be stopped — when all or most of the inert tracer has been egested.

The assimilation efficiencies for various foods and members of a foodweb are pieced together to predict trends in trophic transfer. They are used to complement field surveys and more complex laboratory experiments as described below.

Trophic Transfer

Defining Trophic Position

Inaccuracies in identifying trophic status of species occur in the bioaccumulation literature. For this reason, a few paragraphs will be spent defining some methods for identifying trophic structure.

One way to reduce ambiguity is to conduct a laboratory experiment so that the trophic structure is imposed by design. The disadvantage is the artificial context and, perhaps, unrealistic depiction of trophic dynamics relative to natural communities. Recognizing this flaw, field surveys are conducted to complement or test conclusions from highly structured, laboratory experiments.

In field surveys, the prevalent method for determining trophic status is extraction and abstraction of information from the natural history literature. This is adequate for interpreting results of field surveys only to a degree. Much uncertainty can remain because species in communities have complex feeding strategies that change with age, time, and community composition. Some of this uncertainty can be further reduced with visual observations of species interactions and analysis of gut content (Kling et al., 1992). Regardless, most such renderings and associated quantitation simplify trophic status to a discrete state for specific species, i.e., primary producer, primary consumer, secondary consumer, and so forth. This would be an inadequate description for species that feed at several levels.

Another, more accurate method of quantifying trophic status has become widely available during the last 15 years. **Isotopic discrimination**[4] of light elements such as C, N, and S occurs during trophic transfers, providing an opportunity for quantifying trophic status in natural communities (Fry, 1988; Hesslein et al., 1991). Isotopic discrimination tends to reduce the amount of lighter isotopes (^{12}C, ^{14}N, ^{32}S) in organisms relative to the heavier isotopes (^{13}C, ^{15}N, ^{34}S) during trophic exchange. The lighter isotopes are eliminated from the organisms more readily than the heavy isotopes. Carbon isotopic changes with trophic transfer are

---

[4]Isotopic discrimination is the rate of or extent to which participation in some biological or chemical process depends on the mass of the isotope. Isotopic discrimination or the isotope effect results from differences in the kinetic energy of associated molecules with slightly different masses because they contain different isotopes of an element, e.g., ^{14}N instead of ^{15}N. Differences can also result from distinct vibrational and rotational qualities of molecules (Wang et al., 1975). Discrimination between isotopes is measured as a **discrimination ratio** with the ratio of 1 if no discrimination is occurring.

measurable but smaller in magnitude than those of nitrogen isotopes (Hesslein et al., 1991; Cabana and Rasmussen, 1994). Carbon isotopic ratios vary more within a trophic level too, making them less effective indicators of trophic status than nitrogen isotopes (Hesslein et al., 1991). However, carbon isotopes do provide information on carbon sources (C3 and C4 photosynthetic pathways) as well as trophic structure. Sulfur isotopic ratio may change only slightly (Hesslein et al., 1991) or not at all (Fry, 1988) with trophic exchange but change significantly with sulfur source, making it a better indicator of sulfur source than trophic status. For example, Hesslein et al. (1991) used this technique to identify food bases for broad whitefish (*Coregonus nasus*) and lake whitefish (*Coregonus clupeaformis*). Changes in stable nitrogen isotope composition seem to be the best indicator of trophic status, e.g., Rau (1981), Fry (1988), Hesslein et al. (1991), and Kling et al. (1992).

Changes in nitrogen isotopes (^{14}N and ^{15}N) in organisms within a trophic web are quantified relative to the isotopic ratio of air (Equation 5.1).

$$\delta^{15}N = 1,000 \cdot \left\{ \frac{(^{15}N_{sample})/(^{14}N_{sample})}{(^{15}N_{air})/(^{14}N_{air})} - 1 \right\} \qquad (5.1)$$

The units for $\delta^{15}N$ are ‰ ("per mill" or "per millage"). The $\delta^{15}N$ increases with each trophic exchange because the lighter isotope (^{14}N) is more readily excreted. Isotopic discrimination of nitrogen is not associated with uptake, metabolic breakdown to amino acids, or deamination (Minagawa and Wada, 1984). Nor is there any evidence of discrimination in the urea cycle or in the formation of uric acid: the discrimination is relatively independent of the diverse processes of urine formation (Minagawa and Wada, 1984).

Changes in $\delta^{15}N$ at each trophic exchange can range widely. Minawaga and Wada (1984) describe a range from 1.3 to 5.3 ‰ per trophic exchange. However, the average increase is 3.4 ‰ per trophic exchange (Minagawa and Wada, 1984; Cabana and Rasmussen, 1994). The $\delta^{15}N$ may also change with animal age (Minagawa and Wada, 1984). Despite these considerations, $\delta^{15}N$ can readily reveal trophic structure of diverse field communities (Figure 5.1) and enhance interpretation of contaminant movement in ecological communities. Indeed, it may also be used with human populations to identify diet, e.g., $\delta^{15}N$ for human populations with different but unquantified amounts of diet coming from marine and terrestrial systems (Scheninger et al., 1983). Finally, this approach has the advantage of quantifying intermediate trophic positions. The simplistic assignment of a species to a single trophic level can be avoided. It can be augmented with a dual isotope ($\delta^{15}N$ and $\delta^{13}C$) approach to also imply major vegetable sources of biomass (Fry, 1991) to members of the ecological community.

Estimating Trophic Transfer

Perhaps the simplest way to quantify biomagnification is to divide the contaminant concentration at trophic level n (C_n) by that at the next lowest trophic

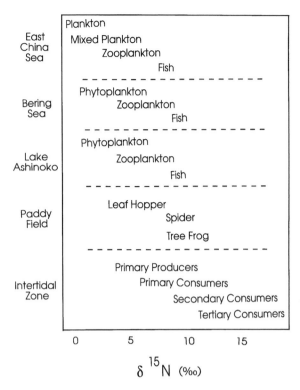

FIGURE 5.1. The change in δ ^{15}N (‰) with increase in trophic level for the diverse communities of the East China Sea (an oligotrophic sea with significant amounts of nitrogen fixation at the time of phytoplankton sampling which brought the δ ^{15}N to near atmospheric levels), the Bering Sea (North Atlantic Ocean), Lake Ashinoko (freshwater lake), a paddy field (terrestrial system) in Konosu, Japan, and an intertidal zone. (A composite and modification of Figures 1 and 2 in Minagawa and Wada [1984].)

level (C_{n-1}) (e.g., Bruggeman et al., 1981; Laskowski, 1991). This **biomagnification factor (B)** may include individual organisms of a known or assumed trophic status.

$$B = \frac{C_n}{C_{n-1}} \tag{5.2}$$

Such a biomagnification factor is based on the assumption that concentrations have reached steady state in the sampled individuals and the capacity is the same for the sampled individuals (e.g., individuals have the same lipid content if lipophilic compounds are being studied). This B can also be expressed in terms of the rate constant-based bioaccumulation model.

$$B = \frac{C_n}{C_{n-1}} = \frac{\alpha\, f}{k_e} \tag{5.3}$$

where α = assimilation efficiency for the ingested species, f = the feeding rate (mass of food · mass of individual^{-1} · time^{-1}), and k_e = the elimination rate constant (Bruggeman et al., 1981). Note that assimilation efficiency often diminishes as feeding rate increases (Clark and Mackay, 1991) so this parameterization is conditional.

Biomagnification factors may be estimated for samples of many organisms from two trophic levels by using a body mass-weighted mean concentration for the two trophic levels,

$$B' = \frac{\left(\sum\limits_{i=1}^{x} C_{n.i}\, w_{n.i}\right)\left(\sum\limits_{j=1}^{z} w_{n-1.j}\right)}{\left(\sum\limits_{j=1}^{z} C_{n-1.j}\, w_{n-1.j}\right)\left(\sum\limits_{i=1}^{x} w_{n.i}\right)} \qquad (5.4)$$

where w = the weight of individuals sampled from the n or n − 1 trophic levels.

Definitions for some bioaccumulation indices acknowledge that water may also contribute to differences in concentrations measured for individuals at different trophic levels. This is necessary in field surveys that do not isolate water and food sources to individuals. The **bioaccumulation factor (BF)**[5] is such an index that has the same form as Equation 5.2 except the source isn't necessarily food alone.

$$BF = \frac{C_{organism}}{C_{source}} \qquad (5.5)$$

where $C_{organism}$ = concentration in the organism resulting from uptake from food and water sources, and $C_{source(s)}$ = concentration in the reference source of contaminant. For lipophilic organic compounds, concentrations may be expressed as a mass per mass of lipid basis. If the BF is greater than 1, biomagnification may be occurring. If it is less than 1, trophic dilution is suggested although other factors such as allometric processes or growth dilution may be contributing to the changes. Experimental designs may employ tandem exposures of two subsets of individuals to either contaminant in water alone or contaminant in both food and water. The differences in the treatment results (BCF derived from the water alone treatment and BF derived from the food plus water treatment) can then be used to assess the significance of trophic exchange relative to accumulation from water alone (Bruggeman et al., 1981). Assuming that uptake from water is significant at the primary producer level only and a distinct layering of trophic levels with minimal overlap, the transfer up the trophic structure can be calculated with knowledge of the concentration at the lowest trophic level and estimates of assimilation efficiencies and feeding rates at each trophic level (Ramade, 1987; Newman, 1995).

Commonly, the concentrations at trophic levels are referenced to that of the ultimate or lowest defined source of contaminant. For example, a **concentration factor (CF)** may be estimated for all trophic levels with the concentration in the water used in the denominator, CF = C_n/C_{water}. Reichle and Van Hook (1970) took

---

[5]Note that bioaccumulation factor designated BSAF is used in a more restricted sense in this book to mean the ratio of concentration in an organism and concentration in sediments (Chapter 3). Also, BAF is the same as BF throughout the literature. For the sake of clarity only, BF will be used here to designate the ratio derived under the assumption that both water and food sources may be contributing to body concentrations of contaminants.

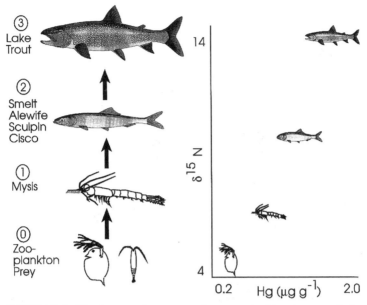

FIGURE 5.2. The increase in mercury from zooplankton to a zooplanktivorous shrimp (*Mysis relicta*) to pelagic forage fish to lake trout (*Salvelinus namaycush*). The trophic structure is quantified with δ ^{15}N: its value increasing roughly 3.4‰ at each trophic exchange. A clear relationship is evident between mercury concentration and trophic status of individuals, indicating biomagnification of mercury. Constructed using data and Figure 2 of Cabana and Rasmussen (1994).

this approach to estimate concentration factors for radionuclides in a terrestrial foodchain with concentrations in plant leaves as the reference concentration. The increase or decrease in concentration relative to that of the source is expressed as a multiple of the source concentration.

The methods described to this point require assignment of species to discrete trophic levels, but this is often an unrealistic simplification of trophic dynamics. It is more effective to quantify trophic transfer by relating concentrations in members of the community to corresponding δ ^{15}N values. Figure 5.2 depicts the results of such an exercise documenting the increase in mercury from zooplankton → a zooplanktivorous shrimp (*Mysis relicta*) → pelagic forage fish → lake trout (*Salvelinus namaycush*) of Canadian lakes (Cabana and Rasmussen, 1994; Cabana et al., 1994). In a survey of seven Canadian shield lakes with various trophic structures leading to lake trout (Figure 5.3), data for mercury concentration and δ ^{15}N fell between the predicted values for a discrete trophic structure. The diagram on the left side of Figure 5.3. shows the expected values for δ ^{15}N based on three possible, discrete structures. (Structure is denoted by the number of levels above zooplankton that the lake trout occupy.)

Regression models can be developed if a continuous variable such as δ ^{15}N is used to quantify trophic transfer. Figures 5.2 and 5.3 suggest that a simple linear

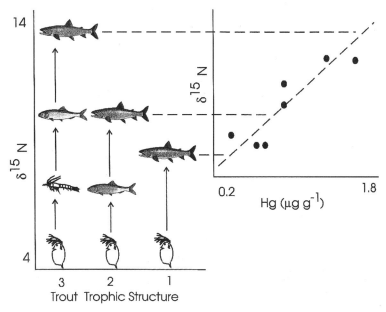

FIGURE 5.3. Correlation between mercury concentration and trophic structure ($\delta\,^{15}$N) for Canadian lakes with different trophic structure relative to lake trout (*S. namaycush*). In some lakes, shrimp and/or forage fish are missing, giving rise to various trophic structures relative to lake trout. Using a discrete assignment of trophic structure, 3 = same structure as shown in Figure 5.2 (zooplankton $\rightarrow$ shrimp $\rightarrow$ forage fish $\rightarrow$ lake trout), 2 = shortened structure without shrimp (zooplankton $\rightarrow$ forage fish $\rightarrow$ lake trout), and 1 = a simple structure with both shrimp and forage fish being insignificant. (Forage fish were predominantly alewife (*Alosa pseudoharengus*), cisco (*Coregonus artedii*), whitefish (*Coregonus clupeaformis, Prosopium cyclindraceum*), sculpin (*Myoxocephalus thompsoni*) and smelt (*Osmerus mordax*).) The panel on the upper right shows the actual mercury concentrations and $\delta\,^{15}$Ns for lake trout from seven Canadian lakes. Note that trout from these lakes are shifted slightly to positions intermediate between the discrete trophic structure levels. (Composite and modification of Figures 2 and 3 of Cabana and Rasmussen [1994].)

regression technique could be used to model the increase in mercury with trophic transfer. Studies of bioaccumulation of organic compounds suggested to Broman et al. (1992) and Rolff et al. (1993) that an exponential model may be more appropriate in some cases.

$$C = a\,e^{b\,(\delta^{15}N)} \tag{5.6}$$

where C = concentration in the organism within the food web, and a and b are estimated parameters. The b is the **biomagnification power**. A positive b indicates a proportional increase in concentration with increase in position within a trophic web (biomagnification) and a negative b indicates a proportional decrease in concentration with increase in position (trophic dilution). Broman et al. (1992) found that decreases and increases in concentrations of various polychlorinated dibenzo-p-dioxins and -dibenzofurans in a pelagic foodweb could be modeled with this relationship.

INORGANIC CONTAMINANTS

Metals and Metalloids

Assimilation studies of metals and metalloids have used the twin tracer technique and single isotope differences between ingested and egested element. Bricelj et al. (1984) fed ^{14}C/ ^{51}Cr labeled algae to clams (*Mercenaria mercenaria*) for 30 to 45 minutes and then measured the amount of both tracers in feces after the clams were transferred to water containing unlabeled algae. The amounts lost after nearly total recovery of the notionally inert ^{51}Cr label in the feces were used to estimate ^{14}C assimilation. Chromium was assimilated in very small amounts but carbon assimilation was approximately 80%. With a similar design (^{60}Co used as an inert tracer for ^{65}Zn assimilation), assimilation of zinc from tissue of the macroalga, *Laminaria digitata,* fed to the amphipod, *Orchestia gammarellus,* was estimated to be nearly 100% (Weeks and Rainbow, 1990). Reinfelder and Fisher (1991) used this design to determine assimilation efficiencies for a sequence of elements when calanoid copepods were fed labeled diatoms. For ^{14}C assimilation, ^{51}Cr was used as the inert tracer. For ^{75}Se assimilation, ^{241}Am was the inert tracer in the twin tracer technique. Assimilation of other elements (^{110m}Ag, ^{109}Cd, ^{32}P, ^{35}S, and ^{65}Zn) were calculated without an inert isotope marker by dividing the amount retained after complete (estimated) gut evacuation by the amount ingested. Assimilation was estimated for sulfur and zinc using algae from cultures in the log phase or stationary phases of growth (Figure 5.4). Given the short residence time in the gut, the elemental content of the algal cytosol was assumed to reflect that available for "liquid" digestion by the zooplankton. This assumption was supported by the clear correlation between assimilation efficiencies for the various elements and the fraction of each present in the algal cytoplasm. Based on these results and earlier work with thorium, lead, uranium and radium (Fisher et al., 1987), Reinfelder and Fisher (1991) suggested that Class B metals and those borderline metals that have greater affinities for sulfur than nitrogen or oxygen (e.g., cadmium, silver, and zinc) will be present in the cytoplasm in higher proportions than Class A metals (e.g., americium, plutonium, and thorium) that have greater affinities for oxygen than nitrogen and sulfur. Assimilation efficiencies are generally higher for the Class B and borderline metals than Class A metals because of their higher availability in the cytoplasm for "liquid" digestion during rapid passage through the zooplankton gut.

Additional factors can influence assimilation efficiencies of metals in other feeding interactions. As mentioned, molluscs can sequester metals in granules and lower their potential for damage. The winkle, *Littorina littorea,* does so with zinc. When tissue of this gastropod is fed to the predatory gastropod, *Nassarius reticula-*

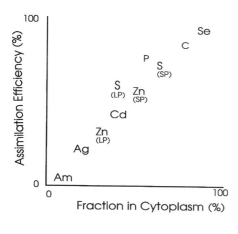

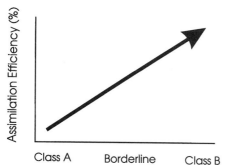

FIGURE 5.4. Assimilation efficiencies of elements contained in diatoms (*Thalassiosira pseudonana*) fed to zooplankton (*Acartia tonsa*, *A. hudsonica*, or *Temora longicornis*) plotted against the amount of each element associated with the cytoplasm of the diatoms (top panel). SP = data derived using stationary phase (senescent) diatom cultures and LP = results derived from diatom cultures in the log phase of growth. (Modified from Figure 1 of Reinfelder and Fisher [1991].) The bottom panel depicts the general trend in assimilation efficiency hypothesized relative to the Class B – borderline – Class A classification of metals (See Chapter 3).

tus, only a minor proportion of the zinc in the associated granules is available for assimilation in the predator (Nott and Nicoladou, 1993). In contrast, there are considerable decreases in magnesium, phosphorus, and potassium in granules during passage through the predator's gut. The detoxification of zinc in the winkle tissue by incorporation into granules decreases the assimilation efficiency of zinc during predation. The assimilation of other metals associated with invertebrate intracellular granules probably is influenced also. Despite the widespread detoxification of metals via sequestration in granules, much work remains to be done regarding their influence on the availability of various classes of metals.

Trophic transfer studies suggest that biomagnification may be more the exception than the rule for metals and metalloids. Wren et al. (1983) studied a series of elements (aluminum, barium, beryllium, boron, calcium, cadmium, cobalt, iron, lead, mercury, magnesium, manganese, molybdenum, nickel, phosphorus, sulfur, strontium, titanium, vanadium, and zinc) at various trophic levels of a Canadian Precambrian Shield lake and found evidence of biomagnification for only mercury. Cushing and Watson (1971) found no evidence for zinc biomagnification in a food chain of water → periphyton → carp. Terhaar et al. (1977) found that mercury, but not silver, showed evidence of biomagnification in an aquatic food chain. Using *in situ* enclosures placed in an aquatic system, Gächter and Geiger (1979) found no evidence for biomagnification of inorganic mercury, copper, cadmium, lead, or zinc. They speculated that alkylated mercury, but not inorganic mercury, increased with increasing trophic status of a species. Numerous authors report biomagnification of mercury in aquatic systems (Wren and MacCrimmon, 1986; Cabana et al., 1994; Kidd et al., 1995) although Huckabee et al. (1979) cautioned that, even with mercury, other factors correlated with trophic status of an organism may confound the identification of biomagnification. Mance (1987) reviewed the literature and found only a few cases suggesting metal or metalloid biomagnification in aquatic systems. In addition to mercury, there was an occasional report of arsenic increase in food webs.

Similarly, there appear to be few clear examples of biomagnification of metals or metalloids in terrestrial systems. Beyer (1986) and Laskowski (1991) concluded independently that biomagnification in terrestrial systems is more the exception than rule. Zinc, an essential and internally regulated metal, may increase with trophic status in terrestrial ecosystems deficient in this element (Beyer, 1986). Wu et al. (1995) suggested that biomagnification occurred in a selenium-contaminated area of California. Selenium concentration factors relative to water extractable soil concentrations were noted to increase in a soil → plant → grasshopper (*Dissosteria pictipennis*) → praying mantis (*Litaneutria minor*) trophic sequence. However, it is difficult to determine the accuracy of this conclusion because there wasn't a consistent pattern of increase with each transfer and covariates such as species longevity and allometric effects were not considered. Only the grasshopper to praying mantis concentration factor indicated an increase in selenium concentration and, according to Table 9 of Wu et al. (1995), the increase was quite variable among sample sites.

Radionuclides

This separate consideration of radionuclides is admittedly arbitrary and results in some overlap with the metals and metalloids discussed above. However, the unique sources and effects of radioactive contaminants provide some justification for this separation and will facilitate a more comfortable transition in later chapters.

Assimilation efficiencies of radionuclides vary widely, e.g., 80% for ^{134}Cs (algae → carp) to nearly 0% for the radionuclide ^{144}Ce (food → fish) which becomes unavailable after being incorporated into structural tissue of prey (Reichle et al., 1970). Amphipods fed brine shrimp had assimilation efficiencies of 6.2, 9.4, and 55% for ^{144}Ce, ^{46}Sc, and ^{65}Zn, respectively. Cesium-137 and ^{134}Cs, widespread fission products with long half-lives (circa 30 y for ^{137}Cs and 2 y for ^{134}Cs), have high assimilation efficiencies (65 to 94%) from a wide range of food sources in terrestrial and aquatic systems. Similarly, ^{47}Ca assimilation efficiencies are quite high in diverse systems (69 to 98%). Almost 100% of ^{86}Rb and ^{187}W were assimilated by grazers of plant foliage (Reichle et al., 1970). Reichle et al. (1970) and Blaylock (1982) summarize this type of data for radionuclides in aquatic and terrestrial systems.

Some (radionuclides of nitrogen, phosphorus, potassium, and sodium) increase with passage up the food web while others (^{47}Ca) are diluted (Reichle and Van Hook, 1970). The increase or decrease is a consequence of their availability in the environment relative to the physiological need for each. Davis and Foster

(1958) observed that the most common trend for many radionuclides was accumulation to relatively high concentrations at the lowest trophic level (primary producers) with subsequent diminution at each transfer thereafter. For example, ^{32}P displayed this behavior in aquatic systems examined by Kahn and Turgeon (1984). Polikarpov (1966) summarized and described processes resulting in concentration factors for many radionuclides in aquatic biota.

Essential elements in short supply and their radioactive analogs tend to increase during trophic exchange (Reichle and Van Hook, 1970). (An elemental **analog** is an element that behaves like, but not necessarily identical to, another element in biological processes, e.g., cesium is an analog of potassium and strontium is an analog of calcium.) However, ^{90}Sr does not increase with trophic exchange because it is incorporated into bone or other structural tissues and, in so doing, has a very low bioavailability to predators. The consequence is trophic dilution for ^{90}Sr (Woodwell, 1967). Reichle et al. (1970) identified **calcium sinks** such as arthropod cuticles as a mechanism for trophic dilution for calcium and its analogs in terrestrial communities.

The radiocesium isotopes, ^{134}Cs and ^{137}Cs, are potassium analogs that have received much deserved attention due to their release from fission-related processes and their relatively long half-lives. Radiocesium and potassium are taken up with similar, high efficiencies but radiocesium tends to be eliminated more slowly than potassium (McNeill and Trojan, 1960).[6] Consequently, the potential exists for a net increase in cesium with each trophic exchange. Radiocesium (^{137}Cs) from fallout did biomagnify from plant tissue to mule deer (*Odocoileus h. hemionus*) to cougar (*Felis concolor hippolestes*), suggesting that biomagnification could also increase ^{137}Cs activities in humans who consume tainted mule deer (Pendleton et al., 1964). There was a 3.4-fold increase of ^{137}Cs from deer to cougar. Similar increases have been described for atmospheric fallout-related ^{137}Cs in a food chain leading from lichens to caribou to Alaskan Eskimos, wolves, and foxes (French, 1967; Woodwell, 1967). Further evidence for the increase in ^{137}Cs levels with trophic transfer is the observation that ^{137}Cs activities are often higher in piscivorous fish than nonpiscivorous fish (Blaylock, 1982; Rowan and Rasmussen, 1994).

ORGANIC COMPOUNDS

Organic compounds exhibit a wide range of patterns relative to trophic transfer but, in general, biomagnification of compounds not subject to metabolism seems to be predicted by the log K_{ow}. Connolly and Pedersen (1988) suggested that biomagnification is possible if log K_{ow} values were above approximately 4: Gobas et al. (1993)

---

[6]Like isotopic discrimination described earlier, a discrimination factor can be estimated for analogs. Discrimination between an element and its analog may be measured as a **discrimination factor or ratio** (e.g., $[Cs]_{food}/[K]_{food}$ divided by $[Cs]_{body}/[K]_{body}$). The discrimination ratio for cesium and potassium in humans is approximately 0.33 (McNeill and Trojan, 1960).

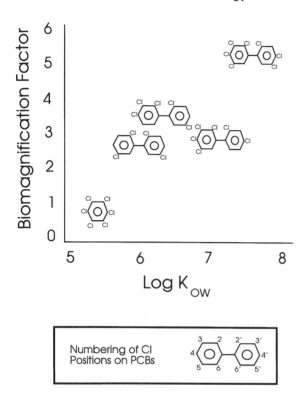

FIGURE 5.5. The increase in biomagnification factor (concentration in white bass/concentration in emerald shiner) with increasing log K_{ow} values for hexachlorobenzene (bottom left) and four polychlorinated biphenyls (PCBs), 2,2',5,5'-tetrachlorobiphenyl, 2,2',3,4,5'-pentachlorophenyl, 2,2',3,4,4',5-hexachlorobiphenyl, and 2,2',3,4,4',5,5'-heptachlorobiphenyl. (Modified from Figure 2 in Russell et al. [1995].) Note that a key to the numbering of chloride atoms attached to the biphenyl rings is provided at the bottom of this figure. The prefixes, tetra-, penta-, hexa-, and hepta- refer to the total number of chloride atoms in the PCB molecule.

suggested log K_{ow} values greater than 6 as an approximate upper limit for biomagnification. Thomann (1989) indicated a range from 5 to 7 for compounds prone to biomagnification. Above 7, biomagnification is hampered by low assimilation efficiencies and BCF for lower trophic levels such as phytoplankton. Below 5, decreased uptake and increased elimination rates will limit the capacity for biomagnification. Russell et al. (1995) found that PCBs with log K_{ow} values of 6.1 or above were subject to biomagnification when white bass (*Morone chrysops*) consumed emerald shiners (*Notropis atherinoides*) (Figure 5.5).

Polychlorinated biphenyls often exhibit biomagnification although Clayton et al. (1977) argued otherwise for zooplankton. Using nitrogen isotopes to quantify trophic status, PCB concentrations in lake trout from 83 Canadian lakes were shown to be determined by biomagnification processes and species lipid content (Rasmussen et al., 1990). Using the lake trout trophic classification discussed earlier for mercury, concentrations of PCBs were shown to increase 3.5 fold for every trophic exchange.[7] Lake Michigan foodweb studies also demonstrated PCB biomagnification (Evans et al., 1991). Clark and Mackay (1991) found the PCB,

---

[7]Interestingly, these authors make the point that fisheries management strategies commonly thought to enhance sports fishing can do damage instead. Forage species are added to enhance the trophic base of a lake and produce more trout. This is seen as a positive action. But, each addition to the trophic structure enhances accumulation of PCBs in trout. If the addition of forage species brings trout tissue concentration above a certain level, a fishing warning or ban can result. The forage fish stocking may damage the sports fishery in some cases.

2,2',3,3',4,4',5,6'-octachlorophenyl to biomagnify in guppies (*Poecilia reticulata*) exposed in the laboratory. Bruggen et al. (1991) fed PCBs to goldfish (*Carassius auratus*) and found that biomagnification occurred, seemingly as a consequence of water solubility-correlated elimination rates.

DDT, DDE, DDD, or the sum of these pesticides may increase with each trophic transfer in field communities (Evans et al., 1991; Kidd et al., 1995). (These pesticides may be studied individually or together as "Σ DDT.") Toxaphene may display biomagnification (Sanborn et al., 1976; Kidd et al., 1995) but to a lesser extent than DDT (Evans et al., 1991). Dibenzo-p-dioxins and -dibenzofurans also were found to increase with trophic position in a Northern Baltic Sea food web (Broman et al., 1992).

All isomers and congeners[8] do not behave similarly during trophic exchanges. For example, congeners of PCBs behaved distinctly in Arctic trophic levels, Arctic cod (*Boreogadus saida*) → ringed seal (*Phoca hispida*) → polar bear (*Urus maritimus*) (Muir et al., 1988). The tri- and tetrachloro PCBs were the dominant congeners in cod. Penta- and hexachloro PCB congeners dominated in the ringed seal, and hexa- and heptochloro PCBs were the dominant congeners in polar bears. Older seals tended to have more of the highly chlorinated PCBs than young seals. The reason for this age difference was likely a combination of the slower elimination of the more highly chlorinated congeners, and the difference in diet between young and adult seals. Young seals tend to eat more amphipods than do adults. In studies of trophic transfer of dibenzo-p-dioxins and -dibenzofurans in a Baltic food web, the total concentration of these contaminants did not increase but the most toxic isomers did exhibit biomagnification to high levels in eider ducks (*Somateria mollissima*) (Broman et al., 1992; Rolff et al., 1993). This study underscores the importance of considering individual as well as summed concentrations of isomers or congeners. Catfish (*Ictalurus punctatus*) preferentially accumulate *cis* to *trans* chlordane because the *cis* isomer appears most resistant to metabolism (Murphy and Gooch, 1995). Likely for this same reason, the *cis* form of chlordane increased relative to the *trans* form in a freshwater trophic chain (Sanborn et al., 1976).

SUMMARY

In this chapter, the transfer of contaminants during trophic interactions was discussed. Methods of estimating such transfer were provided for quantal and continuous trophic structures. Conditions conducive to and examples of biomagnification and trophic dilution were outlined for metals, metalloids, radionuclides and

---

[8]PCBs are manufactured and used as mixtures of chlorinated biphenyls with different numbers and positions of associated Cl atoms. The individual PCBs which share a common form but have different numbers of Cl atoms at different positions are called **congeners**. The numbering of Cl atom positions on the biphenyl common structure of PCB congeners is shown in Figure 5.5.

organic compounds. This chapter ends Section Two of the book which provides description of mechanisms leading to bioaccumulation of contaminants. The next section of the book details the consequences of such bioaccumulation. Effects will be discussed from the biochemical to the global level of ecological organization.

Selected Readings

Broman, D., C. Näf, C. Rolff, Y. Zebühr, B. Fry, and J. Hobbie. Using ratios of stable nitrogen to estimate bioaccumulation and flux of polychlorinated dibenzo-p-dioxins (PCDDs) and dibenzofurans (PCDFs) in two food chains from the Northern Baltic. *Environ. Toxicol. Chem.* 11, pp. 331–345, 1992.

Cabana, G. and J.B. Rasmussen. Modelling food chain structure and contaminant bioaccumulation using stable nitrogen isotopes. *Nature (LOND)* 372, pp. 255–257, 1994.

McNeill, K.G. and O.A.D. Trojan. The cesium-potassium discrimination ratio. *Health Phys.* 4, pp. 109–112, 1960.

Murphy, D.L. and J.W. Gooch. Accumulation of *cis* and *trans* chlordane by channel catfish during dietary exposure. *Arch. Environ. Contam. Toxicol.* 29, pp. 297–301, 1995.

Petersen, B.J. and B. Fry. Stable isotopes in ecosystem studies. *Ann. Rev. Ecol. Syst.* 18, pp. 293–320, 1987.

Reinfelder, J.R. and N.S. Fisher. The assimilation of elements ingested by marine copepods. *Science (WASH DC)* 251, pp. 794–796, 1991.

Rowan, D.J. and J.B. Rasmussen. Bioacumulation of radiocesium by fish: the influence of physicochemical factors and trophic structure. *Can. Fish. Aquat. Sci.* 51, pp. 2388–2410, 1994.

Weeks, J.M. and P.S. Rainbow. A dual-labelling technique to measure the relative assimilation efficiencies of invertebrates taking up trace metals from food. *Functional Ecol.* 4, pp. 711–717, 1990.

Molecular Effects and Biomarkers

Pollutants affect biological systems at many levels, but all chemical pollutants must initially act by changing structural and/or functional properties of molecules essential to cellular activities.

JAGOE (1996)

. . . if effects on the ecosystem are to be predicted and understood, it is necessary to identify the effects of a toxicant on lower levels of biological organization, such as the subcellular level. Specific and sensitive biochemical methods can, therefore, serve as early warning indicators and adverse effects on the ecosystem can be avoided by taking protective measures.

HAUX AND FÖRLIN (1988)

INTRODUCTION

Two important concepts relative to biochemical effects of toxicants are exemplified by the above quotes. First, with the exception of a few agents such as asbestos fibers that inflict physical damage directly to cells and tissues, all toxicant effects begin as interactions with biomolecules. Effects then cascade through the biochemical → subcellular → cellular → tissue → organ → individual → population → community → ecosytem → landscape → biosphere levels of organization. Consequently, an understanding of effects at the biochemical level may provide some insight into the root cause of effects seen at next few higher levels. Also, by understanding biochemical mechanisms, we can better predict effects of untested contaminants based on similarity of biochemical mode of action to well understood contaminants. Further, if several contaminants are present, specific biochemical changes can provide valuable clues about which contaminant is having an effect. In some histochemical methods, tissue localization of biochemical changes can also provide information relative to exposure at target organs or sites. For example, high Phase I enzyme activity (e.g., monooxygenase activity) in a specific tissue may be measured after animals are exposed in the laboratory to a toxicant[1] known

---

[1]A compound that is converted to a carcinogen is a **procarcinogen.** For example, 2-acetylaminofluorene is a procarcinogen that is converted to a carcinogen by Phase I enzymes (N-hydroxylation by monooxygenases) (Stegeman and Hahn, 1994). Similarly, cytochrome P-450 monooxygenases activate benzo[a]pyrene to potent carcinogens. Remember from Chapter 3 that these are cases of activation.

to be metabolized to a potent carcinogen. This finding may be linked to results of surveys indicating high incidence of cancers in the same tissue of individuals from a contaminated site. So, understanding the biochemical mode of action enhances our grasp of causal structure and our ability to predict effects at higher levels of biological organization.

Second, a technical advantage is gained by understanding toxicant effects on biomolecules and molecular responses to contaminants. Changes in biomolecules or suites of biomolecules can indicate exposure to bioavailable contaminants in field situations. The biochemical quality that is changing may be used as a biomarker. As discussed in Chapter 2, a biomarker is a biochemical, physiological, morphological, or histological quality used to imply exposure to or effect of a toxicant. Biomarkers indicate that sufficient toxicant was available for enough time to elicit a response or effect (Melancon, 1995). Because biochemical changes generally are detectable before adverse effects are seen at higher levels of biological organization, the biochemical marker approach is often an early warning or proactive tool. This is a great advantage because responses at higher levels such as the ecosystem are usually measurable only after significant or permanent damage has occurred. Biochemical markers are also useful to monitor the shift back to a normal state after cleanup of a contaminated site. Regardless of their proactive or retroactive utility, the ecological realism (ability to accurately reflect an ecologically meaningful effect or response) is lower for biomarkers than for indicators based on higher level changes such as species richness or reproductive failure.

A biochemical change should have the following qualities to be useful as a biomarker. First, it should be measurable before any adverse, significant consequences occur at higher levels of biological organization (Haux and Förlin, 1988; Campbell and Tessier, 1996). This quality enhances its value as a proactive tool. Second, measurement of the ideal biomarker should be rapid, inexpensive, and sufficiently easy so as to be amenable to widespread use by ecotoxicologists. Third, its measurement should accommodate standard quality control/quality assurance practices. Fourth, the ideal biomarker should be specific to a single toxicant or class of toxicants (Haux and Förlin, 1988; Campbell and Tessier, 1996), although nonspecific biomarkers have value too. Fifth, a clear concentration-effect relationship must exist for the toxicant and biomarker (Haux and Förlin, 1988). Sixth, the ideal biomarker should be applicable to a broad range of **sentinel species** (feral, caged, or endemic species used in measuring and indicating the level of contaminant effect during a biomonitoring exercise) so that it may have the widest possible application (Sanders, 1990). Seventh, established linkage of biomarker changes with some toxicant-related decrease in individual fitness is desirable (Sanders, 1990) but not always necessary. This enhances ecological relevance in discussions of the biomarker change. Finally, the system should be sufficiently well understood so that other qualities of the organism or its environment that influence the biomarker can be accommodated in the experimental design and data interpretation (Campbell and Tessier, 1996). For example, ambient temperature, age and sex can influence monooxygenase activity and should be considered in the design of biomonitoring studies using this biomarker (Kleinow et al., 1987).

The remainder of this chapter examines biochemical aspects of ecotoxicology with the dual goals of understanding the mode of action of toxicants and describing current molecular biomarkers. Enzyme activities, conjugates, and products of Phase I and II breakdown of organic compounds are discussed first. Focus then shifts to metallothioneins and stress proteins. Enzymes and products associated with oxidative stress, toxicant effects on nucleic acids and symptoms of enzyme dysfunction are detailed. Aspects of these topics are explored again during discussions of cancers (Chapter 7).

ORGANIC COMPOUND DETOXIFICATION

Phase I

Biochemical shifts associated with **cytochrome P-450 monooxygenase** induction are often employed to suggest a response to organic contaminants such as polycyclic aromatic hydrocarbons (PAHs) (Van Veld et al., 1990; Wirgin et al., 1994), chlorinated hydrocarbons (Walker et al., 1987), organic compounds in pulp and paper mill effluents (Soimasuo et al., 1995), polychlorinated biphenyls (PCBs) (Brumley et al., 1995), hydrocarbons (Goksøyr and Förlin, 1992), dioxins (Goksøyr and Förlin, 1992), and dibenzofurans (Goksøyr and Förlin, 1992). These monooxygenases are involved in the metabolism of a wide range of xenobiotics as well as fatty acids, cholesterol, and steroid hormones. Although most attention is focused on the cytochrome P-450 monooxygenases, the cytochrome b_5 and NADH-cytochrome b_5 reductase system is also important in xenobiotic metabolism (Di Giulio et al., 1995).

The cytochrome P-450 monooxygenases are hemoproteins associated with membranes, especially in the endoplasmic reticulum (ER). They are often assayed in the microsomal fraction, a cell fraction composed of membrane vesicles (microsomes) and derived from the ER during routine separations. They have molecular weights of 45 to 60 kDa (Goksøyr and Förlin, 1992). The name P-450 is derived from P for pigment and 450 nm, the wavelength at which they have maximum light absorption when bound to CO (Haux and Förlin, 1988; Di Giulio et al., 1995). They are most often associated with the conversion,

$$RH + NADPH + O_2 + H^+ \rightarrow ROH + NADP^+ + H_2O \qquad (6.1)$$

where RH is an organic compound undergoing hydroxylation. Cytochrome P-450 monooxygenases also catalyze epoxidation, (N-, O-, and S-) dealkylation, oxidative deamination, (S-, P-, and N-) oxidation, desulfuration reactions, and oxidative and reductive dehalogenation (Stegeman and Hahn, 1994; Di Giulio et al., 1995). The P-450 complex is composed of two enzymes (cytochrome P-450 and NADPH-cytochrome P-450 reductase), NADPH, and molecular oxygen. Both enzymes are associated with a membrane. This system is often called the **mixed function oxidase** or MFO system (Di Giulio et al., 1995).

There is a diverse nomenclature that must be understood in order to make sense of studies of cytochrome P-450 monooxygenases. The genes coding for cytochrome P-450 monooxygenases are grouped into gene families (circa 27 families) and subfamilies based on similarities in DNA sequences. Names of genes and gene products are based on the root CYP (**Cy**tochrome **P**-450). Numbers and letters are added to the CYP root to designate the particular family (e.g., CYP1 or CYP2), subfamily (e.g., CYP1A or CYP1B), and gene (e.g., CYP1A1 or CYP1A2) (Goksøyr and Förlin, 1992). A standard notation is used for the protein, mRNA, and DNA associated with any particular gene. The protein and mRNA are designated as detailed above, e.g., CYP1A1. Sometimes the protein can be designated as P-450 1A1 instead. The DNA designation is italicized, e.g., *CYP1A1* (Goksøyr and Förlin, 1992). This is important to keep in mind because protein, mRNA, and DNA are all employed in studies of these important Phase I enzymes. Proteins are often assayed with specific antibodies and the mRNA with DNA probes. For example, Haasch et al. (1993) measured immunoreactive CYP1A1 protein and hybridizable CYP1A1 mRNA to suggest liver cytochrome P-450 response in fish to PAH and PCB contamination.

Additional nomenclature appears if enzymatic activities are quantified. Activity may be measured by aryl hydrocarbon hydroxylase hydroxylation of benzo[a]pyrene. Results are expressed as **AHH** (**a**ryl **h**ydrocarbon **h**ydroxylase) activity. Another common assay involves O-deethylation of ethoxyresorufin by ethoxyresorufin O-deethylase. Activity is expressed as **EROD** (**e**thoxy**r**esorufin **O**-**d**eethylase) activity (Goksøyr and Förlin, 1992). Cytochrome P-450 monooxygenase activities measured as AHH or EROD are common biomarkers reflecting response (induction) to significant amounts of bioavailable xenobiotic. For example, Soimasuo et al. (1995) used EROD activity in whitefish (*Coregonus lavaretus*) liver to measure CYP1A response to pulp and paper mill effluent (Figure 6.1). Haasch et al. (1993) used EROD activity in tandem with assays of CYP1A1 protein and mRNA to monitor fish (catfish, *Ictalurus punctatus*; largemouth bass, *Micropterus salmoides*; and killifish, *Fundulus heteroclitus*) response to PAH and PCB exposure. Tomcod (*Microgadus tomcod*) response to xenobiotics was monitored along the North American coast with CYP1A mRNA (Wirgin et al., 1994). Other enzyme assays that also reflect P-450 activity include ethoxycoumarin O-deethylase (ECOD), aflatoxin B1 2,3-epoxidase (AFBI), lauric acid ω-1 hydrolase (LA), testosterone hydroxylase (TH), and phenanthrene hydroxylase (AH) assays (Goksøyr and Förlin, 1992; Brumley et al., 1995).

Induction of cytochrome P-450 is influenced by numerous factors including ambient temperature (Kleinow et al., 1987), the particular toxicant and species being studied (Haux and Förlin, 1988), body weight (Parke, 1981), animal sex (Haux and Förlin, 1988), hormone titers (Haux and Förlin, 1988), and tissue oxygen tension (Parke, 1981). Regardless of these modifying factors, cytochrome P-450 induction has proved to be a reliable biomarker. As discussed already, it was used by Haasch et al. (1993), Wirgin et al. (1994), and Soimasuo et al. (1995) as an indicator for significant exposure to a variety of contaminants. Van Veld et al. (1990) measured cytochrome P-450 protein (immunoassay) and activity (EROD) in spot (*Leiostomus xanthurus*) to indicate PAH contamination in the Chesapeake Bay region. Brumley et al. (1995) used induction of these same biomarkers in sand flatheads (*Platycephalus bassensis*) as indicators of response to PCB exposure.

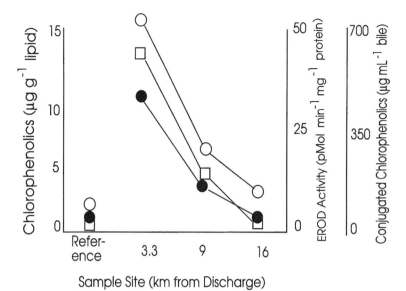

FIGURE 6.1.   The response of biomarkers in juvenile whitefish (*Coregonus lavaretus*) to pulp and paper mill effluents. The concentration of total chlorophenolics in the gut lipids of this sentinel species (●) were elevated in whitefish taken from 3.3 km below the discharge relative to the reference samples and rapidly decreased with distance from the discharge. Similarly, two biomarkers, EROD activity (○) and conjugated chlorophenolics in the bile (□), were highest near the discharge and decreased with distance from the paper and pulp mill effluent. (Generated by combining information from Figures 3, 4, and 5 of Soimasuo et al. [1995]).

Phase II

Phase II enzymes may also be induced by xenobiotics and used in biomarker studies. **Glutathione S-transferase** (GST) which attaches **glutathione** (GSH, a tripeptide made of cysteine, glutamate, and glycine) to the xenobiotic or its metabolites is one candidate enzyme. Another is **sulfotransferase** which conjugates sulfate with the organic compound. **Uridinediphospho glucuronosyltransferase** (UDP-glucouronosyltransferase, UDP-GT) is another. This enzyme catalyzes the transfer of **glucuronic acid** from uridine diphosphate glucuronic acid to electrophilic xenobiotics or their metabolites (Di Giulio et al., 1995). It also binds covalently with electrophilic compounds such as PAHs (George, 1994).

Both the induced activities of these enzymes and concentrations of conjugated products have been explored as biomarkers although Di Giulio et al. (1995) suggested that these enzymes were less valuable biomarkers than Phase I enzymes. As discussed above, Soimasuo et al. (1995) examined GST and UDP-GT induction, concentrations of conjugates in bile, and EROD in whitefish. Induction of EROD was clearly demonstrated for these fish during exposures to pulp and paper mill effluent; however, induction of GST and UDP-GT was not as clear. Conjugated metabolites were sensitive biomarkers that had an obvious trend with distance from discharge (Figure 6.1). Activity of UDP-GT, and conjugates with glucuronic acid and sulfate did increase after laboratory exposure of sand flatheads to PCBs, suggesting to Brumley et al. (1995) that these qualities may be acceptable biomarkers of PCB exposure.

METALLOTHIONEINS

Metallothioneins were first described by Margoshes and Vallee (1957) as cadmium-binding proteins in horse kidneys. Now known to be produced in bacteria, invertebrates, and vertebrates, they are found in highest concentrations in the liver, hepatopancreas, kidneys, gills, and intestines (Roesijadi, 1992).

Metallothioneins function in the uptake, internal compartmentalization, sequestration, and excretion of essential (Cu, Zn) and nonessential (Ag, Cd, Hg) metals. Their involvement in the normal homeostasis of essential metals results in basal levels of metallothionein being present in the absence of toxic metal exposure. Fluctuations in metallothionein levels associated with processes such as molting and reproduction (Roesijadi, 1992), or fluctuations in substances such as glucocorticoids (Karin and Herschman, 1981) can also occur. Reflecting their roles in metal detoxification and sequestration, they are induced by elevated levels of the above metals. This increases the capacity of the organism to bind and effectively sequester toxic metals away from molecular sites of toxic action. Enhanced levels of metallothioneins have been linked to enhanced fitness during metal exposure of individuals (Bouquegneau, 1979; Hobson and Birge, 1989; Sanders et al., 1983). Further, enhanced capacity to produce metallothionein and the associated lessening of toxic metal effects have been linked to population adaptation to chronic metal exposure. For example, Maroni et al. (1987) detected metallothionein gene duplication in *Drosophila melanogaster* populations, leading to enhanced production of metallothionein and higher survival during copper or cadmium exposure.

These cytosolic proteins have molecular weights of roughly 6 to 7 kDa and about 25% of their amino acids are cysteine. They have no aromatic amino acids or histidine (Hamilton and Mehrle, 1986). They have high heat stability, a quality often used during their isolation (Winge and Brouwer, 1986). Poorly characterized proteins or proteins not conforming precisely with the above qualities have been called **metallothionein-like proteins** (Hamilton and Mehrle, 1986). Early in the study of metal-binding proteins of nonmammalian species, proteins were characterized from incompletely purified preparations, resulting in confusion. Roesijadi (1992) recently produced the following, more general list of metallothionein qualities that seem to fit all relevant biomolecules that were properly characterized:

1. low molecular weight with a high metal content,
2. a high cysteine content and no aromatic amino acids or histidine,
3. a unique amino acid sequence, especially regarding cysteine placement,
4. metal-thiolate clusters.

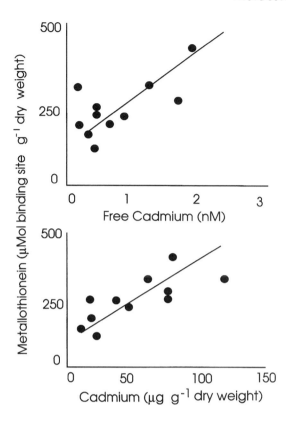

FIGURE 6.2. The relationships between free cadmium (Cd^{2+}) concentration in lake water (sediment-water interface) and metallothionein concentration in the freshwater mussel, *Anodonta grandis* (top panel), and cadmium concentration in the entire mussel and metallothionein concentration in this mussel (bottom panel). Each point represents samples from a different lake in the Rouyn-Noranda mining area of Quebec, Canada. (Modified from Figures 3 and 4 of Couillard et al. [1993]).

Roesijadi includes phytochelatin[2] as a metallothionein in this scheme. As discussed in Chapter 3, phytochelatins are inducible, metal-binding peptides isolated from plants.

Metallothioneins have been used as specific biomarkers for metal contamination under a variety of exposure scenarios. For example, metallothionein levels in whelk (*Bullia digitalis*) consuming metal-tainted grasses reflected site contamination (Hennig, 1986). Hepatic metallothionein levels in juvenile trout (*Salmo gairdneri*) (Roch et al., 1982) and metallothionein in the freshwater mussel (*Anodonta grandis*) (Couillard et al., 1993; Figure 6.2) were effective biomarkers of metal contamination.

Such biomarker studies can be extended even further based on the spillover hypothesis (Campbell and Tessier, 1996; Hamilton and Mehrle, 1986). Under the assumption that binding by metallothionein sequesters toxic metals away from sites of adverse action, the **spillover hypothesis** states that toxic effects will begin to emerge after exceeding the capacity of metallothionein to bind metals. The unbound metals then "spill over" to interact at sites of adverse action. The amount of metal in cells in excess of that bound to metallothionein is correlated with some

---

[2]If free of metals, phytochelatin has the form of (γ-glutamic acid-cysteine)$_n$-glycine with n being 3, 5, 6, or 7 (Grill et al., 1985).

measure of adverse effect. One example of such use of the spillover hypothesis is the study by Klaverkamp et al. (1991) of white sucker (*Catostomus commersoni*) inhabiting lakes around the Flin Flon smelter (Manitoba, Canada).

STRESS PROTEINS

The **cellular stress response** is an "orchestrated induction of key proteins that form the basis for the cell's protein protection and recyling system" (Sanders and Dyer, 1994).[3] A cellular stress response can be elicited under the influence of heat, anoxia, some metals, some xenobiotics, ethanol, sodium arsenate, or UV radiation (Craig, 1985; Thomas, 1990; Di Giulio et al., 1995). Stress-induced proteins were first studied after organisms experienced abrupt changes (5 to 15°C) in temperature. These **heat shock proteins (hsp)** were distinguished from one another according to their molecular weights. Respectively, there are 90, 70, 60, 16-24, and 7 kDa groupings designated as hsp90, hsp70, hsp60, low molecular weight (LMW), and ubiquitin. These proteins are now known to be induced by other stressors and, consequently, have been renamed **stress proteins** to reflect this more general role. The terminology based on heat stress has been modified. Now the stress protein groupings are also called stress90, stress70, chaperon 60 (cpn60), LMW, and ubiquitin. The term **chaperon**, used collectively for stress90, stress70, and cpn60, reflects their role of associating with and directing the proper folding and coming together of proteins. They also protect proteins from denaturation and aggregation, and enhance refolding of damaged protein to a functional conformation. Their final role is in the transport of proteins to their intracellular location where they then are folded to a functional conformation (Craig, 1993).

Physical and chemical agents that have significant **proteotoxicity** (toxicity due to protein damage) (Hightower, 1991) induce stress proteins. Enhanced production of stress90, stress70, cpn60, LMW, and ubiquitin is initiated by denatured protein to protect, repair, or vector for breakdown the proteins of the cell. Some stress proteins are always present (e.g., stress70 [Welch, 1990]) while others (LMW) appear only under stressful conditions (Di Giulio et al., 1995). Stress90 is present at high concentrations under unstressed conditions and is induced to even higher levels by stress. Stress70 is present at lower concentrations during normal conditions and is induced by stress. Cpn60, which is found in the mitochondria and facilitates protein movement and folding, is present at low levels under normal conditions and is inducible. Because stress70 and cpn60 are inducible by stress, are highly conserved proteins, and are not normally present at high levels as is stress90, they are excellent candidates as biomarkers (Sanders, 1990). The

---

[3]This is one of the most often discussed characteristics of the cellular stress response; however, Sanders (1990) also discussed as part of this response the induction of **glucose regulated proteins** (grps) under low glucose or oxygen conditions. The grps are structurally similar to hsp, are present at basal levels in unstressed cells, and are induced in glucose- or oxygen-deficient cells exposed to toxicants which modify calcium metabolism, e.g., lead (Sanders, 1990).

LMW proteins are more variable in structure and in inducibility among species: these qualities detract from the utility of LMW as biomarkers. Ubiquitin is induced by stressors and its structure is very conservative evolutionarily, making it a potential biomarker (Sanders, 1990).

Stress proteins recognize and bind to exposed regions of denatured proteins which are rich in hydrophobic peptides (Agard, 1993). With the aid of stress70 and stress90, denatured or aggregated proteins are unfolded and refolded properly to restore their function. Proteins damaged beyond repair are bound by ubiquitin which helps move them to lysosomes for final breakdown (Di Giulio et al., 1995).

Increased concentrations of the various stress proteins are used as biomarkers of general cellular stress response. For example, Sanders and Martin (1993) correlated general contamination with cpn60 and stress70 in archived tissues of marine species. Because different stressors induce the various stress proteins to different degrees, Sanders and Dyer (1994) suggested that the patterns of stress protein induction could be used to suggest the particular toxicant inducing the response. Patterns from field samples can be compared to those obtained with single candidate toxicants in the laboratory. They referred to this approach as **stress protein fingerprinting**. Because the stress proteins have evolved in a very conservative manner and the cellular stress response is so universal, Sanders and Dyer (1994) suggested that this approach has more universal application than many other biomarker methods.

OXIDATIVE STRESS AND ANTIOXIDANT RESPONSE

Combining reduction of molecular oxygen with energy generation during aerobic metabolism (Figure 6.3) creates the potential for **oxidative stress**, i.e., damage to biomolecules from free oxyradicals. Oxyradical-generating compounds such as hydrogen peroxide (H_2O_2) are also produced by aerobic metabolism and other processes, contributing to oxidative stress. Free radicals[4] such as the superoxide radical ($O_2^\bullet$) and hydroxyl radical ($^\bullet OH$) can damage proteins, lipids, DNA, and other biomolecules. Oxyradicals are produced during electron transport reactions in mitochondria and microsomes, photosynthetic electron transport, phagocytosis, and normal catalysis by prostaglandin synthase, guanyl cyclase, and glucose oxidase (Di Giulio et al., 1989).

Organisms cope with this situation in two ways. They produce antioxidants that react with oxyradicals. These antioxidants include vitamin E, vitamin C, β-carotene, catecholamines, glutathione, and uric acid (Winston and Di Giulio, 1991). Notice here that glutathione has a dual role as a substrate for Phase II conjugation and as an antioxidant. In addition to these antioxidants, enzymes which reduce the amount of oxyradicals present at any instant are involved in avoiding oxidative

---

[4]A **free radical** is a molecule having an unshared electron. The unshared electron is usually designated by a dot, •. Free radicals are extremely reactive. **Oxyradicals** are free radicals with an unshared electron of oxygen, e.g., $RO^\bullet$ where R is some oxygen-containing compound.

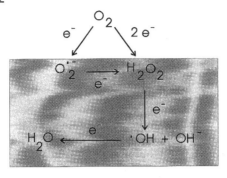

O$_2$ Reduction to Water

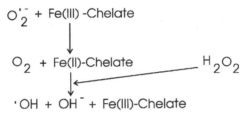

Catalyzed Haber-Weiss Reaction

$O_2^{\cdot-}$ + Fe(III) -Chelate

O$_2$ + Fe(II)-Chelate   H$_2$O$_2$

$^{\cdot}$OH + OH$^-$ + Fe(III)-Chelate

FIGURE 6.3. Reduction of O$_2$ to water during aerobic respiration (O$_2$ + 4e$^-$ → H$_2$O). Molecular oxygen can be converted to the superoxide radical (O$_2^{\cdot-}$) by adding one electron (e$^-$) or to hydrogen peroxide (H$_2$O$_2$) by adding 2 electrons. The superoxide radical can be converted to hydrogen peroxide by the addition of another electron. Hydrogen peroxide can then be reduced to the hydroxyl radical ($^{\cdot}$OH), producing a hydroxyl anion (OH$^-$). Water is produced with the addition of a fourth electron. Two highly reactive, free radicals are generated along this reaction sequence. Hydrogen peroxide is also produced and, through the Haber-Weiss reaction (O$_2^{\cdot}$ + H$_2$O$_2$ → $^{\cdot}$OH + OH$^-$ as shown in the top panel in a box), generates oxyradicals. The catalyzed Haber-Weiss reaction is a greatly accelerated Haber-Weiss reaction catalyzed by metal chelates as shown in the bottom panel.

damage. **Superoxide dismutase** (SOD) decreases the amount of superoxide radical in the cell by catalyzing reaction 6.2. **Catalase** (CAT) and **glutathione peroxidase** reduce levels of hydrogen peroxide via reactions 6.3 and 6.4, respectively.

$$2 + O_2^{\cdot} + 2\,H^+ \rightarrow H_2O_2 + O_2 \tag{6.2}$$

$$2 + H_2O_2 \rightarrow H_2O + O_2 \tag{6.3}$$

$$2\ Reduced\ Glutathione + H_2O_2 \rightarrow Oxidized\ Glutathione + H_2O \tag{6.4}$$

Xenobiotics can cause oxidative damage indirectly by interfering with these mechanisms of coping with oxidative stress. They may also participate directly in reactions leading to oxidative stress. Xenobiotics can form oxyradicals such as **alkoxyradicals** ($RO^{\bullet}$) and **peroxyradicals** ($ROO^{\bullet}$) and, in so doing, cause damage (Di Giulio et al., 1995). For example, carbon tetrachloride can undergo the following reaction to generate a trichloromethyl radical, $CCL_4 + e^- \rightarrow CCl_3^{\bullet} + Cl^-$. This reaction, involving the NADPH-cytochrome P-450 system, contributes to carbon tetrachloride damage (necrosis, cancers) to the human liver where the free radical reacts with lipids, proteins and DNA (Slater, 1984). Xenobiotics may also form free radicals during interactions with oxyradicals. For example, promethazine reacts with the hydroxyl radical, produces OH^-, and becomes a free radical itself. Polycyclic aromatic hydrocarbons (PAHs) are activated by biotransformation to their free radical forms (Slater, 1984). Still other contaminants (quinones, aromatic nitro compounds, aromatic hydroxylamines, bipyridyls such as paraquat and diquat, and some chelated metals [Di Giulio et al., 1989]) are reduced to radicals and then undergo **redox cycling** to produce superoxide radicals from molecular oxygen. The initial reduction of the contaminant to a free radical may be facilitated by one of several reductases (Di Giulio et al., 1989). Because the contaminant enters the redox cycle as a free radical but exits in its original form at the end of a cycle, it is available to recycle many times and produce considerable amounts of oxyradicals.

Because contaminants can interfere with the normal mechanisms of coping with oxidative stress and can contribute free radical concentrations themselves, considerable effort has been spent in studying contaminant-related changes in oxidative stress. Elevated levels of superoxide dismutase, catalase, or glutathione peroxidase in exposed individuals may suggest elevated oxidative stress. Changes in these enzymes may be observed for individuals in contaminated sites (Roberts et al., 1987; Regoli and Principato, 1995) or those exposed to contaminants in the laboratory (Víg and Nemcsók, 1989). Antioxidant pools may also be examined. For example, mussels exposed to metals (Regoli and Principato, 1995) or paraquat (Wenning et al., 1988) show significant changes in glutathione concentrations.

Increased concentrations of free radicals can cause membrane dysfunction due to **lipid peroxidation** (oxidation of polyunsaturated lipids). **Malondialdehyde**, a breakdown product from lipid peroxidation, is indicative of oxidative damage of lipids from a variety of toxicants (Thomas, 1990). For example, Atlantic croaker (*Micropogonias undulatus*) exposed to PCBs (Aroclor 1254) or cadmium had elevated concentrations of microsomal malondialdehyde (Wofford and Thomas, 1988).

Free radicals can form covalent bonds with a variety of biomolecules. Free radicals covalently bind to membrane components such as enzymes and receptors to change their structures and functions (Slater, 1984). Free radicals cause damage by reacting with sulfhydryl groups of proteins and other biomolecules and, in so doing, influence their function (Slater, 1984). A final, important reason exists for the interest paid to contaminant-influenced oxidative stress. The formation of free radicals near DNA can lead to mutations and, as a consequence, increased risk

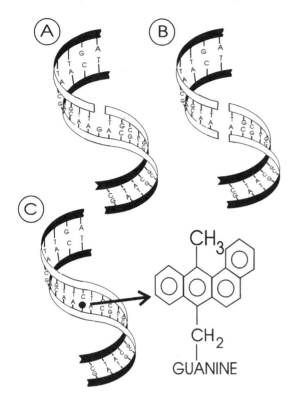

FIGURE 6.4. Various types of damage can occur to DNA due to toxicants. Toxicants or free radicals may produce single (A) or double (B) strand breaks. Xenobiotics or their metabolites may react with bases to form adducts (C). Here, metabolism of 7,12-dimethyl-benz[a]anthracene leads to covalent bonding of its metabolite to guanine to form a DNA adduct. Interactions such as those with free radicals can also modify bases, i.e., oxidize bases such as thymine and guanine to thymine glycol and 8-hydroxyguanine, respectively (Di Giulio et al., 1995).

of cancer or other genotoxic consequences.[5] Malins (1993) found DNA (guanine and adenine) lesions induced by the hydroxyl radical in fish exposed to carcinogens. He suggested that these types of lesions in fish and humans lead to increased misreading of the DNA template and increased cancer risk.

DNA MODIFICATION

As suggested in the last paragraph, toxicants and their products can be discussed in terms of **genotoxicity**, i.e., the damage by a physical or chemical agent to genetic materials such as chromosomes or DNA. At the molecular level, agents such as free radicals produce breaks in one or both strands of the DNA molecule (Figure 6.4). Oxyradicals also oxidize bases. Xenobiotics and their metabolites may bind

---

[5] The formation of free radicals is key to understanding radiation effects too. Free radicals formed from water and molecular oxygen during irradiation cause damage to cells. However, free radicals can also be used to our advantage in radiation treatments of cancers. Damage to cancer cells can be enhanced during radiation treatment by administering a radiosensitizer such as derivatives of nitro imidazoles (Slater, 1984). **Radiosensitizers** enhance the production of free radicals in the area receiving radiation which leads to more effective destruction of cancer cells.

covalently to a base or, less frequently, another portion of the DNA molecule to form an **adduct** (Shugart, 1995).

Metals can bind to phosphate groups and heterocyclic bases of DNA and, in so doing, change the stability and normal functioning of the DNA (Eichhorn et al., 1970). Magnesium binds to phosphate groups in the DNA backbone and stabilizes the DNA structure, but copper binds between bases, competes with the normal hydrogen binding, and destabilizes the DNA structure (Eichhorn, 1975). Because the matching of specific base pairs is a matter of degree of attraction for complementary versus noncomplementary pairs, the modification of hydrogen bonding by metals can contribute to base mispairing. For example, mercury forms strong crosslinks between the strands of the DNA molecule. Normal DNA repair mechanisms are overwhelmed if alterations occur excessively often and high mutation rates can result.

The expected diploid chromosome number (2N) can be disrupted due to chromosome breakage and result in a deviation from the usual number of chromosomes (**aneuploidy**) or structural aberrations in chromosomes. Agents that cause chromosome damage in living cells are classified as **clastogenic**. All of these genotoxic effects can have **mutagenic** (causing mutations), **carcinogenic** (causing cancers), and **teratogenic** (causing developmental malformations) consequences (Jones and Parry, 1992).

Various methods are available to assess genotoxicity. Chromosome damage may be determined visually. Aneuploidy and other clastogenic effects may be quantified using flow cytometry after staining the DNA in cells with a fluorescent dye. With flow cytometry, fluorescence is used to measure the amount of DNA in individual cells of a sample as each cell flows through an excitation light beam. The distribution of DNA concentrations in the population of cells is examined for significant numbers of cells with atypical amounts of DNA (Shugart, 1995). Clastogenic activity may also be reflected in the number of **micronuclei** (membrane bound masses of chromatin) in cells. The presence of many micronuclei suggests damage to the cell's ability to divide properly (Jones and Parry, 1992).

DNA adducts may be assayed using a ^{32}P-labeling method (Jones and Parry, 1992). Deoxyribonucleoside 3'-monophosphates are produced by hydrolysis of the DNA molecule. The DNA molecule is broken down into deoxyribonucleoside 3'-monophosphates and then labeled with ^{32}P. Most of the deoxyribonucleoside 3'-monophosphates contain one of the four normal bases in DNA but some will have modified bases covalently bound to an adduct. The labeled deoxyribonucleoside 3'-monophosphates are separated chromatographically to the four normal bisphosphates of adenosine, cytidine, guanosine, and thymidine, plus bases with adducts. The amount of radioactivity associated with the adducts reflects the number of DNA adducts and the potential for genotoxic effects. The occurrence of adducts has been correlated with cancer risk (Gaylor et al., 1992).

The amount of DNA breakage may also be estimated and correlated with exposure to genotoxic contaminants. Shugart (1988) described an alkaline unwinding assay to estimate the degree of single-strand breakage. After a specified time of exposure to alkaline conditions, the DNA strands from samples unwind to different degrees depending on the relative amounts of double- and single-stranded DNA in the sample. Differences in fluorescence intensity of single- and double-stranded

DNA are used to measure the relative amounts of each in the samples. Identical amounts of isolated DNA from control and exposed individuals are placed under alkaline conditions, allowed a set time to unwind, and then the fraction of DNA that is double stranded in each sample is estimated via fluorescence. This fraction is used to imply the relative amount of single-strand breaks in the various samples. For example, this fraction dropped within ten days of bluegill sunfish (*Lepomis macrochirus*) exposure to benzo[a]pyrene (Shugart, 1988), suggesting an increase in single-strand breaks with exposure. Meyers-Schöne et al. (1993) correlated environmental contamination with single-strand breaks in DNA of freshwater turtles using this alkaline unwinding assay.

ENZYME DYSFUNCTION AND SUBSTRATE POOL SHIFTS

Contaminants can have significant effects on enzymes and these effects are used routinely as biomarkers. Metals can influence protein-mediated catalysis, transport, and gas exchange by modifying protein structure (secondary, tertiary or quaternary) and stability (Ulmer, 1970). Metals bind to a wide range of electron donor groups of proteins such as imidazole, sulfhydryl, hydroxyl, carboxyl, amino, quanidinium and peptide groups (Eichhorn, 1975). Changes in secondary and tertiary structure can lead to lowered or elevated enzyme activities. Normally, metals stabilize quaternary structure of many proteins: substitution of another metal for the usual stabilizing metal can interfere with the coming together of peptide chains to form stable and functional oligomers, e.g., multimeric enzymes. Some enzymes are stabilized by metals (e.g., lysozyme by Mg^{2+}) and substitution of other metals for these stabilizing metals can enhance denaturation (Ulmer, 1970). Metals are also present at active sites of biomolecules such as carboxypeptidase, alkaline phosphatase, carbonic anhydrase, cytochrome c, and hemoglobin (Eichhorn, 1975). Displacement of the appropriate metal by another can change the functioning of these proteins.

Several steps in the synthesis of heme (Figure 6.5) are modified by toxicants, and associated enzyme activities and substrate pools are used as biomarkers to reflect sublethal poisoning. **Porphyrins** are produced as intermediates in heme synthesis. Porphyrins with four to eight carboxyl groups tend to be produced in excess and are excreted in urine. The relative amounts of each of these excreted porphyrins tends to be consistent among individuals. However, mercury poisoning interferes with normal heme synthesis, promotes oxidation of reduced porphyrins, and shifts porphyrin pools in the urine as a consequence. For example, rats exposed to mercury have increased levels of porphyrins with four and five carboxyl groups (Woods et al., 1993). Woods et al. (1993) used such shifts to indicate sublethal exposure to mercury in dentists who use mercury in silver-mercury amalgam fillings. Male dentists categorized as having either no mercury, or greater than 20 µg L^{-1} of mercury, in their urine had pentacarboxylporphyrin concentrations of 0.76 and 3.07 µg L^{-1}, respectively.

Polyhalogenated aromatic hydrocarbons such as PCBs may also interfere

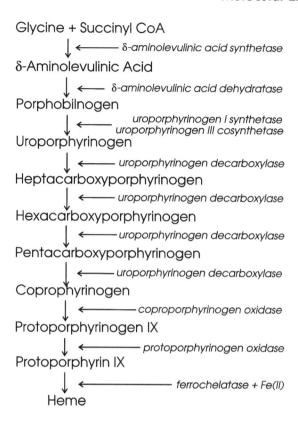

Glycine + Succinyl CoA

δ-Aminolevulinic Acid

↓ ⟵ δ-aminolevulinic acid synthetase

Porphobilnogen

↓ ⟵ δ-aminolevulinic acid dehydratase

Uroporphyrinogen

↓ ⟵ uroporphyrinogen I synthetase / uroporphyrinogen III cosynthetase

Heptacarboxyporphyrinogen

↓ ⟵ uroporphyrinogen decarboxylase

Hexacarboxyporphyrinogen

↓ ⟵ uroporphyrinogen decarboxylase

Pentacarboxyporphyrinogen

↓ ⟵ uroporphyrinogen decarboxylase

Coprophyrinogen

↓ ⟵ uroporphyrinogen decarboxylase

Protoporphyrinogen IX

↓ ⟵ coproporphyrinogen oxidase

Protoporphyrin IX

↓ ⟵ protoporphyrinogen oxidase

Heme

↓ ⟵ ferrochelatase + Fe(II)

FIGURE 6.5. Steps in the synthesis of heme. Enzymes catalyzing each conversion are italicized and placed to the right of the reaction sequence. (Modified from Figure 1 in Woods et al., 1993).

with heme synthesis, possibly by inhibiting uroporphyrinogen decarboxylase or by P-450 generation of oxyradicals that oxidize porphyrinogens (Peakall, 1992). Hepatic porphyrins have been used as biomarkers of polyhalogenated aromatic hydrocarbon exposure for species such as pike (*Esox lucius*) (Koss et al., 1986) and herring gulls (*Larus argentatus*) (Fox et al., 1988)

Other biomarkers can indicate effect to the respiratory pigments of animals. Cadmium exposure of flounder (*Pleuronectes flesus*) depressed blood hematocrit, hemoglobin titers, and red blood cell counts (Johansson-Sjöbeck and Larsson, 1978). Lead depresses the activity of δ-**aminolevulinic acid dehydratase** (δ-ALAD or ALAD), an enzyme catalyzing the conversion of δ-aminolevulinic acid to porphobilnogen during heme synthesis (see Figure 6.5). Exposure of rainbow trout (*Salmo gairdneri*) to lead lowered ALAD activities and caused anemia (Johansson-Sjöbeck and Larsson, 1979). Fish exposed to lead-containing mine drainage also had low blood ALAD activities (Dwyer et al., 1988; Schmitt et al., 1993). This effect of lead can be modified by the presence of metals such as cadmium and zinc (Berglind, 1986; Schmitt et al., 1993).

Because of their importance in animal osmoregulation and cell water regulation, ATPases (adenosine triphosphatases) have been the subject of considerable study (Thomas, 1990). Their presence on gill surfaces makes them particularly vulnerable to contact with toxicants. Yap et al. (1971) exposed bluegill sunfish (*Lepomis machrochirus*) to PCBs and noted depressed Mg^{2+}- and Na^+,K^+-ATPase activities. Rock crab (*Cancer irroratus*) gills exposed to cadmium or lead also showed depressed Na^+, K^+-ATPase activity.

SUMMARY

Understanding molecular effects enhances our ability to assign causal linkage to effects at higher levels of organization and to predict effects of untested chemicals based on similar molecular interactions with biomolecules. It also provides biomarkers as tools to proactively measure effects in field situations. Some biomarkers are quite specific (lead's effect on ALAD activity), specific to a class of toxicants (metallothioneins), or general (stress proteins). All have value if the temptation to attach too much ecological relevance to a biochemical response is avoided.

Selected Readings

Di Giulio, R.T., W.H. Benson, B.M. Sanders, and P.A. Van Veld, Biochemical Mechanisms: Metabolism, Adaptation, and Toxicity, in *Fundamentals of Aquatic Toxicology, Effects, Environmental Fate, and Risk Assessment*, 2nd ed., Rand, G.M., Ed., Taylor & Francis, Washington, DC, 1995.

Di Giulio, R.T., P.C. Washburn, R.J. Wenning, G.W. Winston, and C.S. Jewell. Biochemical responses in aquatic animals: a review of determinants of oxidative stress. *Environ. Toxicol. Chem.* 8, pp. 1103–1123, 1989.

Goksøyr, A. and L. Förlin. The cytochrome P-450 system in fish, aquatic toxicology and environmental monitoring. *Aquat. Toxicol (AMST)*. 22, pp. 287–312, 1992.

McCarthy, J.F. and L.R. Shugart (Eds.). *Biomarkers of Environmental Contamination*. Lewis Publishers, Chelsea, MI, 1990, p. 457.

Peakall, D. *Animal Biomarkers as Pollution Indicators*. Chapman & Hall, London, 1992, p.291.

Shugart, L.R. Environmental Genotoxicology, in *Fundamentals of Aquatic Toxicology, Effects, Environmental Fate, and Risk Assessment*, 2nd ed., Rand, G.M., Ed., Taylor & Francis, Washington, DC, 1995.

Shugart, L.R. Molecular Markers to Toxic Agents, in *Ecotoxicology: A Hierarchical Treatment*, Newman, M.C. and C.H. Jagoe, Eds., CRC Press, Inc., Boca Raton, FL, 1996.

Thomas, P. Molecular and biochemical responses of fish to stressors and their potential use in environmental monitoring. *Amer. Fish. Soc. Symp.* 8, pp. 9–28, 1990.

Cells, Tissues, and Organs

... the problem took the form of habitat pollution → DDE accumulation in prey species →
DDE in predators → decline in brood size → potential extermination.
The same phenomenon can [be written as] lipid soluble toxicant →
bioaccumulation in organisms with poor detoxification systems ... → vulnerable
target organ [shell gland] → inhibition of membrane-bound ATPases ... →
potential extermination. Ecologists would claim a decline in population
recruitment, biochemists an inhibition of membrane enzymes.
SIMKISS (1996)

INTRODUCTION

Two superficially opposing views of how to deal with hierarchical subjects exist. At one extreme, the reductionist approach attempts to understand the behavior of the simplest units at the lowest levels of organization, and uses this understanding to explain phenomena at all higher levels. By analogy, a clock is understood completely if one understands the workings of all the clock parts. In apparent contrast, the holistic approach holds that, because unique properties emerge at higher levels and complex interactions among parts beget complex dynamics, an understanding of higher order processes is much more useful than building a causal structure from the lowest or most fundamental level to the highest. In reality, both approaches are successful only when used in a mutually supportive fashion. Alone, neither works beyond a limited scope. Prediction is limited if one doesn't understand the components of a system but, instead, only describes phenomena at the highest level of organization. Limited prediction makes it impossible to move effectively toward the scientific goal of ecotoxicology. The reductionist approach alone is also impractical because our limited knowledge makes prediction of some emergent properties impossible based solely on mechanisms at lower levels. Also causal structure does not always proceed up the artificial, hierarchical structure of biological organization. Effects at the tissue level can be influenced by physiological (one level up) in addition to biochemical (one level down) mechanisms. Arguably, the physiological causes could be traced back to their biochemical mechanisms. However, our incomplete understanding of cascading events, the network of interlinked cause-effect phenomena, and the uncertainty associated with the magnitude of each phenomenon in the causal web condemns such an exercise to failure beyond a limited number of links.

In some cases, causal structure can be defined clearly for several levels of organization. One such causal sequence from the biochemical → cell → individual that can be quickly evoked is activation of a xenobiotic by monooxygenases → production of oxyradicals and consequent DNA damage → increased risk of cancer in the liver → increased chance of death to an individual. However, extrapolation of this sequence to the population level would be difficult, and to the level of an ecological community level, highly speculative. Causal linkage fails almost immediately for other cases such as that between ALAD activity in red blood cells and individual fitness. Future work will eventually allow clearer definition for some of these linkages, but it is unlikely that clear linkage will ever be made in all instances.

Success in ecotoxicology is enhanced by keeping holistic goals and limitations in mind while applying reductionistic methods to questions. Regardless of the level being examined, one should use the reductionist method of studying the "simplest system you think has the properties you are interested in" (Levinthal quoted in Platt [1964]). This should be done in full anticipation that important properties may emerge at higher levels.

As discussed in Chapter 1 and reflected again in the quote above, conceptual coherency is maintained by understanding that ". . . processes at one level take their mechanisms from the level below and find their consequences at the level above . . . Recognizing this principle makes it clear that there are no truly 'fundamental' explanations,[1] and makes it possible to move smoothly up and down the levels of the hierarchical system without falling into the traps of naive reductionism or pseudo-scientific holism" (Caswell, 1996; also Bartholomew, 1964). Thus coherency is maintained in a conceptual "relay race" in which all legs (levels of biological organization) are equally crucial to achieving the goal (Newman, 1995). Attempts to fulfill the goal of ecotoxicology using either the reductionist or holistic approaches alone are equally absurd attempts to "swallow the ocean." Whether attempted in one impossibly large gulp or in an impossibly large number of small gulps, the ocean will not be swallowed.

GENERAL CYTOTOXICITY AND HISTOPATHOLOGY

The study of effects at the cell, tissue, and organ level of organization is invaluable for several reasons. In the context of the "relay race" just described, these effects are consequences of biochemical mechanics that provide interpretative power for effects to individuals. They integrate damage done at the molecular level (Hinton

---

[1]A related incongruity arises as a consequence of not fully appreciating this context for ecotoxicological study. Often, statements are made that a specific level of organization is the most important to understanding ecotoxicological phenomena. Such statements are reminiscent of the Ptolemaic theory (the Earth is the center of the universe) and reflect a false paradigm which I will call the **Ptolemaic incongruity**. This incongruity that any particular level of biological organization holds the central role in the science of ecotoxicology leads to confusion, narrowness of vision, and wasteful intransigence.

and Laurén, 1990a). These biomarkers can be used as an early warning system for potential effects at the level of individual and, sometimes, population. **Histopathology**, the study of change in cells and tissues associated with communicable or noncommunicable disease, provides a cost-effective way to verify toxicant effect as well as exposure (Hinton, 1994).

There are two disadvantages to histopathological biomarkers. First, the normal histology and variations in normal histology with season, diet, reproductive cycle, and other processes may be poorly understood for sentinel species (Hinton and Laurén, 1990a). Much more descriptive work (normal science) is required in this area to improve the effectiveness of histopathological biomarkers. Second, although methods now exist to routinely quantify effects, most histopathological studies are qualitative and quantitation is needlessly neglected (Jagoe, 1996). Both of these limitations can be overcome by more normal science and a change in emphasis to quantitative methods.

Necrosis

A wide range of **lesions** (pathological alterations of cells, tissues, or organs) indicate exposure to toxicants and suggest mechanisms of action. Often, lesions are associated with a specific **target organ** as a result of preferential toxicant transport to, accumulation in, or activation within that organ. **Cytotoxicity** (toxicity causing cell death[2]) may be reflected in a tissue or target organ as **necrosis** (cell death from disease or injury). Pyknosis, the most obvious evidence of necrosis, involves the cell nucleus. With **pyknosis (pycnosis)**, the distribution of chromatin in the nucleus changes with the material condensing into a strongly staining mass. The nucleus stains intensely basophilic and becomes irregular in shape (Sparks, 1972). Sometimes the nucleus disintegrates **(karyolysis)**. The cytoplasm of necrotic cells tends to be more acidophilic[3] (eosinophilic) than that of viable cells. Mitochondria swell and more cytoplasmic granules may appear. Necrosis may also be indicated by displacement or separation of the cell from its normal location in a tissue (Meyers and Hendricks, 1985), e.g., cell sloughing from gill epithelium or arterial walls.

Different types of necrosis are characteristic of various insults. **Coagulation necrosis** is characterized by extensive coagulation of cytoplasmic protein, making the cell appear opaque. The cell outline and position within its tissue are retained

---

[2]In everyday use, terms such as "death" lack the clear distinctions that are important in ecotoxicology. The distinction between cell death and somatic death is a good example. Cell death or necrosis occurs in living as well as dead individuals. Death of an individual is **somatic death**. As we have already seen, stress is another term requiring unusual attention in its use. In the last chapter, we discussed cellular and oxidative stress. Later, we will discuss Selyean (individual) and ecosystem stresses. Each describes a different phenomena.

[3]An **acidophilic component** of a cell (e.g., the cytoplasm) is one that is readily stained by an acidic dye and a **basophilic component** (e.g., the nucleus) is one that is readily stained by a basic dye. In general preparations such as those shown in this chapter, hematoxylin and eosin are used to stain the nucleus and cytoplasm, respectively.

for some time after cell death (Sparks, 1972; Hinton, 1994). Coagulation necrosis occurs in the alimentary tract of mammals after ingestion of phenol (Sparks, 1972). It can also result from acute exposure to inorganic mercury because mercury denatures and precipitates proteins (Sparks, 1972). Renal failure may result from accumulation and eventual spillover of metallothionein-associated metals in kidneys (the target organ) and subsequent coagulation necrosis (Hinton and Laurén, 1990a) of kidney cells. This type of necrosis may also occur with an abrupt cessation of blood flow and, for this reason, Hinton and Laurén (1990b) suggest that it is a good biomarker for cytotoxicity. **Liquefactive (cytolytic) necrosis** occurs with rapid breakdown of the cell as a consequence of the release of cellular enzymes. Many cells undergoing liquefactive necrosis in a tissue, especially a tissue with much enzymatic activity, can produce fluid-filled, necrotic spaces in the tissue (Meyers and Hendricks, 1985). Hinton and Laurén (1990b) suggest that liquefactive necrosis is less useful as a biomarker than coagulation necrosis because liquefactive necrosis is often associated with infection. Several other forms of necrosis have been described. With **caseous necrosis**, cells disintegrate to form a mass of fat and protein. **Gangrenous necrosis** is a combination of coagulation and liquefactive necrosis (Sparks, 1972) often resulting from puncture (with associated lack of blood supply to the damaged tissue[4]) and subsequent infection. Meyers and Hendricks (1985) describe **fat necrosis** which involves deposits of saponified fats in dead fat cells. **Zenker's necrosis** occurs only in skeletal muscle and is similar to coagulation necrosis. All reflect cell death but coagulation necrosis seems to best reflect toxicant effect.

Inflammation

Inflammation can be useful as a biomarker of toxicant effect. It is a response to cell injury or necrosis that isolates and destroys the offending agent or damaged cells (Sparks, 1972). As such, inflammation often is associated with necrosis. For example, hepatic necrosis due to toxicant action can be accompanied by inflammation (Hinton and Laurén, 1990b). Inflammation continues through to lesion healing. The net result of inflammation is healed tissue with damaged cells being replaced by functioning cells (La Via and Hill, 1971).

The four **cardinal signs of inflammation** are heat, redness, swelling, and pain although heat is irrelevant for poikilotherms. Blood vessels in the damaged region dilate to increase blood flow to the area. This causes redness and heat. Pain is caused by the pressure of tissue swelling. Swelling is a result of fluid from the blood passing through blood vessel walls and into tissues of the inflamed area. Leucocytes can leave the blood vessels and enter the area of damaged tissue. Consequently, the infiltration of such cells (neutrophilic granulocytes or neutrophils, mononuclear cells, and lymphocytes) also indicates inflammation. Later in the process, small blood vessels begin to form and connective tissue begins to grow in

---

[4]The resulting inadequacy of blood supply to surrounding tissues is called **ischemia**.

a mass called the **granulation tissue** (La Via and Hall, 1971). Scar tissue may be formed due to fibroblast and collagen infiltration of the damaged tissue (Sparks, 1972). If inflammation continues for too long as a consequence of chronic damage or infection, dense collections of collagenous connective tissue form and can produce tissue dysfunction. Consequently, such accumulations can reflect chronic inflammation.

Although the inflammatory response is common to all metazoan phyla, details vary. Students interested in studying nonmammalian species are advised to consult books describing the pathology and immunology of invertebrates and lower vertebrates (e.g., Sparks, 1972; Cooper, 1976) rather than relaying solely on mammalian pathology books.

Other General Effects

Several other general effects of toxicants are observed in cells and tissues. Swelling of mitochondria has been correlated with metal contamination (Aloj Totaro et al., 1986) or poisoning (Squibb and Fowler, 1981). **Lipofuscin** accumulation is another biomarker of toxicant effect. Lipofuscin is a degradation product of lipid peroxidation that accumulates in cell vacuoles called **residual bodies** (La Via and Hall, 1971). Lipofuscin accumulates with age in some cell types (e.g., neurons) giving it the name, **age pigment** (La Via and Hall, 1971; Aloj Totaro et al., 1986). It also increases with exposure to some toxicants. For example, lipofuscin granules in squid (*Torpedo marmorata*) neurons increased with copper exposure (Aloj Totaro et al., 1985; 1986), reflecting enhanced lipid peroxidation due to oxyradical generation by copper (see Chapter 6). Interestingly, the portion of the nervous system with the most lipofuscin accumulation (the electric lobe) had the lowest activities of superoxide dismutase of all central nervous system tissues examined. Metal-rich granules (Figure 7.1) are another biomarker that have been discussed previously. They are found in almost all animal phyla (Simkiss, 1981). Their presence in unusually high densities in tissues such as the hepatopancreas, arthropod midgut, and kidney suggests a detoxifying response to metal exposure (Brown, 1978; Hopkin and Nott, 1979; Simkiss and Taylor, 1981; Mason et al., 1984; Hopkin et al., 1989).

GENE AND CHROMOSOME DAMAGE

Contaminants can produce changes in DNA and such changes can affect cells and tissues. Such changes can increase both somatic and genetic risk. **Somatic risk** is that of an adverse effect to the exposed individual resulting from genetic damage to somatic cells (e.g., damage leading to cancer) and **genetic risk** is the risk to the progeny of an exposed individual as a consequence of heritable genetic damage (e.g., damage to germ cells or gametes leading to a nonviable fetus or offspring with a birth defect). These effects are manifest in a variety of cellular structures or

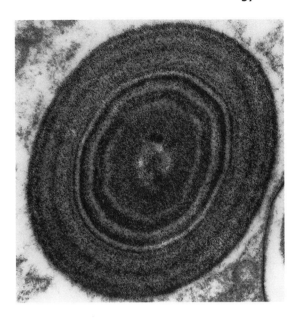

FIGURE 7.1. A transmission electron micrograph of a Type A granule (cross section) from the digestive gland of the spider, (*Dysdera crocata*). Note the characteristic concentric layering in the granule. The granule section is approximately 2 μm in diameter. (Courtesy of S.P. Hopkin, University of Reading, with permission of Chapman & Hall.)

processes (Figure 7.2). Several effects are routinely used as biomarkers and provide a mechanism for consequences at the individual level such as mutation or cancer.

Toxicant-mediated chromatid breakage can be correlated with the incidence of **sister chromatid exchange (SCE)**[5] of DNA (Figure 7.2A). Dixon and Clarke (1982) showed a clear dose-SCE response for mutagen[6] (mitomycin C) exposure of the blue mussel (*Mytilus edulis*). A similar relationship was found for *Mytilus edulis* larvae exposed to mutagens (Harrison and Jones, 1982). Consequently, sister chromatid exchange has been proposed as an assay of DNA damage (Tucker et al., 1993). In this assay, one chromatid in the pair is stained with 5-bromodeoxyuridine. Then, after two cycles of cell division, tissue samples are examined for DNA exchange between chromatids of labeled segments. With no exchange, chromatids would remain either completely stained or unstained after cell division. When exchange between chromatids has occurred, chromatids with both unstained and stained segments would be produced. The number of SCE per metaphase or SCE per chromosome is used as a measure of DNA damage.

Structural **chromosomal aberrations** include breakage and loss of segments of DNA, or chromosomal rearrangements. Chromosomal breaks involve double-strand breaks as shown in Figure 7.2B (left). Fragments of chromosomes may also be apparent. Also visible may be "gaps" in chromosomes, small discontinuities in

---

[5]Before cells divide, the DNA in each chromosome is duplicated to produce two identical chromatids. As the chromosomes condense prior to division, each appears as a pair of **chromatids** connected at a common centromere. At the metaphase plate, **sister chromatids** come together with those of the other homologous chromosome to form a **tetrad**. The four sister chromatids (2 associated with each chromosome) may exchange segments of homologous DNA by the breaking and rejoining with crossing over of DNA segments.

[6]A **mutagen** is a physical or chemical entity capable of producing mutations.

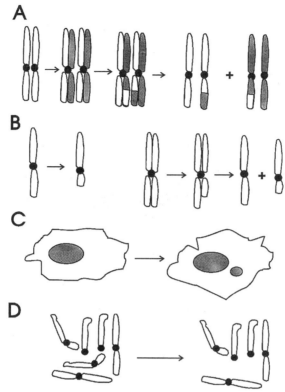

FIGURE 7.2. Damage to genetic materials can be assessed with a variety of cellular qualities. Increased rates of sister chromatid exchange (A) can reflect the rate of DNA damage although the exchange itself is not injurious. Beginning as a pair of homologous chromosomes (a homolog), four chromatids are formed. Sister chromatids in this tetrad (four chromatids paired as two homologs at metaphase) may exchange material. In the sister chromatid exchange assay, DNA is stained in the first round of cell division with 5-bromodeoxyuridine and the cells then synthesize unstained DNA in subsequent divisions. This results in stained and unstained chromatids that may exchange DNA in future cell divisions. This exchange is seen as chromatids with both stained and unstained segments. The rate of exchange is correlated with the frequency of DNA breakage. Chromosomes may be damaged (B, left) producing a chromosomal aberration. Viewed under the light microscope, these aberrations may appear as breaks or gaps in chromosomes. If only one chromatid in a homolog is broken, a chromatid aberration occurs (B, right). With a chromatid aberration, only one chromosome in one of the two daughter cells would be damaged. Failure of mitotic processes can produce micronuclei (C) or aneuploidy, a deviation from the usual ploidy (e.g., the usual 2N diploid number of chromosomes) (D).

the chromosome. A **chromatid aberration** occurs if only one strand (chromatid) is broken (Figure 7.2B, right).

There have been numerous studies using aberrations as biomarkers. Chromosomal aberrations in bone marrow cells were higher in mice (*Peromyscus leucopus*) and cotton rats (*Sigmodon hispidus*) from a petroleum, heavy metal and

PCB-contaminated site than in mice and rats from uncontaminated sites (McBee et al., 1987). Similarly, chromosomal aberrations in blood lymphocytes of petroleum refinery workers (0.023 to 0.037 breaks per lymphocyte) were elevated relative to those of control populations (0.015 to 0.021 breaks per lymphocyte) (Khalil, 1995). Lead poisoning increased the incidence of chromosomal aberrations in mouse bone marrow cells (Forni, 1980). Chromosomal and chromatid aberrations in workers at a lead oxide factory were elevated by sublethal lead exposure (18.7% versus 5.1% abnormal metaphases per lymphocyte for factory workers and controls, respectively) (Forni, 1980).

Anomolies during cell division, such as chromosome damage and/or spindle dysfunction, can produce micronuclei, nuclear segments isolated in the cytoplasm from the nucleus (Nikinmaa, 1992) (Figure 7.2C). Red blood cells of fish exposed to toxicants have been reported to have elevated numbers of micronuclei (Nikinmaa, 1992). Soft shell clams (*Mya arenaria*) exposed to PCB contaminated sediments of the New Bedford Harbor (Massachusetts) had a three-fold higher incidence of micronuclei in blood cells than clams from a clean site (Martha's Vineyard) (Dopp et al., 1996). Further, the number of leukemic cells per mL of hemolymph was correlated with the number of blood cells with micronuclei, suggesting a common mechanism between production of micronuclei and severity of leukemia in clams from contaminated sites.

Failure of chromosomes to properly segregate during cell division can result in aneuploidy (Figure 7.2D). With a chromosome preparation, such a condition would appear as an atypical number of chromosomes in the cell, e.g., 2N-1 or 2N+1 or more chromosomes. As discussed in Chapter 6, flow cytometry has been used to quantify aneuploidy in cells taken from individuals exposed to genotoxic agents. For example, Lamb et al. (1991) used flow cytometric methods to document aneuploidy in the red blood cells of turtles (*Trachemys scripta*) from a reservoir with elevated levels of radiocesium (^{137}Cs) and radiostrontium (^{90}Sr).

CANCER

Normal cells have the capacity to multiply and increase in tissues. This increase in the numbers of cells in a tissue or organ is called **hyperplasia**. Hyperplasia in response to a variety of normal stimuli such as in the tissue repair process described above is called **physiologic hyperplasia** (La Via and Hill, 1971). There is also pathologic or neoplastic hyperplasia. Sometimes, excessive hyperplasia occurs during response to injury or irritation, resulting in a type of pathologic neoplasia called **compensatory hyperplasia**. **Neoplastic hyperplasia** results from a hereditary change in a cell such that it no longer responds properly to chemical signals that would normally control cell growth, providing mechanism to cancerous growth. Such **neoplasia** is "hyperplasia which is caused, at least in part, by an intrinsic heritable abnormality in the involved cells" (La Via and Hill, 1971). Neoplasia that tend to remain differentiated in their morphology and grow slowly are not as invasive of tissues as other neoplasia and are termed benign. Those which take on undifferentiated forms, tend to grow rapidly, and invade other tissues are called malignant. Malignant cancers are more life-threatening than benign cancers. Often pieces of the original cancerous growth can dislodge, move to another tissue via the circulatory or lymphatic system and establish other foci of cancerous growth. This process (**metastasis**) leads to the spread of a malignant cancer from the site of origin throughout the body.

Cancer is a result of heritable change in cells. A chemical (e.g., numerous carcinogens), physical (e.g., ionizing radiation), or biological (e.g., a retrovirus) agent modifies a gene or its normal relation to other genes which results in neoplasia. This may occur through point mutations, deletions, additions, rearrangements as described above, or by gene insertion by a virus. A retrovirus may insert DNA into a cell chromosome and transform the cell so it responds improperly to growth regulating signals. A chemical or physical agent may interact with the cell's genome such that a gene involved with the normal growth and differentiation of cells (**proto-oncogene**) is changed to an **oncogene**, a gene causing cancer. Change to an oncogene results in loss of normal growth dynamics and/or differentiation of the cell. Cancer results from the inappropriate activation or activity of a gene that would normally be involved in cellular growth and/or differentiation (Moolgavkar, 1986). Other genes called **suppressor genes** may function normally to suppress cell growth and may inhibit abnormal growth. Like proto-oncogenes, suppressor genes are thought to function in the process of growth and differentiation in normal cells; the proto-oncogenes are associated with stimulatory and suppressor genes with inhibitory aspects of normal cell growth. Cancer can develop if some agent or event affects the functioning of the suppressor gene.

Some agents initiate the neoplastic process and some promote the development of the tumor (La Via and Hill, 1971). This has led to the classification of two groups of agents, **initiators** which convert normal cells to latent tumor cells and **promoters** which enhance the growth and continued expansion of latent tumor cells (Figure 7.3). Cancers are thought to pass through an initiation stage and then a promotional stage. Lesions with consequent cell proliferation (e.g., those associated with asbestos fiber in the lung), high levels of hormones with associated hyperplasia (**hormonal oncogenesis** [La Via and Hill, 1971]), and chemical agents can act to promote cancer. Often, inappropriate inactivation of suppressor genes is also associated with the promotion process (Aust, 1991), resulting in unregulated growth (Van Beneden and Ostrander, 1994). After initiation and promotion, the final step of carcinogenesis occurs, cancer progression. **Cancer progression** is the change in the biological attributes of neoplastic cells over time that lead to a malignancy (La Via and Hill, 1971). Obviously, there can be a long **latent (or latency) period** between exposure to a carcinogenic agent and the clinical appearance of cancer. This makes assignment of cause and effect very difficult because the exposure may have occurred years prior to the appearance of the effect.

The latent period and multistage process leading to cancer also makes prediction of the exact shape of the dose-cancer response curve difficult. Two theories exist and are supported by different data. The **threshold theory** assumes that there is no effect below a certain low dose (e.g., Downs and Frankowski, 1982; Cohen, 1990). Above the threshold, the slope of the response versus dose curve increases rapidly. The dose-response curve takes on the appearance of a hockey stick (Figure

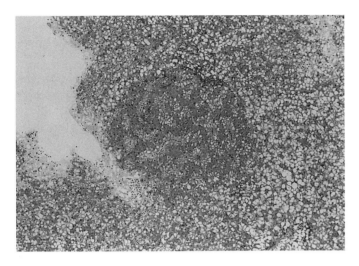

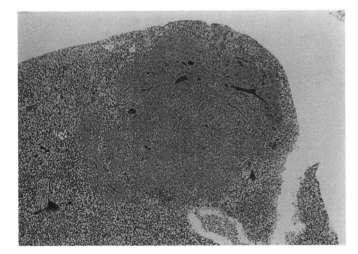

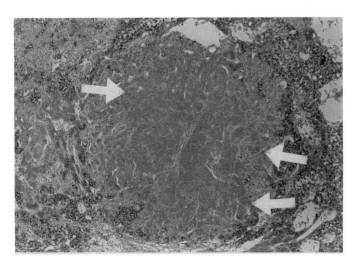

FIGURE 7.3. The progression of neoplasms in the medaka (*Oryias latipes*) initiated with the carcinogen, diethylnitrosamine. Fish had more rapid development of foci, and shorter times between exposure and realized effect (latent period) if they were fed the promoter, 17-β-estradiol in their diet. The top micrograph shows a basophilic focus of cellular alteration in the liver 25 weeks after exposure to diethylnitrosamine and provision of a diet containing estradiol (Hematoxylin and eosin, X150, Courtesy of Janis Brencher Cooke, University of California - Davis). The middle micrograph shows a solid basophilic adenoma in medaka liver exposed as already described for the top micrograph. (Hematoxylin and eosin, X75, Courtesy of Janis Brencher Cooke, University of California-Davis). The bottom micrograph shows a hepatocellular carcinoma in medaka liver at 12 weeks of exposure. Note the distinct architecture of the cancer relative to the surrounding tissue. Arrows indicate the locations of numerous mitotic figures in the carcinoma. (Hematoxylin and eosin, X125, Courtesy of Swee, The University of California-Davis).

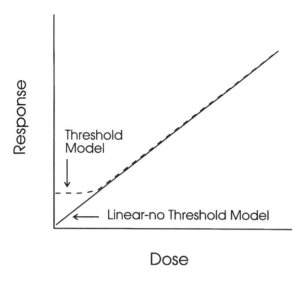

FIGURE 7.4. The linear-no threshold and threshold theories describe two dose-response models differing from each other at lower doses. The linear-no threshold model is described by a straight line and the threshold model is described by a "hockey stick" curve.

7.4). The **linear-no threshold theory** is based on several radiation-induced cancer studies where no apparent threshold exists for effect. This theory assumes that any lack of cancers below a certain dose reflects our inability to measure low incidences of cancers at these exposure levels. This dose-response model describes a straight line with no threshold.

Chemical and physical agents can act by initiation or promotion of the process leading to cancer. Relevant mutagenic mechanisms are those described previously for alterations of DNA and chromosomes. Further, agents that inhibit DNA repair may also increase the probably of cancer. Differences in DNA **repair fidelity** (accuracy in repairing and returning the DNA to its original state after damage) associated with various agents can also result in differences in carcinogenicity. For example, Robison et al. (1984) suggested that chromium as chromate, nickel, and mercury cause damage to DNA but that there is a difference in the damage and its repair. Generally, mercury produces more single-strand breaks while nickel and chromium cause more protein-DNA crosslinking. Because of differences in the fidelity of repair between these two types of damage, nickel and chromium (lower repair fidelity) are more carcinogenic than mercury (higher repair fidelity).

The presence of tumors and various tumor cell qualities associated with gene products are used as biomarkers for environmental carcinogens. In the study of micronuclei in soft shell clams discussed above (Dopp et al., 1996), leukemic cells were also measured using an antibody assay which recognized surface alterations of the cells. Enzymatic and histochemical alterations associated with cancer cells may also be used (e.g., Moore and Myers, 1994). More often, the incidence of cancers in populations is correlated with the level of contaminant. For example, liver neoplasms in sole (*Parophyrs vetulus*) from Puget Sound were correlated to sediment contamination by PAHs (Stein et al., 1990). The incidence of cervical neoplasia in Czech women increased two years (latent period) after the Chernobyl release of radionuclides (Borovec, 1995).

GILLS AS AN EXAMPLE

Gills are often damaged or changed by exposure to toxicants and will be used as an example for integrating many of the nongenetic cellular effects described in this chapter. (The consequences of genetic damage are already illustrated relative to cancer.) Gills also provide good examples of unique changes not easily predicted from biochemical mechanisms alone. The consequences of gill changes to the individual are also easily demonstrated.

The fish gill (Figure 7.5, top panel) is composed of **primary lamellae** (filaments) which extend outward at right angles from the branchial arches. On the dorsal and ventral sides of each primary lamellae are parallel rows of **secondary** (respiratory) **lamellae** which are the major sites of gas exchange. The epithelium covering the lamellae is a double layer of cells with intercellular lymphoid spaces between the two cell layers. **Chloride cells** (Figure 7.6) which function in ion regulation are found predominantly on the epithelium of primary lamellae but also on the secondary lamellae.

Upon exposure to diverse toxicants as metals, detergents, elevated H^+, or nitrophenols, the outer epithelium of the secondary lamellae often lifts away and leaves an enlarged, fluid-filled space between itself and the inner epithelial layer of cells (Skidmore and Tovell, 1972; Mallatt, 1985; Evans, 1987). Granulocyte densities in this space may rise, suggesting inflammation. With acute exposure to agents such as zinc, chloride cells may begin to separate from the epithelium or ruptures may occur in the outer epithelial layer. Cells may appear swollen with distended mitochondria, i.e., necrosis.

The number of chloride cells on the primary lamellae increased with the exposure of rainbow trout (*Salmo gairdneri*) to dehydroabretic acid, a component of Kraft mill effluent (Tuurala and Soivio, 1982). Increased number of chloride cells also occurred in mosquitofish (*Gambusia holbrooki*) exposed to inorganic mercury (Jagoe et al., 1996). Also, the size of chloride cells of mosquitofish gills increased with inorganic mercury exposure. This increase is an example of **hypertrophy**, an increase in cell size (and function) resulting from an increase in the mass of cellular components often as a compensatory response (La Via and Hill, 1971; Meyers and Hendricks, 1985). The hyperplasia and hypertrophy of chloride cells is thought to be a compensatory response to ion imbalance associated with gill damage (Jagoe et al., 1996). Individually or together, hyperplasia and epithelial lifting often produce a fusion of secondary lamellae with a consequent reduction in the capacity for gaseous exchange across the gill surface. Exposure of salmon fry to aluminum (Figure 7.5, bottom) produced extensive fusion of lamellae. The hyperplasia and hypertrophy of chloride cells of the secondary lamellae caused similar fusion of lamellae as the spaces between primary lamellae were filled with cells from the primary lamellae (Figure 7.6, bottom). The consequences at the physio-

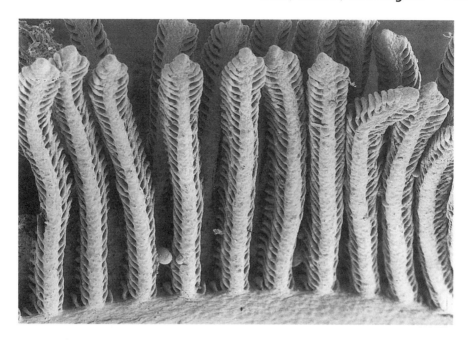

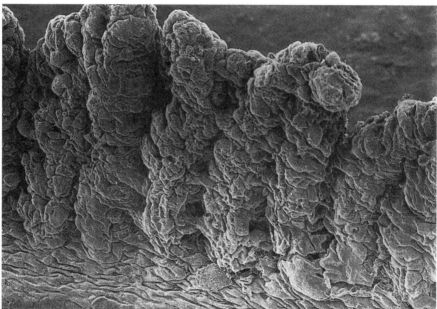

FIGURE 7.5. Electron micrographs of gills from Atlantic salmon (*Salmo salar*) fry. The top panel shows the normal gill morphology with primary lamellae extending (vertically here) from the branchial arch. Perpendicular to and on both sides of the main axis of each primary lamellae are the secondary lamellae. The bottom micrograph shows the gills of salmon fry after 30 d exposure to 300 μg L⁻¹ of aluminum. Note the extensive fusion of the secondary lamellae. (Courtesy of C.H. Jagoe, University of Georgia.)

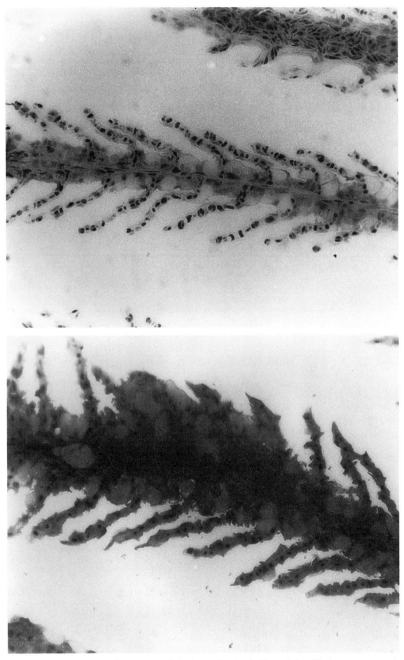

FIGURE 7.6. Micrographs of mosquitofish (*Gambusia holbrooki*) gills in cross section. The top panel shows the normal gill with secondary lamellae extending outward from the primary lamellae. Note the large chloride cells on the primary lamellae between the secondary lamellae. After exposure for 14 d to 60 μg L^{-1} of inorganic mercury (bottom panel; see Jagoe et al., 1996), the primary lamellar epithelium began filling in the spaces between adjacent secondary lamellae. The chloride cells (large cells with lightly stained cytoplasm) were involved to a large extent in this hyperplasia, becoming larger (hypotrophy) and more abundant on the primary lamellae. The secondary lamellae appeared to shorten or disappear as a consequence. Although reported elsewhere in response to gill irritation with toxicants (e.g., Tuurala and Soivio, 1982), no necrosis, separation of epithelium from the secondary lamellae, or inflammation were noted. (Toluidine blue; distance between secondary lamellae at their base is circa 20 μm; Courtesy of C.H. Jagoe, University of Georgia.)

logical and individual levels were a reduction in oxygen exchange across gill surfaces and death at lethal exposures.

SUMMARY

In this chapter, effects at the cellular, tissue, and organ level from exposure to toxicants were described. They included necrosis, inflammation, and specific changes including increased numbers of lipofuscin deposits and metal-laden granules. Damage to genes and chromosomes was described along with several associated assays for damage. Cancer development as a consequence of exposure to chemical and physical agents was discussed briefly and linked to damage due to somatic mutations. Cancer is a clear example of consequences to individuals of cellular damage. Finally, gill response to toxicants was used as an example integrating several of the cellular changes discussed in the chapter. This example linked changes at the cellular level to consequences at the physiological (ion imbalance and decreased oxygen diffusion) and individual (somatic death) level of biological organization.

Selected Readings

Hinton, D.E. Cells, Cellular Responses, and Their Markers in Chronic Toxicity of Fishes, in *Aquatic Toxicology. Molecular, Biochemical and Cellular Perspectives*, Malins, D.C. and G.K. Ostrander, Eds., CRC Press, Inc., Boca Raton, FL, 1994.

Li, A.P. and R.H. Heflich. *Genetic Toxicology*. CRC Press, Inc., Boca Raton, FL, 1991, p. 493.

Malins, D.C. and G.K. Ostrander. *Aquatic Toxicology. Molecular, Biochemical, and Cellular Perspectives*. CRC Press, Inc., Boca Raton, FL, 1994, p. 539.

Meyers, T.R. and J.D. Hendricks. Histopathology, in *Fundamentals of Aquatic Toxicology. Methods and Applications*, Rand, G.M. and S.R. Petrocelli, Eds., Hemisphere Publishing Corp., Washington, DC, 1985.

Sublethal Effects to Individuals

*Our notions of law and harmony are commonly confined to those instances which we detect;
but the harmony which results from a far greater number of seemingly conflicting, but really
concurring laws, which we have not detected, is still more wonderful.*

THOREAU (1854)

GENERAL

In the last chapter, we discussed effects at the cellular to organ levels including cell death. In the next chapter, death of individuals (somatic death) will be detailed. Sandwiched between these two categories of toxicant effects are sublethal effects to individuals. The emphasis in the last chapter was on use of sublethal effects to indicate mode of action. The emphasis here will be on adverse impacts. Although these sublethal effects are often more difficult to measure than lethal effects, they are as likely, or more likely, to be important in determining the ultimate consequences of many pollution scenarios.

Sublethal effects are effects occurring at concentrations (or doses) below those producing direct somatic death (Rand, 1985). They are most often recognized as a change of some important physiological process, growth, reproduction, behavior, development, or a similar quality. Nearly always, they are adverse or putatively adverse effects, lowering an individual's fitness.

The concept of sublethal effects was first developed with a pharmacological or mammalian toxicological emphasis. Although useful and widely applied in ecotoxicology, its meaning is more ambiguous. Some sublethal effects will have lethal consequences in an ecological context, i.e., an arena in which the individual must successfully compete with other species, avoid predation, and cope with multiple stressors. For example, an individual may be able to survive a sublethal exposure but have reduced ability to evade predators or effectively forage for food. Thus, the result of a sublethal exposure may be death in a natural setting. The concept of **ecological mortality or death** is used to describe the toxicant-related diminution of fitness of an individual functioning in an ecosystem context that is of a magnitude sufficient to be equivalent to somatic death (Newman, 1995). It is important to remain open to the very real possibility of ecological mortality at exposures that are judged sublethal based on laboratory assays. Also, a sublethal effect that results in an individual that is incapable of producing viable offspring could be considered a lethal effect because the individual's fitness (ability to contribute offspring to the next generation) could be equivalent to that of a dead individual (Rand and Petrocelli, 1985). On the other hand, complex behaviors such as avoidance of contaminated areas of the habitat may ameliorate sublethal consequences of contamination.

SELYEAN STRESS

When dealing with sublethal effects, the concept of stress is frequently invoked. The implication is that sublethal exposure causes individuals to function suboptimally. The concept of stress to individuals must be clarified here because it does not always have the same meaning to different people. When applied in a medical or scientific sense to individuals, stress has a precise meaning that does not correspond with its general meaning. This confusion of the scientific/medical and general meaning of stress in the ecotoxicology literature has led to many problems.

Hans Selye formulated the medical concept of stress as applied to individuals about 60 years ago. As a medical student, he noted a nonspecific response of the human body when extraordinary demands were made upon it (Selye, 1973). He referred to this nonspecific response as stress and defined it as "the state manifested by a specific syndrome which consists of all the nonspecifically induced changes within a biological system" (Selye, 1956).[1] Selyean stress is a specific suite of changes constituting the body's attempt to reestablish or maintain homeostasis while under the influence of a stressor (Adams, 1990).

The **General Adaptation Syndrome (GAS)** associated with Selyean stress has three phases: the alarm reaction, adaptation or resistance, and exhaustion phases (Figure 8.1). All phases of the GAS serve to resist deviation from, or to regain, homeostasis (Newman, 1995). In the alarm component, blood pressure and heart rate increase as short-term responses to compensate for an immediate stressor. Because these short-term, energy-intensive mechanisms cannot be maintained indefinitely, the organism may enter the second stage if stress persists. Responses that enhance tissue level compensation such as enlargement of the adrenal cortex are typical of the adaptation stage; however, adverse responses such as shrinkage of the thymus, lymph nodes, and spleen, and appearance of gastric ulcers can also occur for mammals stressed to this phase (Selye, 1973). If exposure to enough stressor continues for sufficient time, the individual's ability to resist change is exceeded and it enters an exhaustion phase. In this last stage, the individual slowly fails in its efforts to compensate for the effects of the stressor and eventually dies if exposure continues.

Selyean stress is a specific syndrome. Phenomena such as necrotic damage

---

[1] We will further qualify stress to an individual as **Selyean stress** so as to distinguish it from other applications of the term in this book and throughout the literature.

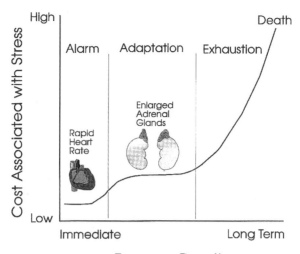

FIGURE 8.1. The three stages of the General Adaptation Syndrome (GAS). In the initial alarm phase, immediate responses to stress include increases in blood pressure and heart pumping rate. Cells of the adrenal cortex also discharge their granules into the blood. With long durations of exposure, adrenal enlargement occurs and cells of the adrenal cortex may become rich in granules again (Selye, 1973). Finally, after sufficient exposure, the ability of the body to compensate for the stressor's effects is exceeded and the individual slowly becomes exhausted. If the stress continues, the individual will die.

or metallothionein induction were not included in the stress concept as originally proposed by Selye. The term "Selyean stress" should not be used for these effects. Necrosis constitutes damage, not a response to a stressor and metallothionein induction is a specific response to a class of chemicals, i.e., cadmium, copper, mercury, silver, and zinc. Induction of cytochrome P-450 monooxygenase by a PAH would not be part of a stress response, but the stress protein response to a stressor as described in Chapter 6 seems to fit into the stress concept. It is important to keep clear these distinctions among Selyean stress, damage, specific responses to, and effects of toxicants. In the remainder of this chapter, most of the sublethal effects described will not be examples of Selyean stress.

GROWTH

Growth is often chosen as the response variable to measure sublethal effects. Not only is it easy to quantify, growth integrates a suite of biochemical and physiological effects (Mehrle and Mayer, 1985) into one quality that is often linked to individual fitness. For example, growth is reduced for fish under acidic conditions and for green tree frog (*Hyla cinerea*) tadpoles exposed to a combination of low pH and elevated aluminum concentrations such as is often found in water bodies impacted by acid rain (Jung and Jagoe, 1995). Woltering et al. (1978) measured reduced growth of largemouth bass (*Micropterus salmoides*) exposed to ammonia and attributed the reduced growth to decreased feeding rate. Growth was reduced for redbreast sunfish (*Lepomis auritus*) inhabiting a stream contaminated with mixed waste (Adams et al., 1992). In contrast, white sucker (*Catostomus commersoni*) from a lake contaminated with metals had higher growth rates but reduced longevity relative to white suckers from a nearby, uncontaminated lake (McFarlane and Frazin, 1978). Some toxicants reduce growth in fish but differences in body size between exposed and nonexposed individuals may disappear later due to compensatory growth (Sprague, 1971).

The dose-response relationship for toxicant-influenced growth often conforms to a threshold model (Figure 7.4), but this is not always the case. Sometimes, a stimulatory effect is exhibited with exposure to low, subinhibitory levels of toxicants or physical agents. Such **hormesis** is not usually a toxicant-specific response. Instead, it is a general phenomenon that appears during exposure to a variety of stressors. For example, peppermint (*Mentha piperita*) exposed to a series of concentrations of the growth inhibitor, phosfon, grew fastest at the lower concentrations of 2.5×10^{-12} to 2.5×10^{-8} M (Calabrese et al., 1987). Faster growth of peppermint at these concentrations than in the control group yields a biphasic dose-response. The **biphasic dose-effect model** is shaped like the threshold model in Figure 7.4, but the curve dips down from the controls before increasing with dose. In this case, the downward dip would reflect enhanced growth of plants exposed to low concentrations but, in other cases, it might reflect decreased mortality or increased fecundity at low concentrations relative to controls. Stebbing (1982) and Sagan (1987) suggest that the mechanism for hormesis may be regulatory overcompensation or an overresponse of organisms after subinhibitory challenge. Regardless, hormesis has been documented for survival, growth, seed germination, cancer incidence, antibody titer, and fertility in individuals exposed to a variety of agents including toxicants and radiation (Doust et al., 1994; Robohm, 1986; Stebbing, 1982; Sagan, 1987; Wolff, 1989).

Interest in hormesis is not restricted to nonhuman species. Indeed, hormesis is a foundation concept of **homeopathic medicine**, a branch of medicine not given full attention by many physicians in North America but practiced to varying degrees throughout the world (Figure 8.2). Homeopathic medicine, founded by the Samuel Hahnemann, is based on the **law of similars** (a drug that induces symptoms similar to those of the disease will aid the body in defending itself by stimulating the body's natural responses). By stimulating these responses (symptoms), a drug is thought to enhance the processes that the individual uses in resisting the disease.

FIGURE 8.2. A pharmacy display window in Barcelona, Spain advertising homeopathic medication ("homeopatia" = homeopathy). Because homeopathic medicine does not focus on the causal mechanisms of disease, it is judged to be an untenable practice by many professionals. Perhaps the concept of hormesis has not been incorporated as fully as warranted into the science of ecotoxicology because of its association with this controversial branch of medicine.

DEVELOPMENT

Developmental Toxicity and Teratology

Some contaminants can adversely impact the developing fetus or embryo. The *in utero* effect of mercury on humans was mentioned briefly in our discussion of the Minamata disease. *In utero* effects of mercury include macrocephali (abnormally large head), asymmetrical skull, depressed optical region of the skull, lowered IQ, poor muscular coordination, hearing loss or impairment, poor speech, poor walking skills, and mental retardation (Khera, 1979). Any physical or chemical agent, such as mercury, that is capable of causing developmental malformations is called a **teratogen**. **Teratology** is the science of fetal or embryonic abnormal development. Some of the qualities mentioned above for mercury involve functional deficiencies, e.g., lowered IQ, that may not be considered in classic teratology. Also, some contaminants slow growth of the developing organism but do not produce an anatomical abnormality. The broader term, **developmental toxicity** is used to include altered growth and functional deficiencies in addition to teratogenic effects (Weis and Weis, 1989a).

Teratogenic effects often have a threshold: a critical amount or concentration of teratogen is needed before an effect is manifest (Weis and Weis, 1989a). Although some teratogens, such as the infamous thalidomide,[2] are specific in their action, most teratogenic contaminants are believed to be relatively nonspecific. According to **Karnofsky's law**, any agent will be teratogenic if it is present at concentrations or intensities producing cell toxicity (Bantle, 1995). Teratogens act by disrupting mitosis, interfering with transcription and translation, disturbing metabolism, and producing nutritional deficits (Weis and Weis, 1987). Consequences of these disruptions include abnormal cell interactions, migration and growth, and excessive or inadequate cell death. In general, effects early in development tend to be more deleterious than those occurring later because early damage affects cells that will go on to differentiate and to become involved in a wider range of organs and tissues (Bantle, 1995).

Most developmental toxicology focuses on effects occurring during or after exposure of the egg to contaminants. Exposure may occur across the placenta, from contaminants deposited in the yolk, from egg exposure before fertilization, between egg shedding and elevation of the chorion, or after elevation of the chorion (Weis and Weis, 1989a). However, evidence suggests that some congenital diseases and birth defects in humans may also be linked to male exposure to contaminants (Gardner et al., 1990; Stone, 1992). For example, men working at the Sellafield nuclear fuel processing plant in England had a higher incidence of children with leukemia than control groups (Gardner et al., 1990), notionally due to a chromosomal translocation which activated proto-oncogenes (Evans, 1990; Kondo, 1993). Previously, such **male-mediated toxicity** (disease and birth defects produced by a father's exposure to a physical or chemical agent) had been judged to be insignificant based on epidemiological studies of atomic bomb survivors of Hiroshima. However, Stone (1992) reports that males in some professions (painters, mechanics, and farmers) may be at higher risk of fathering children with teratogenic problems.

A wide range of developmental problems has been described for organisms exposed to contaminants. For fish, the most common are those of the skeletal system and associated musculature, circulatory system, optical system, and retardation of growth (Weis and Weis, 1989a). Skeletal system problems include **scoliosis** (the lateral curvature of the spine) and **lordosis** (the extreme, forward curvature of the spine) (Figure 8.3). Slowing of growth may increase the time a developing individual remains in a critical stage and, consequently, the likelihood of a problem becoming manifest (Weis and Weis, 1989a). Birds exposed to contaminants produced embryos with a variety of eye, limb, beak, heart, and brain abnormalities (selenium exposure) (Ohlendorf et al., 1986) or had increased incidence of egg failure (mercury, selenium, DDE and DDT exposure) (Henny and Herron, 1989). Re-

---

[2]In the early 1960s, the sedative, thalidomide was given to pregnant European women to treat nausea. It soon became apparent that abnormalities occurred if it was taken during a critical two-week period of active limb morphogenesis (La Via and Hill, 1971). Roughly 10,000 children were born with severely malformed limbs. This event lead to revision of drug testing procedures. Although it has become symbolic of the tragic consequences of drug use prior to extensive testing, thalidomide is prescribed more judiciously today as an immunosuppressant and is being considered for treatment of some symptoms of AIDS.

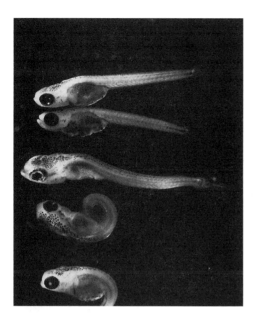

FIGURE 8.3. Fish (*Fundulus heteroclitus*) exposed to no (top three individuals) or 10 mg L^{-1} of Pb^{2+} (bottom two individuals) during development. The effect threshold is typically 1 mg L^{-1} of Pb^{2+}. Note the failure of the exposed individuals to uncurl after hatching. (Courtesy of P. Weis, UMDNJ - New Jersey Medical School).

cently, Weis and Weis (1995) demonstrated that **behavioral teratology** (behavioral abnormalities in otherwise normal appearing individuals arising after exposure to an agent as an embryo) can also be important. They showed that exposure of mummichog (*Fundulus heteroclitus*) embryos to methylmercury lowered the ability of this fish to capture prey after hatching.

There are standard assays for measuring developmental effects. The widely-accepted **FETAX** (frog embryo teratogenesis assay) uses embryos of the clawed frog (*Xenopus laevis*). Although the test species is not native to North America or Europe, the convenience of producing ample amounts of eggs and sperm, well-established procedures for the assay (e.g., Bantle and Sabourin, 1991; ASTM, 1993), and the large database for this species make it an appealing tool (Bantle, 1995). In the assay, eggs are exposed to different concentrations of the contaminant for a set time, e.g., 96 h. The proportions of exposed eggs showing mortality and the proportion of living embryos with developmental abnormalities are scored for each treatment concentration. Using formal methods described in the next chapter, the concentrations producing 50% mortality of eggs (LC50) and producing 50% of embryos with abnormalities (EC50 or TC50) are calculated. (EC stands for effective concentration and TC stands for teratogenic concentration.) A **teratogenic index** (TI) is calculated as the LC50 divided by the EC50. It reflects the developmental hazard of a contaminant. Higher values indicate increased developmental hazard of the tested contaminant. Bantle (1995) believes that TI values less than 1.5 indicate little developmental hazard.

Sexual Characteristics

Estrogenic chemicals (xenobiotic estrogens or xenoestrogens) mimic estrogen and can cause changes in the sexual characteristics of individuals. Like estrogen, these chemicals regulate the activity of estrogen-responsive genes by binding to estrogen receptors (Jobling et al., 1996). They disrupt the hormonal systems, affecting sex organ development, behavior, and fertility. For example, male sea gulls may ignore nesting colonies, and females may pair and nest together as a consequence of modified behavior by estrogenic chemicals (Hunt and Hunt, 1977; Luoma, 1992). Indeed, gull egg treatment with DDT results in feminization of the reproductive system of hatched males (Fry and Toone, 1981). Estrogenic chemicals include xenobiotics such as DDT, DDE, dioxin, PCB, and alkylphenols (e.g., p-nonyl-phenol and the surfactant, alkylphenol polyethoxylate). Xenobiotic estrogens can bind to hormone receptors and induce an abnormal, elevated response. Alternatively, some estrogenic chemicals with minimal estrogenic activity bind to the receptor and block the normal hormone's action as a consequence (McLachlan, 1993). Kelce et al. (1995) found that some estrogenic chemicals such as DDE can also block androgen receptor-mediated processes, and in doing so, act as **androgen receptor antagonists**.

Bergeron et al. (1994) provided a straightforward demonstration of PCBs acting as xenobiotic estrogens. Eggs of the red-eared slider turtle (*Trachemys scripta*) were exposed to either: no PCB, different concentrations of PCBs, or the hormone, estradiol-17β (Figure 8.4). Because this species displays temperature-dependent sex determination, researchers could manipulate the sexes of hatchlings by con-

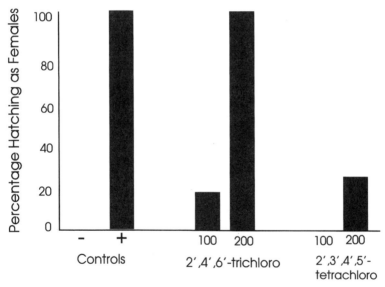

FIGURE 8.4. The percentage of hatchling turtles that are females after exposure to PCBs. Eggs were incubated at 26 to 28°C which will cause all hatchlings to be males (negative control). Eggs in the negative control treatment (-) were spotted on their surfaces only with the solvent used for the other treatments (95% ethanol) and those in the positive control treatment (+) were spotted with the hormone, estradiol-17β. The two PCBs spotted onto eggs were 2′, 4′, 6′-trichloro-4-biphenylol (labeled 2′, 4′, 6′-trichloro) and 2′, 3′,4′, 5′- tetrachloro-4-biphenylol (labeled 2′,3′,4′,5′-tetrachloro). Both were dosed at 100 and 200 μg per egg as indicated above the PBC labels. (Modified from Figure 1 of Bergeron et al., 1994.)

trolling incubation temperatures. In the experiment, eggs were incubated at a temperature (26 to 28°C) that would produce all males. The application of 200 µg of 2′, 4′, 6′-trichloro-4-biphenylol to the eggshell surface resulted in 100% of the hatchlings emerging as females despite the incubation temperature. The estrogenic activity of a second PCB (2′, 3′,4′, 5′-tetrachloro-4-biphenylol) was much lower than that of 2′, 4′, 6′-trichloro-4-biphenylol.

Masculinization of females can also result from exposure to contaminants. **Imposex** (the imposition of male characteristics on females, e.g., a penis or vas deferens) occurred for mosquitofish (*Gambusia* sp.) exposed to Kraft mill effluent (Howell et al., 1980). Females developed gonopodia, a modified anal fin which functions as an intromittent organ in males, and displayed male reproductive behavior. Bortone et al. (1989) suggested that mosquitofish exposed to Kraft mill effluent take in sterols (stigmastanol and β-sitosterol) that have been modified by *Mycobacterium smegmatis* to compounds having androgenic effects. Tributyltin (TBT) compounds used in antifouling paints have also been implicated in high incidence of imposex in marine snails inhabiting coastal regions of North America and England (Bryan and Gibbs, 1991; Saavedra Alvarez, 1990). In areas of heavy boat traffic, the occurrence of a large proportion of reproductively incompetent individuals resulted in decimated populations of important whelk species and consequent shifts in the composition of the ecological community (Bryan and Gibbs, 1991).

Developmental Stability

Developmental stability, the capacity of an organism to develop into a consistent phenotype in an environment, may also be influenced by, and used to suggest impact of, contaminants. Developmental stability has been shown to be correlated with fitness within a particular setting. Individuals that fail to achieve a consistent phenotype often show lower survivorship or reduced reproductive output. As will be seen, developmental stability is easily and inexpensively measured in the field or laboratory, and is relatively universal in its applicability (Zakharov, 1990; Graham et al., 1993a). Thus, developmental stability has great advantage as an indicator of contaminant effects.

Developmental stability is often examined by measuring deviations from perfect form. For bilaterally symmetrical organisms, this is often calculated as deviations from perfect symmetry. Bilateral characters, such as the lengths of the right and left claws of a crab, are measured and the difference calculated, $d = \text{Length}_{right} - \text{Length}_{left}$. Usually many characters are measured or counted. Deviations from perfect bilateral symmetry (**fluctuating asymmetry or FA**) measured within a population are thought to reflect perturbations of normal developmental processes. If the mean of the d values from a population is 0 and the distribution of d is normal, the variance of d within the population is a measure of fluctuating asymmetry. Using the notation of Zakharov (1990), the mean (M_d) and standard deviation (σ_d) of d are the following,

$$M_d = \frac{\sum d_i}{n} \tag{8.1}$$

$$\sigma_d = \sqrt{\frac{\sum(d - M_d)^2}{n-1}} \tag{8.2}$$

where n = the number of individuals measured.

Bilateral organisms can also exhibit directional asymmetry and antisymmetry. **Directional asymmetry** exists if the mean is not 0.[3] For example, measurement of d for weights of left and right arms of humans would display directional asymmetry because most humans are right-handed. **Antisymmetry** is indicated if the distribution of d is bimodal. Graham et al. (1993b) give the example of male fiddler crab claws as antisymmetry: some males have larger right claws but others have larger left claws. Although directional asymmetry and antisymmetry may be used to measure contaminant effects on developmental stability, fluctuating asymmetry is most often used.

Minimal fluctuating asymmetry is expected under optimal (benign) conditions. As conditions become increasingly different from some optimal range, as in the case of increasing concentrations of a contaminant, fluctuating asymmetry is expected to increase. This increase in fluctuating asymmetry reflects movement away from developmental stability.

Several field and laboratory studies have shown the value of measuring fluctuating asymmetry to assess toxicant effects. For example, Graham et al. (1993a) measured traits of morning glory (*Convolvulus arvensis*) growing at various distances from an ammonia plant in the Ukraine and found that fluctuating asymmetry was highest for leaf samples taken nearest the factory. In a laboratory study in which flies (*Drosophila melanogaster*) were fed either lead or benzene, fluctuating asymmetry increased with exposure to these toxicants, suggesting a diminished ability to maintain a consistent phenotype (Graham et al., 1993c).

REPRODUCTION

Lowered fitness of individuals due to reproductive impairment is arguably among the most useful sublethal effects measured by ecotoxicologists (Sprague, 1971, 1976). In the first chapter, reproductive failure traced to DDT/DDE inhibition of Ca-dependent ATPase in the eggshell gland (Kolaja and Hinton, 1979) and consequent eggshell thinning was discussed as one of the earliest events leading to our increased awareness of pesticide impacts to nontarget species. In this chapter, we discussed adverse impacts to the sea gull reproductive system due to an estrogenic xenobiotic and population failure of whelks experiencing high incidences of imposex as a consequence of TBT exposure. Reproduction remains one of the most frequently measured qualities in both field and laboratory studies of contaminants.

The following field studies further illustrate the value of measuring reproductive variables. White suckers (*Catostomus commersoni*) from a lake near the

---

[3]According to Zakharov (1990), a simple t-test can be used to determine if M_d is significantly different from 0. The test with the appropriate t value assesses whether $tM_d = M_d/(\sigma_d/\sqrt{n})$.

Flin Flon metal smelters (Manitoba, Canada) had smaller eggs, and lower egg and larval survival than suckers sampled from a clean lake (McFarlane and Franzin, 1978). *Catostomus commersoni* had a high incidence of reproductive failure in a lake contaminated with metals (Munkittrick and Dixon, 1988); however, this failure was attributed to modification of the food base for the sucker, not to a direct effect of the metals. Western mosquitofish (*Gambusia affinis*) taken from a selenium-contaminated water body showed lowered fry survival and more stillborn fry than mosquitofish from a clean site (Saiki and Ogle, 1995). Starry flounder (*Platichthys stellatus*) from a highly urbanized region of the San Francisco Bay had more previtellogenic oocytes and lower embryo success than flounder sampled from a less urbanized region (Spies et al., 1989). Purple sea urchin (*Strongylocentrotus purpuratus*) egg fertilization was diminished by oil production effluent (Krause, 1994).

Numerous laboratory experiments demonstrate the impact of toxicants on reproductive performance and augment field studies such as those just mentioned. In one such study, reproductive impairment by metals was estimated by using a target 16% reduction in the number of young produced per female *Daphnia magna* (Biesinger and Christensen, 1972). Mosquitofish (*G. holbrooki*) in mercury-spiked mesocosms had low numbers of late stage embryos (Mulvey et al., 1995). Standard laboratory methods also exist for measuring various reproductive qualities as affected by contaminants. Weber et al. (1989) outline these methods for reproduction of zooplankton species and fathead minnow (*Pimephales promelas*). Chapman (1995) describes a standard method for quantifying fertilization success of purple urchin eggs.

All of these methods are useful for predicting effects on field populations if combined with an adequate understanding of the reproductive biology and ecology of the species in question. In the absence of such information, effectiveness of identifying consequences of the sublethal effect is uncertain. For example, a delay in the onset of reproduction may have trivial consequences for one species but catastrophic consequences for another (Newman, 1995).

PHYSIOLOGY

Deviations from homeostasis associated with sublethal exposure often reflect physiological alterations. Such physiological alterations may be used to infer a mode of action of the toxicant as well as document a lowered capacity to maintain homeostasis or normal functioning (Brouwer et al., 1990). For example, acetylcholinesterase inhibitors[4] such as many organophosphate and carbamate insecti-

---

[4]**Acetylcholinesterase inhibitors** are compounds, such as some insecticides, that inhibit the normal functioning of acetylcholinesterase, an enzyme which breaks down the neurotransmitter, acetylcholine. After release from the presynaptic neuron, acetylcholine diffuses across the synapse to bind at a receptor on the postsynaptic neuron (or on the neuromuscular junction), resulting in nerve impulse transmission. At the receptor site, it must then be hydrolyzed by acetylcholinesterase to choline and acetic acid in order to facilitate subsequent normal transmission of nerve impulses.

cides affect feeding, respiration, swimming, and social interactions of nontarget as well as target species by impairing the senses and neuromuscular activities (Mehrle and Mayer, 1985). Some toxicants, like acetylcholinesterase inhibitors, may be straightforward in their mode of action but others may involve a complex series of causal linkages. For example, contaminant-modified functioning of the endocrine system may impact numerous processes under hormonal control (Brouwer et al., 1990).

Some physiological effects involve threshold concentrations. Ammonia toxicity to rainbow trout (*Oncorhynchus mykiss*) is a good example of a physiological effect with thresholds at both ends of a concentration gradient (Lloyd and Orr, 1969). Above a certain threshold concentration, ammonia causes increased water flux into trout that is counterbalanced by an increase in urine production. This effect becomes lethal above a threshold reflecting the maximum rate of urine production.

Physiological changes often studied for nonhuman animals include impaired performance (e.g., swimming speed or stamina), respiration, excretion, ion regulation, osmoregulation, and bioenergetics (e.g., food conversion efficiency) (Sprague, 1971). Infrequently, immunological capabilities or disease resistance are examined (Anderson, 1990) although these responses have considerable potential for understanding effects of toxicants. Common, sublethal effects to plants include water status, stomatal function, root growth, respiration, transpiration, nitrogen or carbon fixation, chlorophyll content and photosynthesis (Baker and Walker, 1989). Most of these sublethal effects are assumed to compromise important physiological functions for which diminished efficiency implies lowered fitness.

Studies of toxicant effects on respiration may describe oxygen consumption directly under normal, resting, or maximum exertion regimes. Also used is the **scope of activity or metabolic scope**, the difference between the rates of oxygen consumption under maximum and minimum activity levels. The scope of activity reflects the respiratory capacity or amount of energy available for the diverse demands on or activities of an organism. The oxygen consumption rate may be combined with the nitrogen (ammonia) excretion rate (O:N ratio) to suggest the relative dependence of respiration on carbohydrate and lipid resources versus the deamination of amino acids. For example, white mullet *(Mugil curema)* exposed to benzene and shrimp (*Macrobranchium carcinus*) exposed to metals had significant shifts in the O:N ratio (Correa, 1987; Correa and Garcia, 1990).

More complicated indicators of metabolism may be needed to address specific problems. For example, energetic analysis may reveal a disruption of energy balance associated with toxicant exposure (Dillon and Lynch, 1981). The **scope of growth** (P = production) may be calculated from an energy budget of exposed individuals; it is the amount of energy taken into the organism in its food (A) minus the energy used for respiration (R) and excretion (U): P = A - R - U (Cockerham and Shane, 1994). It reflects the energy available for growth and production of young. For organisms exposed to toxicants, energy is expended in response to the toxicant and the prediction is that scope of growth will be decreased.

The **adenylate energy charge (AEC)** (Equation 8.3) can also reflect the balance of energy transfer between catabolic and anabolic processes.

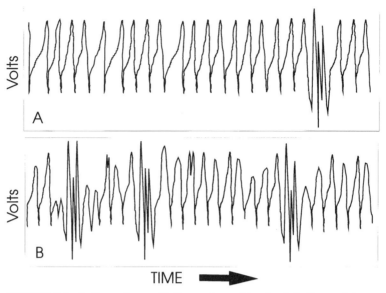

FIGURE 8.5. Patterns for ventilation and coughing for fish. In panel A, ventilation (small evenly-spaced peaks) and coughing (strong, rapid cluster of peaks toward the right hand side) are traced through time for a control individual. Ventilation frequency increases and amplitude decreases, and cough frequency often increases with exposure to contaminants (Panel B).

$$AEC = \frac{ATP + 1/2 \ ADP}{ATP + ADP + AMP} \tag{8.3}$$

where ATP, ADP, and AMP = concentrations of adenosine tri-, di- and monophosphate, respectively. Giesy et al. (1981) reported a drop in AEC in crayfish (*Procambarus pubescens*) and freshwater shrimp (*Palaeomonetes paludosis*) exposed to cadmium, and suggested a diminished energy status for these exposed crustaceans.

Sublethal effects on respiratory activity or respiratory organs may be determined by examining movements associated with respiratory organs, e.g., gills. In a common assay, fish are placed into a chamber receiving a solution of toxicant or an effluent suspected of containing a toxicant. The chamber is equipped with electrodes that measure the change in low voltage electrical fields produced by respiratory movement. In the absence of contaminant, changes in potential over time (Figure 8.5, Panel A) show regular and uniform respiratory movements with occasional **coughs or gill purges**. The coughs are abrupt, periodic reversals of water flow over the gills that function to dislodge and eliminate excess mucus from the gill surfaces. Cough frequency may increase during exposure to various contaminants, often because the irritant causes excess mucus production on the respiratory surfaces (Figure 8.5, Panel B). Bluegill sunfish (*Lepomis macrochirus*) exposed to heavy metals or chlorinated hydrocarbons (Bishop and McIntosh, 1981; Diamond et al., 1990) and brook trout (*Salvelinus fontinalis*) exposed to very low concentrations of methyl or inorganic mercury (Drummond et al., 1974) had increased cough rates. The frequency and amplitude of the signals from ventilation also are sensitive indicators of sublethal effect. Ventilation frequency increased with exposures to cadmium (Bishop and McIntosh, 1981), copper (Thompson et al., 1983), zinc (Thompson et al., 1983), chlorinated hydrocarbons (Diamond et al., 1990), and other organic compounds (Kaiser et al., 1995). Diamond et al. (1990) and Kaiser et al. (1995) demonstrated that the amplitude of the ventilation signal can also decrease with exposure to sublethal concentrations of toxicants.

Osmoregulation and ion regulation capacities can also decline during contaminant exposure. Atlantic salmon (*Salmo salar*) in ammonia-spiked seawater experienced an increase in plasma osmolality (Knoph and Olsen, 1994). Eel (*Anguilla rostrata*) osmoregulation was disrupted by DDT inhibition of Na^+, K^+- and Mg^{2+}-ATPase activity in the intestine where the flux of water is linked with the flux of these ions. Ion balance was disrupted in Atlantic salmon exposed to ammonia, flounder (*Platichthys flesus*) exposed to cadmium, and rainbow trout (*O. mykiss*) exposed to copper (Larsson et al., 1981; Laurén and McDonald, 1985; Knoph and Olsen, 1994). Acid conditions, alone or in combination with elevated aluminum concentrations, also altered ion regulation by fish (Fromm, 1980; Witters, 1986).

A wide range of physiological qualities are also measured for plants under the influence of contaminants. Metals (cobalt, nickel, and zinc) modify the water balance, stomatal closure, and leaf orientation of bean (*Phaseolus vulgaris*) seedlings (Rauser and Dumbroff, 1981). Exposure to PCBs (Doust et al., 1994) and heavy metals (Baker and Walker, 1989) can reduce photosynthetic activity of plants. Air pollutants such as ozone and sulfur dioxide (SO_2) cause leaf **chlorosis**, a blanching of green color due to the lack of production or the destruction of chlorophyll (Landis and Yu, 1995). Heavy metals can affect plant respiration, and carbon and nitrogen fixation in addition to inhibiting photosynthesis (Baker and Walker, 1989).

BEHAVIOR

Animal activities are studied in **behavioral toxicology**, the science of abnormal behaviors produced by exposure to chemical or physical agents. In addition to those already discussed relative to ventilation movement and teratology, behavioral abnormalities include changes in preference or avoidance, activity level, feeding, performance, learning, predation, competition, reproduction, and a variety of social interactions such as aggression or mutual grooming (Table 8.1) (Rand, 1985; Henry and Atchison, 1991). Most often, these effects are measured in a laboratory setting but some studies measure *in situ* changes in behavior (e.g., Gray, 1990). Unfortunately, behavioral effects are underutilized in assessments of risk for three reasons (Giattina and Garton, 1983): (1) it is difficult to objectively score some behaviors; thus, leaving open the possibility of biased information; (2) considerable variability can exist in behavioral data, and (3) it is often difficult to extrapolate accurately from highly-structured laboratory experiments of behavior to behavior in field situations. However, the first two problems can be minimized by careful de-

TABLE 8.1.   Behaviors Commonly Used to Reflect Sublethal Effects of Contaminants

Behavior	Examples
1. Preference or Avoidance	Change in response to light, temperature, salinity, or current. May avoid or move toward a stimulus differently after toxicant exposure. Salmon exposed to DDT shift their preferred water temperature from 19.1 to 23.4°C (Ogilvie and Miller, 1976).
2. Activity Level	Fatigue (lethargy) of workers with chronic lead poisoning (Bornschein and Kuang, 1990). Hyperactivity of fiddler crabs (*Uca pugilator*) exposed to tributyltin (Weis and Perlmutter, 1987) or Arctic char exposed to chlorine (Jones and Hara, 1988).
3. Feeding	Cessation of or diminished feeding by fish after exposure (Jones and Hara, 1988); Deviation from predictions of optimal foraging theory (Sandheinrich and Atchison, 1990).
4. Performance	Ability to swim against a current or maintain proper orientation to a current (rheotaxis) (Little and Finger, 1990). Critical swimming speed of fish lowered by exposure (Schreck, 1990).
5. Learning	Memory impairment of humans exposed to excess metals or metalloids, and memory loss due to mercury poisoning (Bornschein and Kuang, 1990); poorer response and higher error rate of exposed mammals in "lever pulling" learned behavior experiments (Gad, 1982).
6. Respiratory Activity	See examples given in the chapter section on physiological effects of sublethal exposures.
7. Predation	Lowered ability to avoid predator (largemouth bass) by mosquitofish (*G. holbrooki*) exposed to radiation or mercury (Kania and O'Hara, 1974; Goodyear, 1972); suboptimal predator foraging or prey-switching (Atchison et al., 1996).
8. Competition	Zooplankton species grazing and filtration rates modified by toxicants (general details discussed by Atchison et al., 1996)
9. Reproductive Behavior	Decreased libido after occupational lead exposure (Bornschein and Kuang, 1990); masculization of mosquitofish discussed earlier in this chapter.
10. Social Interactions	Grooming in mammals after exposure (Gad, 1982); emotional lability of humans exposed to excess manganese, or increased irritability or depression associated with lead poisoning of humans (Bornschein and Kuang, 1990).

sign and execution of experiments and the third point is no more true of behavioral assays than for many assays used today in risk assessment. Regardless, the results of behavioral studies are most effectively used to assess contaminant impact when combined with results from a suite of other lethal and sublethal effects studies (Atchison et al., 1987).

DETECTING SUBLETHAL EFFECTS

Sublethal and chronic sublethal effects are detected and quantified in several ways. For regulatory testing, data are often generated with experimental designs similar

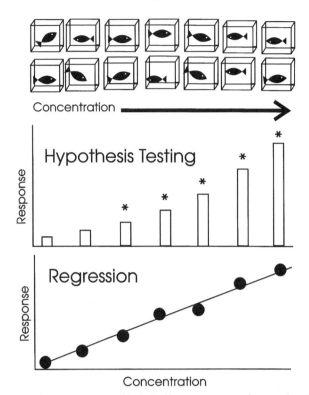

Concentration

Hypothesis Testing

Response

Regression

Response

Concentration

FIGURE 8.6. For regulatory testing, sublethal effects are most often analyzed using hypothesis testing methods. A series of tanks (seven sets of duplicates here) receiving increasing concentrations of contaminant are used to estimate the response for each concentration, including a control set of tanks. Data are then tested to determine if the response at each concentration was significantly different from that of the control. In the middle panel, an asterisk (*) is used to indicate exposure concentrations with effects that are significantly different from the control. Here, the responses of the five highest concentrations were significantly different from the control: the lowest exposure concentration was not significantly different from the control. These data can also be used in regression models to develop a predictive relationship between concentration and effect (bottom panel).

to that shown at the top of Figure 8.6. Effects are measured at one or several time intervals in replicates receiving various concentrations of toxicant. For example, fathead minnow (*Pimephales promelas*) larvae may be exposed to a series of toxicant concentrations (four replicate tanks containing 10 embryos each per exposure concentration) and the growth of the larvae measured after 21 days. (If growth were measured, the response in Figure 8.6 would decrease with increasing concentration.) The results are then analyzed using either a hypothesis testing or regression method.

Hypothesis testing most often involves a one-way analysis of variance (ANOVA) and post-analysis of variance approach (Figure 8.7), although other tests are also applicable. Generally, in **analysis of variance (ANOVA)**, the total variance (total sum of squares) is broken down into the variance among and within treatments, e.g., among the different concentration treatments and within replicates for each concentration treatment. The variance within treatments (mean sum of

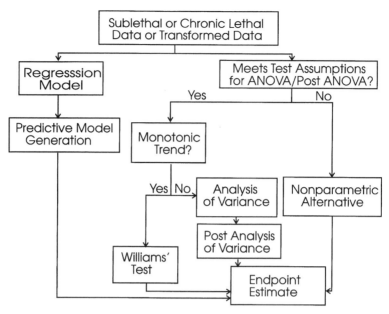

FIGURE 8.7. A flow diagram of methods for analyzing sublethal (and chronic lethal) effects.

squares$_{within}$) is assumed to reflect the sampling or error variance, and that among treatments (sum of squares$_{among}$) is an estimate of error variance plus any additional variance associated with the treatment. If there were no differences among treatment means, then these two measures of variance would be equal. This fact leads to a useful statistic for testing the null hypothesis of equal means among treatments. The ratio of these two variances (F = mean sum of squares$_{among}$/mean sum of squares$_{within}$) is compared to tabulated F statistics to test for significant deviation from the null hypothesis of no difference among treatments.

The ANOVA approach is based on two assumptions: equal variances among treatments and normally distributed data. These two assumptions are assessed before performing ANOVA.[5] Normality is conveniently tested with the **Shapiro-Wilk's test** although graphical methods are also adequate (Newman, 1995). The assumption of homogeneity of variance is tested using **Bartlett's test** for data that, according to some test such as the Shapiro-Wilk's test, already satisfy the assumption of normality. The ANOVA methods are relatively robust to violations of these two formal assumptions (Miller, 1986; Salsburg, 1986). Newman (1995) states that "... probabilities derived from ANOVA are close to the real probabilities if the underlying distribution is at least symmetrical and the variances for the treatments are within three-fold of each other."

Often transformations of the data aid in meeting these assumptions of ANOVA. The most common transformation is the **arcsine square root transforma-**

---

[5]A third and very important assumption of ANOVA is that observations are independent. It is usually satisfied with rigorous experimental design in which subjects are randomly assigned to treatments.

tion which is particularly helpful in meeting the assumption of homogeneous variances for proportions of exposed individuals responding,

$$Transform\ P = \arcsin \sqrt{P}$$

where P = the measured effect, such as the proportion of the exposed organisms responding.

If ANOVA leads to the rejection of the null hypothesis, we know that treatment (i.e., exposure concentration) had a significant influence on mean response but we don't know which means were significantly different. Extending the fathead minnow growth example, we know that growth was different among the different exposure concentrations but we do not know which concentrations had significantly higher or lower growth rates than the others. A series of post-ANOVA methods can identify the specific treatments that are different from each other. Some test for significant difference between all possible pairs with the control (control mean versus each treatment concentration mean). For example (Figure 8.6), six pairs of means are tested because there was a control treatment and six exposure concentration treatments: one pair for each concentration tested against the control. Other methods can be applied, as will be discussed.

If the data or transformations of the data grossly violate one or both of the assumptions of ANOVA, statistical power may be sacrificed by using nonparametric, post-ANOVA tests. **Steel's many-one rank test** and the **Wilcoxon rank sum test with Bonferroni's adjustment** are the most commonly employed (Weber et al., 1989). Although the Steel's many-one rank test is often recommended if there are equal numbers of observations for all treatments (e.g., duplicate tanks of fish for all concentrations including the control treatment) and the Wilcoxon rank sum with Bonferroni adjustment for unequal number of observations (Weber et al., 1989), the slightly more powerful Steel's many-one rank test can also be used in designs with unequal observation numbers (Newman, 1995).

If the data or their transformations are acceptable for ANOVA, several parametric methods are available that have more statistical power than the nonparametric methods just described. **Dunnett's test** or a **t-test with a Bonferroni adjustment** can be applied although Dunnett's test is the slightly more powerful. Newman (1995) suggests that, although widely used, the t-test with a Bonferroni adjustment is also slightly less powerful than the equally convenient **t-test with Dunn-Šidák adjustment**.

If one can assume that a monotonic trend (consistent increase or decrease in response) will occur with increasing toxicant concentration, the even more powerful **Williams' test** can be applied. Unlike the tests described above, the alternate hypothesis is no longer inequality of means between treatment pairs. Instead, the alternate hypothesis becomes "there is a monotonic trend in effect with treatment concentration." The test is done in two steps. In the first, significant deviation from the null hypothesis (equal mean responses among all treatments) is tested. If there is a significant deviation, a last stage of the test is completed in which the lowest concentration having a significantly different mean from the control is identified. Obviously, Williams' test would not be appropriate if one suspected hormesis in the data because the assumption of monotonicity would not be met.

Any of these methods could produce a data summary as depicted in the middle panel of Figure 8.6. A mean response of the control (leftmost bar) is compared to mean responses for the treatment concentrations and those significantly different from the control response are identified. An asterisk was placed above treatment means differing significantly from the control.

With results from analyses of the kind just described, biological effects of various concentrations of chemicals are predicted for sublethal and chronic lethal effects. However, the process of using tests of statistical significance to make projections about biological significance is more difficult than it may first appear. Statistical hypothesis tests only demonstrate that something happened that differed significantly from the null hypothesis: they say nothing about the biological significance of that deviation. Considerable judgment must be applied to statistical results in order to successfully predict ecotoxicological consequences.

To assist in the extrapolation of statistical results to ecotoxicological consequences, a number of descriptive concepts have been developed. The **no observed effect concentration (or level)(NOEC or NOEL)** is the highest test concentration for which there was no statistically significant difference from the control response. To emphasize that the effect is an adverse one, the term may sometimes be expanded to no observed adverse effect concentration or NOAEC. In Figure 8.6, the lowest concentration would be the NOEC (second bar from the left). The **lowest observed effect concentration (or level) (LOEC or LOEL)** is the lowest concentration in a test with a statistically significant difference from the control response (third bar from the left). Again, the word, adverse, may be added to the term. The **maximum acceptable toxicant concentration (MATC)** is "an undetermined concentration within the interval bounded by the NOEC and LOEC that is presumed safe by virtue of the fact that no statistically significant adverse effect was observed" (Weber et al., 1989). Weber et al. (1989) define the term, **safe concentration** to mean "The highest concentration of toxicant that will permit normal propagation of fish and other aquatic life in receiving waters. The concept of a 'safe concentration' is a biological concept, whereas the 'no observed effect concentration' is a statistically defined concentration."

These definitions and their implied applications in assessing risk have several shortcomings. First, the NOEC and LOEC values will depend on the particular concentrations chosen for the experiment: they are tied as much to the experimental design as to any toxicological reality. Next, the process results in higher NOEC and LOEC values if one uses a suboptimal design (low statistical power) or poor technique (high error variance). Consequently, inferior design and technique can be rewarded with higher NOEC and LOEC values than would be estimated with superior design and technique: the concentrations identified as having an effect would be higher with inferior methods. The presumption of a "safe" MATC between the NOEC and LOEC cannot be extended beyond the design, species, and exposure durations used in the specific test without much additional supportive information. The MATC has no statistical confidence interval because the LOEC and NOEC are used to define it. Finally, such data may have dubious predictive value for estimating a safe concentration (". . . that [permitting] propagation of fish and other aquatic life in receiving waters"). A statistically significant reduction in reproduction (e.g., 50%) may be much higher than that which will eventually lead to local extinction of a species population (e.g., 20%). Regardless, these types of data, augmented with supportive data and professional judgment, are used extensively in ecological risk assessments today.[6]

Another approach to analyzing concentration-response data is to use regression methods (Figure 8.6, bottom panel) (Stephan and Rogers, 1985; Hoekstra and Van Ewijk, 1993). Data may be fit to a specific concentration-effect model by least squares or maximum likelihood methods. Concentrations (and their associated confidence intervals) having some biologically significant effect, such as a 10% reduction in fecundity, are calculated via interpolation with the model. The ability to extrapolate downward from results may be another advantage of this approach if one is confident in the shape of the concentration-response model. However, if one incorrectly assumes a linear model when a threshold concentration exists, predictions from regression models will lead to false conclusions. Also, if so much variation exists in the data that a good model cannot be identified, the regression approach becomes compromised and the ANOVA approach may be required (Stephan and Rogers, 1985).

SUMMARY

In this chapter, sublethal effects were described including the general adaptation syndrome and effects to growth, development, reproduction, physiology, and behavior. Many effects overlap these artificial categories, e.g., behavioral teratology. Ambiguity associated with whether or not such "sublethal" effects may result in death in an ecological arena was emphasized. Statistical methods used to detect, model, and predict sublethal response to toxicants were described along with the difficulties of using the associated results to predict ecotoxicological impact or risk. Regardless of the difficulties in prediction, measures of sublethal effects are likely to be as important, or more important, than the measures of acute or chronic lethal effect described in the next chapter to accurately assess the consequences of contamination.

Suggested Readings

Atchison, G.J., M.G. Henry, and M.B. Sandheinrich. Effects of metals on fish behavior: A review. *Environ. Biol. Fishes.* 18, pp. 11–25, 1987.

---

[6]Statistical significance is often used directly to assign ecological significance in ecological risk assessment. This unwise practice will be labelled the **maulstick incongruity**. (A maulstick is a stick used by artists to steady the brushhand while painting.) Just as an inferior painting would be expected from an artist who used the maulstick instead of the brush to apply paint, use of statistical methods alone instead of biological data supported by statistical methods to determine biological significance leads to an inferior decision. The misuse of an otherwise effective tool leads to an inferior product and misinformation.

McLachlan, J.A. Functional toxicology: A new approach to detect biologically active xenobiotics. *Environ. Health Perspect.* 101, pp. 386–387, 1993.

Newman, M.C. Hypothesis Tests for Detection of Chronic Lethal and Sublethal Stress, in *Quantitative Methods in Aquatic Ecotoxicology*, Newman, M.C., CRC Press, Inc., Boca Raton, FL, 1995.

Palmer, A.R. Waltzing with asymmetry. *Bioscience* 46, pp. 518–532, 1996.

Selye, H. The evolution of the stress concept. *Amer. Sci.* 61, pp. 692–699, 1973.

Sprague, J.B. Measurement of pollutant toxicity to fish - III. Sublethal effects and "safe" concentrations. *Water Res.* 5, pp. 245–266, 1971.

Stebbing, A.R.D. Hormesis - The stimulation of growth by low levels of inhibitors. *Sci. Total Environ.* 22, pp. 213–234, 1982.

Weis, J.S. and P. Weis. Effects of environmental pollutants on early fish development. *CRC Critical Reviews in Aquatic Sciences* 1, pp. 45–73, 1989a.

Acute and Chronic
Lethal Effects to Individuals

Pollutants matter because of their effects on populations, and so, indirectly,
on communities too, but pollutants act by their effects on individual organisms.
MORIARTY (1983)

GENERAL

Overview

Methods used by ecotoxicologists to determine lethality often have their origins in mammalian toxicology. In the developing field of ecotoxicology, this transplantation allowed very rapid, initial advancement since established techniques and concepts could be quickly incorporated. It also led to some inconsistencies that must be resolved if ecotoxicology is to progress further. Also, some very worthwhile techniques in mammalian toxicology have yet to be assimilated into ecotoxicology. This chapter will explore these borrowed concepts and techniques, discuss those remaining underexploited, and identify inconsistencies between applications in mammalian toxicology and ecotoxicology.

Acute, Chronic, and Life Stage Lethality

Early in ecotoxicology, a crude distinction was made between acute and chronic lethality. **Acute lethality** refers to death following a short and often intense exposure. The duration of an acute exposure in toxicity testing is generally 96 or fewer hours (Sprague, 1969). Although death is assumed to occur within that short time, there are exceptions in which acute exposure results in death over a long period of time. For example, a short exposure to high concentrations of beryllium may produce an effect which becomes apparent only after a long period has transpired (Casarett and Doull, 1975). **Chronic lethality** refers to death resulting from a more prolonged exposure. By recent convention, a chronic test should be at least 10% of the duration of the species' life span (Suter, 1993), but this is not always the case. Sometimes a test of shorter duration is discussed as a chronic test. The distinction between acute and chronic is often blurred.

Another important distinction can be made for lethality testing based on life stages. An elaborate **life cycle study** may determine lethality, growth, reproduction, development, or other important qualities at all stages of a species life, e.g., Mount and Stephan (1967). **Critical life stage testing** focuses on a particular life stage such as newly-hatched individuals. Often, the most critical life stage is, or is assumed to be, an early life stage, leading to the development of **early life stage (ELS)** tests (McKim, 1985; Weber et al., 1989). The critical life stage approach is based on the sound assumption that protection of the most sensitive stage will ensure protection of all life stages: the most sensitive stage of an individual's life cycle will determine its fate under lethal challenge. A dubious extension of this concept (**weakest link incongruity**) is often made in which one assumes that exposure of field populations to concentrations identified in testing as causing significant mortality at a critical life stage will result in significant impact on the field population. However, loss of individuals from certain life stages may or may not have much bearing on population demographics or the risk of local extinction for species populations.[1] For example, a 10% reduction in the number of larvae during an oyster spawn due to toxic effect may have minimal impact on the likelihood of an oyster population becoming extinct. Very high mortality is expected for the planktonic larvae under normal conditions and oyster populations accommodate widely varying annual recruitment. It is important to keep in mind that the term "critical life stage" refers to the life stage of an individual most sensitive to poisoning, not the life stage most critical to population viability. We will add more to this point in the next chapter.

Test Types

Tests used to quantify lethality vary depending on the medium of concern, i.e., water, sediment, or soil. The nature of the toxicant or toxicant mixture may also influence the test design. For example, exposure solutions must not be aerated for tests involving volatile compounds. For TBT-based antifouling paints, dosing may be done by placing painted surfaces (e.g., discs or rods) into the feed water flows to the different exposure tanks (Bryan and Gibbs, 1991). Different amounts of TBT leach from these surfaces into the tanks, producing a range of exposure concentrations. Exposure duration, test species, resources, and expendable time also influence methods. A chronic exposure of a suite of endemic species may be highly desirable yet impractical. Instead, a representative species that is easily cultured in the laboratory may be exposed for only the most sensitive time in its life cycle. The

---

[1]This situation results from the difference in emphasis between medical/mammalian toxicologists and ecotoxicologists. For a toxicologist dealing with humans, the individual is justifiably the focus of decisions, concepts, and associated methodologies. The ecotoxicologist tends to focus on population and community viability instead. This distracting inconsistency is a consequence of the rapid infusion of methods from mammalian toxicology into ecotoxicology.

results may then be used to imply the risk to endemic species over their entire life cycles.

A series of exposure designs has been established for tests quantifying lethal effects of toxicants in waters. In **static toxicity tests**, individuals are placed into one of a series of exposure concentrations. The exposure water is not changed during the test. The advantage of this design is that it is easy to perform and inexpensive. Also, minimal volumes of toxic solutions are generated (Peltier and Weber, 1985). But toxicant concentrations can change during exposures due to sorption to the container walls and other solid phases, volatilization, bacterial transformation, photolysis, and many other processes. Waste products of the test organisms may build up during the test and oxygen concentrations may drop to undesirable levels. For these reasons, most static tests are used to measure acute lethality, not chronic lethality. A **static-renewal test** can minimize some of these problems. Test solutions are completely or partially replaced with new solutions periodically during exposures, or organisms are periodically transferred to new solutions. A **flow-through test** uses continuous flow or intermittent flow of the toxicant solutions through the exposure tanks. The flow-through design eliminates or greatly minimizes the problems just discussed for static tests. However, flow-through tests produce large volumes of toxicant solutions that must be treated. They also require more time, space, and expense (Peltier and Weber, 1985). Although individual containers of various toxicant concentrations can be used as sources of test waters, often a special apparatus called a **proportional diluter** mixes and delivers a series of dilutions of the contaminant solution to the test tanks. The solution being diluted may be a toxicant solution or an effluent suspected of having an adverse effect on aquatic biota. With effluent testing, lethal effects are expressed as percentages of the total exposure water volume made up of the effluent (e.g., 45% effluent to 55% diluent by volume) resulting in the toxic response.

There are several methods associated with solid phase testing. Organisms may be placed into spiked or contaminated soils. This type of test has been used with important soil invertebrates such as nematodes (Donkin and Dusenbery, 1994) and earthworms (Gibbs et al., 1996). Similarly, sediment toxicity tests may involve spiked or contaminated sediments. The **spiked bioassay approach (SB)** generates a concentration-response model for, or tests hypotheses regarding, effects to individuals placed into sediments spiked with different amounts of toxicant (Giesy and Hoke, 1990). Concentration may be based on total concentration in the sediment, interstitial water concentration, or concentration in some notionally bioavailable fraction of the sediments. Sediment toxicity may also be implied using an elutriate test of a nonbenthic test species. In an **elutriate test**, a nonbenthic species such as *Daphnia magna* is exposed to an elutriate produced by mixing the test sediment with water and then centrifuging the mixture to remove solids from the elutriate (McIntosh, 1991). Exposure to various dilutions of the elutriate allows an amount-response analysis of associated data as described above for effluent tests. The results of such a test would be most appropriately used to assess lethality of plumes produced during sediment dredging activities.

DOSE-RESPONSE

Basis for Dose-Response Models

More as a consequence of the early history of ecotoxicology than through careful comparison to alternatives, most lethality tests in ecotoxicology involve the dose-response[2] approach. This approach and associated quantitative methods were taken directly from mammalian toxicology and remain the cornerstone of toxicity testing in ecotoxicology. The time-response models described later are utilized much less frequently.

With the **concentration-response approach**, a series of toxicant concentrations is delivered to containers as illustrated earlier in Figure 8.6. There are replicate containers for each treatment (concentration) to allow estimation of variation within each treatment and at least one control treatment receiving no toxicant.[3] Individuals are randomly placed into the containers until a predetermined number of individuals is present in each, e.g., 10 fish per tank. Mortality is tallied in each tank as the number dying of the total number exposed after a certain time or series of times such as 24, 48, 72, and 96 h. The paired data (proportion dying, exposure concentration) for the tanks are used to calculate lethality.

The predominant methods for analyzing dose- or concentration-response data are based on the concept of **individual effective dose (IED)** or **individual tolerance**. According to this concept, there exists a smallest dose (or concentration) needed to kill any particular individual and this IED is a characteristic of that individual. This concept and its applicability to dose-response data analysis is illustrated in Figure 9.1. The top panel shows the skewed, log normal distribution of IED values thought to be typical of populations. A sigmoidal dose-response curve would be produced if seven random samples of 10 fish each from this same population were given doses corresponding to the six IED groupings in the top panel and a control dose. This presumptive log normal curve is the basis for the probit method, the most common approach to analyzing dose-response data.

Although the IED concept is presented almost exclusively as the foundation of dose-response models, other concepts are invoked to support various methods of analysis. The log logistic model has been suggested for years as an alternative to the log normal model, leading to a protracted controversy about the relative values of logit versus probit methods. The log logistic model also predicts a sigmoidal curve like that shown in Figure 9.1. The foundation for the logistic model is its

---

[2]To improve readability, dose-response and concentration-response are used interchangeably here. The analyses of these two models are identical: only the delivery of toxicant differs. A dose-response model is associated with a delivered amount of toxicant, e.g., 5 mg of toxicant is injected into the organism. A concentration-response model is associated with exposure to a concentration in the organism's environment, e.g., 5 mg L^{-1} of toxicant is present in the water into which a fish is placed.

[3]A solvent may be used to produce the toxicant solution as required for some organic compounds. In such a case, a "solvent control" treatment is included too and methods for quantifying toxic effect are modified slightly to include this blank in the experimental design. The modifications are made to ensure that any effect of the solvent on response does not unintentionally bias the results.

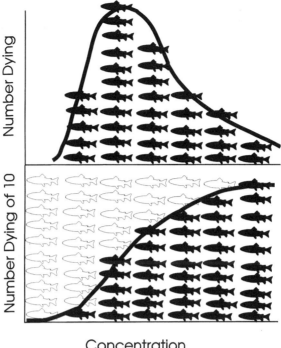

Concentration

FIGURE 9.1. The IED concept and the analysis of dose-response lethality data. The top panel depicts the distribution of individual effective doses among 35 fish taken randomly from a population. Each individual's response was placed into a category (column) based on whether its IED fell into one of six different ranges of IED values. The distribution of such IED values within a population is thought to be log normal as evidenced by an asymmetric curve with a few very tolerant individuals. The bottom panel shows the results of exposing sets of 10 individuals from this same population to a series of doses for a set time. Surviving fish (white) had IED values greater than the exposure dose and dead fish (black) had IED values less than or equal to the exposure dose. The result is a typical, sigmoidal dose-response curve.

linkage to processes such as enzyme kinetics, autocatalysis, and adsorption phenomena (Berkson, 1951).

Berkson (1951) attacked the IED concept and advocated the use of the log logistic model instead of the log normal model. He based his argument on an experimental screening of combat pilots for tolerance to high altitude conditions. Candidate aviators were placed into decompression chambers and their individual tolerances measured. Assuming the IED concept was correct, those failing to exhibit symptoms of the "bends" to a certain critical pressure passed the test and low tolerance individuals were rejected from further consideration as pilots. Berkson broke from the standard test protocol and asked that a set of candidates be retested to see if individual responses remained the same between tests. The results showed poor agreement between repeated tests: the IED concept had failed to explain this dose-response phenomena.

As a consequence, counter argument to the IED concept has been made based on the idea that individuals do not have unique tolerances. The argument is made that the probability of death for any individual is a consequence of a random process or processes that may conform to a log normal, log logistic or another model. Berkson (1951) and Finney (1971) argued that some processes, such as those described in Chapter 7 for cancer risk, are based on probabilities that a specific sequence of events will occur and lead to death or cancer. The distribution of differences in occurrence of the event (i.e., appearance of a clinical cancer or death) is not related to differences among unique individuals: the distribution is a consequence of probabilities associated with events taking place in all individuals. In this model, which particular individual responds quickly ("sensitive") or slowly ("tolerant") is a matter of chance alone. Individual response is a consequence of random events that are described by probability distributions. Certainly, the observation that the log normal model seems to work for tests with microbes and zooplankton composed of cloned individuals casts some doubt on the IED concept as the sole underlying explanation for the log normal model. Gaddum (1953) suggested that a random process in which several "hits" are required to produce death could also form the basis for the log normal model. Remarkably, whether one or both of these concepts is the basis for the majority of dose-response relationships remains untested.

Although this debate may appear trivial, the consequences of repeated exposure of a population are different under these two concepts. This can be illustrated with a thought problem in which all covariates affecting lethal impact such as animal sex, size, and age are identical for all individuals. If the IED concept were correct, survivors of a first exposure would be the most tolerant and the impact on the population of survivors would be less in a subsequent exposure. In the second case where probability of death is the same for all individuals, the survivors will not be inherently more tolerant and the impact of a second exposure would be as large as that of the first.

A Weibull model also provides good fit for the dose-response curve but is rarely used (Christensen, 1984; Christensen and Nyholm, 1984; Newman, 1995; Newman and Dixon, 1996). The Weibull model can describe a multistage process such as that discussed in Chapter 7 for carcinogenesis. When applied, it seems to fit dose-response data as well as the generally-accepted log normal and log logistic models.

Many dose-response relationships for lethality have threshold concentrations below which no discernible increase in mortality occurs. Most of the models described here can and should be modified to include lethal thresholds if required. Further, with chronic exposures, natural or spontaneous mortality may be occurring simultaneously with toxicant-induced mortality. Such spontaneous mortality can also be included in dose-response models.

Fitting Data to Dose-Response Models

Based on the concepts and models just discussed, methods have been developed to analyze dose-response data. Data (proportions, doses, or concentrations) may be used directly or after transformation. Often, the objective of transformations is to make linear the relationship between dose and response. Measurements or transformations of measurements used in the analysis of biological tests are termed **metameters**. Both dose or concentration metameters, and effect metameters may be used for dose-response data. The most common dose or concentration metameter is the logarithm of dose or concentration. Which effect metameter is appropriate depends on whether the log normal, log logistic, or another model is assumed. For example, the log normal model is assumed if the log dose or log concentration transformation is paired with the probit metameter of the proportion dying. The log logistic model is assumed if the log dose or concentration is paired with the logit metameter of the proportion dying.

The probit transformation is derived from the **normal equivalent deviation (NED)**, the proportion dying expressed in terms of standard deviations from the mean of a normal curve. For example, a proportion corresponding with the mean (50% of exposed individuals are dead) would have an NED of 0: a proportion below the mean by one standard deviation (16% of exposed individuals are dead) would have a NED of –1. In introducing the probit method, Bliss (1935) viewed negative NED values as inconvenient and added five to NED values to avoid negative numbers.[4] The resulting metameter is the **probit**. Probit analysis (log normal model) is performed using the log dose or log concentration versus probit of the proportion dead.

$$Probit(P) = NED(P) + 5 \qquad (9.1)$$

where P = proportion of exposed individuals that died by the end of the exposure, and

NED = the normal equivalent deviation.

The **logit** metameter is based on the log logistic model and has the form,

$$Logit(P) = \ln\left[\frac{P}{1-P}\right]. \qquad (9.2)$$

A transformed logit is more commonly employed than that calculated by Equation 9.2 because values of this transformed logit are nearly the same as probit values except for proportions at the extreme ends of the curves.

$$Transformed\ Logit = \left[\frac{Logit(P)}{2}\right] + 5 \qquad (9.3)$$

where logit(P) = logit value estimated by Equation 9.2. Other effect metameters are used much less commonly as discussed in Newman (1995). The Weibull transformation is one metameter that could be used more often (Christensen, 1984;

---

[4]The need to avoid negative numbers can and has been questioned. Analysis with the NED instead of the probit produces the same results. The NED used in this way is often called the **normit** metameter and the associated method, normit analysis.

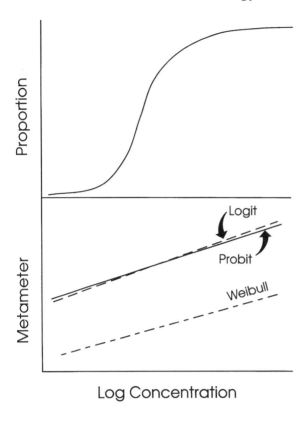

FIGURE 9.2. The typical, sigmoidal curve for concentration-response data (top panel) and lines resulting from the probit, logit, and Weibull transformations (bottom panel).

Christensen and Nyholm, 1984). Equation 9.4 gives the form of the **Weibull metameter**.

$$U(P) = \ln\left(-\ln\left(1 - P\right)\right) \tag{9.4}$$

All of these metameters will generate a straight line for appropriate dose-response data (Figure 9.2). Except at the tails of these lines, the probit and logit metameters have nearly identical values for any set of data.

The **median lethal dose (LD50)** and its associated confidence limits are the most common statistics derived from dose-response models. The LD50 is the dose resulting in death of 50% of exposed individuals by a predetermined time, e.g., 96 hours. Similarly, a **median lethal concentration (LC50)**[5] and its confidence limits are commonly derived from concentration-effect data. It is the concentration resulting in death of 50% of exposed individuals by a predetermined time. Median values were adopted as benchmarks instead of low values (e.g.,1% or 5%) because medians tend to be more consistent and to have narrower confidence intervals

---

[5]For sublethal or ambiguously lethal effects, the term **median effective concentration (EC50)** may be used instead of LC50. It is often applied to species such as invertebrates for which death is difficult to score and events such as cessation of ventilation or general movement are scored. It is also the term used if sublethal events are being analyzed.

than lower percentiles. Median values also have an advantage because models such as the probit and logit produce very similar results toward the median. Although these advantages are quite real, it is important to keep in mind that the median was not chosen because it has any special biological significance. In many cases, concentrations killing lower proportions would be much more meaningful for the ecotoxicologist attempting to determine risk upon toxicant release to the environment.

Many ecotoxicologists estimate values other than the median. Such LCX statistics (e.g., 96 h LC5 = the concentration killing 5% of exposed individuals by 96 h) are more helpful in assessing adverse effects, but they have generally wider confidence intervals and results are more model dependent. This last point is often forgotten because of the misconception, based on our past preoccupation on the median, that all models give similar statistics. Model independence is true for practical purposes at the median but it becomes progressively less so toward the tails of the model. Careful model selection (e.g., probit versus logit analysis) becomes important as attention moves away from the LC50. Methods of comparing candidate models will be discussed toward the end of this section.

Numerous methods for estimating the LC50 are available (Table 9.1). Indeed, interpolation from a simple line produced with different sets of metameters, such as the probit of P versus log of concentration (Figure 9.3), could be used to graphically estimate the LC50. However, the 0% and 100% mortality treatments would not easily be plotted on such a graph, and visual fitting would be subjective. For these reasons, more formal methods are applied for estimation of the LC50, its 95% confidence interval, and the slope of the concentration-effect line. These last two parameters are as important as the LC50 itself. The LC50 has little utility without some measure of confidence in its calculated value. Also, the slope of the line provides valuable information as can be illustrated easily with Figure 9.3. Imagine a second line intersecting the drawn line at the LC50 but give this second line a much steeper slope. Although the LC50 would be the same for both lines, a small change in concentration has much more of an effect with one toxicant (steeper slope) than the other (shallow slope). This is an important piece of information.

The easiest but most subjective method for estimation of an LC50 is the **Litchfield-Wilcoxon method**. It is a semigraphical method in which data are graphed first as in Figure 9.3 and the points on the line corresponding to proportions 16, 50, and 84% mortality used to estimate the LC50 and its 95% confidence interval. The antilogarithm of the log concentration corresponding with the p = 0.50 is taken as the LC50. A slope factor (S) is calculated from the concentrations corresponding with 16% (LC16), 50% (LC50), and 84% (LC84) mortalities.

$$S = \frac{\dfrac{LC84}{LC50} + \dfrac{LC50}{LC16}}{2} \tag{9.5}$$

This S and the total number of animals exposed in treatments within the range of LC16 and LC84 (N′) are used to generate the upper and lower 95% confidence limits.

TABLE 9.1. Established Methods for Estimation of LC50.

Method	Advantages	Disadvantages	References
Litchfield-Wilcoxon	Quick, semigraphical method	Results are dependent on the individual who is fitting the data "by eye."	Litchfield and Wilcoxon, 1949; Stephan, 1977; Peltier and Weber, 1985; Newman, 1995
Maximum Likelihood Estimation (MLE) or χ^2 Fitting of Specific Model	Powerful, parametric method which can use any of a series of possible models	Requires a specific model such as the log normal or log logistic models; iterative method that is tedious to do manually; MLE results are slightly biased; MLE methods may not converge properly.	Armitage and Allen, 1950; Berkson, 1955; Stephan, 1977; Peltier and Weber, 1985; Newman, 1995
Spearman-Karber	A robust, nonparametric method not requiring a specific model; can trim data to minimize the undue influence of extreme values.	Toxicity curve must be symmetrical; not as powerful as parametric methods.	Hamilton et al., 1977; Stephan, 1977; Peltier and Weber, 1985; Newman, 1995
Binomial	Can be used for data with no partial kills.	Estimation of confidence interval ignores sampling error (Newman, 1995).	Stephan, 1977; Peltier and Weber, 1985; Newman, 1995
Moving Average	Easily implemented	Simple equations for this method require specific progression of toxicant concentrations and numbers of replicates.	Stephan, 1977; Peltier and Weber, 1985

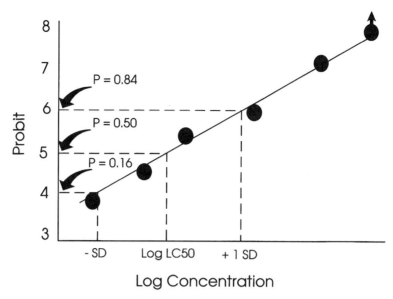

FIGURE 9.3. Based on a log normal model, probit methods can be used to generate a line from concentration-effect data. An estimate of the log LC50 ± one standard deviation can be grossly approximated from such a graph. Note that a point with 100% mortality is often indicated by an arrow attached to the data point. This indicates that 100% mortality would likely have occurred prior to the time endpoint and some lower concentration likely would have produced 100% at exactly that time endpoint. Other, more rigorous methods discussed in the text provide better estimates.

$$f_{LC50} = S^{\frac{2.77}{\sqrt{N'}}} \tag{9.6}$$

$$Upper\ Limit = LC50 \cdot f_{LC50} \tag{9.7}$$

$$Lower\ Limit = LC50/f_{LC50} \tag{9.8}$$

As easy as this method is, it is not a consistent tool because visual fitting of the line will vary among individuals.

Maximum likelihood estimation (MLE) for the log normal, log logistic, or other models is a parametric method which avoids subjectivity but takes on the assumption of a specific model. Probit, logit, and other approaches are most often applied with MLE methods if there are two or more **partial kills** (treatments in which some, but not all, exposed individuals are killed). Maximum likelihood estimation results have very good precision relative to those of other methods. The MLE estimates can be slightly biased for small sample sizes; however, this bias is often in the range of that of the other methods. The MLE method is an iterative process and is most often done with a computer. Occasionally, the MLE method will fail to converge after many iterations or will converge on an inappropriate ("local") solution.

Because various models can be fit to data, questions arise about goodness-of-fit among candidate models. The χ^2 values estimated for each model may be used for this purpose. Since the χ^2 value will decrease as fit improves, the ratio of χ^2 values for the different models will reflect relative goodness-of-fit for each model to the data. If the χ^2 **ratio** is less than 1, the model whose χ^2 value is in the numerator (e.g., χ^2_{probit} for $\chi^2_{probit}/\chi^2_{logit}$) fits the data better than the model whose χ^2 value is in the denominator. More formal tests can be carried out with the χ^2 ratio to determine if the fit for one model is significantly better than that of the other.

If it is difficult or unnecessary to assume a specific model for the dose- or concentration-effect data, the nonparametric **Spearman-Karber method** is available to estimate the LC50. The technique only requires a symmetrical toxicity curve. The technique has many steps but can be done with a hand calculator if necessary. Values at the extreme tails may or may not be trimmed to minimize undue influence of extreme values on the estimate. Trimming rules are provided by Hamilton et al. (1977). If trimming is done, the influence of anomalous values will be reduced; however, the standard error of the estimate will increase.

Two other methods are commonly discussed in ecotoxicology. The **binomial method** allows estimation of the LC50 if there are no partial kills although sampling error is ignored with this method (Newman, 1995). The **moving average method** may be implemented with straightforward equations if the toxicant concentrations are set in a geometric series, and there are equal numbers of individuals exposed in each treatment.

Occasionally, a model cannot be unambiguously fit to data for chronic toxicity tests. An ANOVA design as described previously may then be applied. The lowest concentration at which mortality is significantly higher than the control is determined via hypothesis testing.

Incipiency

Incipiency when applied to lethality of contaminants is the concentration (or dose) at which an increase in toxicant concentration (or dose) begins to produce an increase in the measured effect. It is often measured with the **incipient median lethal concentration (incipient LC50)**, the concentration below which 50% of individuals will live indefinitely relative to the lethal effects of the toxicant. This concentration is also called the asymptotic, ultimate, or threshold LC50 by various authors. It may be determined graphically in a variety of ways. Figure 9.4 shows one approach taken by van den Heuvel et al. (1991) for rainbow trout (*Oncorhynchus mykiss*) exposed to pentachlorophenol. The LC50 values were determined for a series of times and the reciprocals of the LC50 values plotted against times of exposure. Alternatively, a double logarithm plot may be used (Newman, 1995). The concentration at which the curve becomes parallel to the x-axis is an estimate of the incipient LC50. Although widely used and convenient to estimate incipiency,

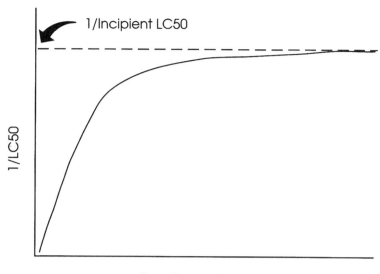

FIGURE 9.4. The incipient lethal concentration is estimated as the point at which the curve of 1/LC50 versus duration of exposure begins to run parallel to the x-axis.

this measure is difficult to assign ecotoxicological significance. The median value has ambiguous meaning relative to the continued viability of an exposed population. Statistical limitations also exist for this graphical approach (Chew and Hamilton, 1985). For example, concentration is set in the design yet treated as an independent variable in subsequent analysis.

Mixture Models

Estimation of lethality is confounded if chemicals are present in mixtures. **Potentiation** may occur if one chemical, not toxic itself at the exposure concentration, enhances the toxicity of a second chemical in a mixture.[6] For example, sublethal concentrations of isopropanol greatly enhance the toxic effects of carbon tetrachloride to the mammalian liver (Klaassen et al., 1987). Potentiation is used to improve the effectiveness of some insecticide formulations. Piperonyl butoxide, added to formulations, potentiates pesticide action by inhibiting their breakdown by the cytochrome P450 system.

Other effects of combined chemicals can be easily illustrated using the concept of **toxic units (TU)**, amounts or concentrations of different toxicants expressed in units of lethality such as units of LD50 or LC50. Very often, the toxic unit is expressed as a fraction of the incipient median lethal concentration. For ex-

---

[6]The term, potentiation, is also used by many authors (e.g., Thompson, 1996) in the context of synergism as discussed below.

ample, if toxic units are based on the incipient LC50, chemical A with an incipient LC50 of 20 mg L^{-1} would be present at 0.5 TU in a 10 mg L^{-1} solution. Similarly, chemical B (LC50 = 100 mg L^{-1}) would be present as 0.5 TU in a 50 mg L^{-1} solution. If the toxicity of two toxicants in combination is **additive**, the simple sum of the toxic units of the two toxicants would equal the actual toxicity measured for the mixture, e.g., 0.5 TU of A + 0.5 TU of B should equal 1.0 TU of effect when combined as a mixture. In the above example, a mixture of 10 mg L^{-1} of A plus 50 mg L^{-1} of B should result in 50% of the exposed individuals dying. If the mixture results in less than 1 TU of lethality, chemicals A and B are said to be less than additive or **antagonistic.** They are **synergistic** if their combined effect is more than additive.

Antagonism can be further broken down based on the underlying mechanism (Klaassen et al., 1987). **Functional antagonism** results from two chemical-eliciting opposite physiological effects and, as a consequence, counterbalancing each other. With **chemical antagonism**, two toxicants react with one another to produce a less toxic product. For example, cyanide and a toxic metal may combine in mixture to form a less toxic complex. **Dispositional antagonism** involves the uptake, movement within the organism, deposition at specific sites, and elimination of the toxicants. The presence of the two toxicants together shifts one or more of these processes to lower the impact of the toxicants on the site(s) of action or target organ(s). This may involve lowered chances of interaction with a target, e.g., less time available to interact or lowered concentration available to interact. For example, ethanol enhances mercury elimination in mammals (Hursh et al., 1980; Khayat and Shaikh, 1982) and could modify (increase or decrease) the toxic effect of mercury as a consequence. The last type, **receptor antagonism**, occurs where two or more toxicants bind to the same receptor and each toxicant blocks the other from fully expressing its toxicity. Klaassen et al. (1987) give the example of using O_2 to counter the effects of carbon monoxide poisoning based on receptor antagonism. Newman and McCloskey (1996a) found evidence of receptor antagonism for binary mixtures of metals effects on response of the **Microtox**® **assay,** a rapid, bacterial assay in which a decrease in bioluminescence is thought to reflect toxic action.

Marking and Dawson (1975) generated an **additive index** for assessing the joint action of toxicants in mixtures. Letting, A_m and B_m = the toxicity (e.g., incipient LC50) of toxicants A and B when present in mixture, and A_i and B_i = toxicity of A and B when they are tested separately,

$$\frac{A_m}{A_i} + \frac{B_m}{B_i} = S \qquad (9.9)$$

The toxicants are antagonistic if S < 1, synergistic if S > 1, or additive if S = 1 (Figure 9.5, "sum of toxic contributions" scale). Although shown here for a binary mixture, several toxicants can be added to Equation 9.9 if desired. Unfortunately, the units are not linear to the right and left sides of 1 on this scale. A change from +1 to +2 is not of the same magnitude as a change of +1 to 0. Units can be made linear by making values to the left of additivity equal to -S+1 and values to the right of additivity equal to 1/S-1.

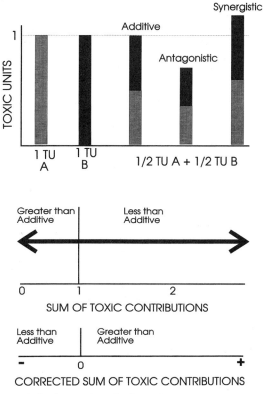

FIGURE 9.5. The combined effect of toxicants may be quantified by expressing toxicant concentrations in mixtures as toxic units (top panel). If the realized effect expressed in terms of TU is less than the calculated sum of TU for both toxicants A and B in a mixture, the chemicals are said to be antagonistic. Their effect is synergistic if the realized effect is greater than the calculated effect based on their individual actions. The two scales include the nonlinear scaling of Marking and Dawson (1975) (middle) and additive index (bottom) which is a linear scaling of combined toxicant effect.

$$Additive\ Index\ (AI) = \frac{1}{S} - 1 \ \ for \ \ S \le 1.0 \qquad (9.10)$$

$$Additive\ Index\ (AI) = -S + 1 \ for \ S \ge 1.0 \qquad (9.11)$$

This additive index (AI) is linear on both sides of additivity (0) (Figure 9.5, "corrected sum of toxic contributions"). Negative and positive numbers indicate less than and greater than additivity, respectively.

The toxic unit and additive index are widely used tools in ecotoxicology. Marking's additive index (Markings and Dawson, 1975; Markings, 1985) is a straightforward means of visualizing combined effects of mixtures. The concept of additivity also forms the foundation for estimation of combined effects of contaminants in solutions or solids. For example, Di Toro et al. (1990) assumed additivity in estimating the combined effects of metals in contaminated sediments and performed mathematical modeling accordingly. The current, extensive use of the toxic unit and additive index is a consequence of their ease of application and historical role in ecotoxicology, not their power to quantify and test for significant deviations from additivity. More thorough analysis of mixtures can be done with the general linear model approach including interaction terms (see Neter et al., 1990). There are now many convenient software packages for personal computers which implement such general linear modeling procedures which should allow more general implementation of these methods.

SURVIVAL TIME

Basis for Time-response Models

The **time-response approach** complements the dose- or concentration-response approach. In the dose- or concentration-response approach, exposure time is held constant although results (numbers dying) are often acquired at a series of time intervals. This approach provides a gross indication of the influence of exposure duration; however, full consideration of temporal dynamics is sacrificed in order to generate estimates of chemical toxicity expressed as amounts or concentrations.

Accurate measurement of the effect of exposure duration is also extremely important to assessing ecotoxicological consequences of contamination. To accomplish this, times-to-death (TTD) are measured in the time-response approach. More data are generated for each exposure tank, resulting in enhanced statistical power. For example, 10 TTD values are generated instead of one proportion from an exposure tank containing 10 fish. Equally advantageous, because the endpoint (TTD) is associated with an individual, important qualities influencing toxic consequences such as individual sex or weight can be included in the analysis. The enhanced power also allows more effective inclusion of other qualities such as temperature, toxicant concentration, or nutritional history. The inclusion of concentration in the survival time approach provides a highly-desirable model that predicts effect as a function of both exposure duration and concentration (Figure 9.6). Finally, as we will discuss in the next chapter, results of time-response analysis can be readily incorporated into demographic analysis.

Despite these clear advantages, ecotoxicologists still underutilize time-response methods. Traditions including those taught to students and formalized in regulations seem to have a strong hold on ecotoxicologists in this area. Kooijman (1981), Chew and Hamilton (1985), Dixon and Newman (1991), Newman and Aplin (1992), Newman et al. (1994), Newman (1995), Roy and Campbell (1995), Newman and Dixon (1996), and Newman and McCloskey (1996b) provide general discussion and examples of time-response methods application in ecotoxicology.

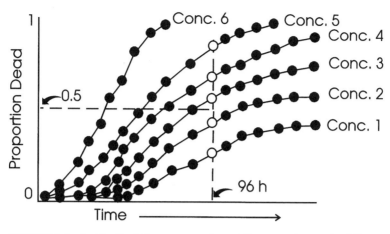

FIGURE 9.6. An idealized time course of mortality occurring at six different concentrations. Note that, if only data from 96 hours were used, only five points would be available to calculate the LC50. In contrast, many more data points (approximately 70 points) would be gathered by 96 hours if times-to-death for individuals were noted instead of the final proportion dying.

Fitting Survival Time Data

The **Litchfield method** is a simple method for analyzing survival time data (Litchfield, 1949). It was developed during the same year as the Litchfield-Wilcoxon method for estimating LC50. The tests are very similar, differing only in minor details. First, the proportion of exposed individuals responding is tabulated with duration of exposure. The probit of the proportion is plotted against log of exposure duration to produce a straight line, e.g., Figure 9.7. (Originally log-probability paper was used but these transformations produce the same results.) A line is fit by eye to these data and the time corresponding to the probit for 50% mortality is the estimated LT50 (**median lethal time.**)[7] To calculate the 95% confidence interval, a slope factor is derived from this line.

$$S = \frac{\dfrac{LT84}{LT50} + \dfrac{LT50}{LT16}}{2} \tag{9.12}$$

where LT16, LT50, and LT84 = time intervals corresponding to when 16, 50, and 84% of exposed individuals were dead, respectively. An f_{LT50} is calculated to estimate the 95% confidence interval for the LT50.

---

[7]As discussed for LC50, a **median effective time** (ET50) may be used instead of LT50 if the effect is sublethal or ambiguously lethal.

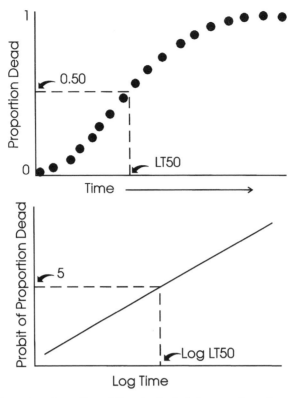

FIGURE 9.7. Linear transformation of time-to-death data assuming a log normal model. A sigmoidal curve is generated if the cumulative proportion of the exposed individuals that have died is plotted against exposure time (top panel). Under the assumption of a log normal model, a straight line is produced (bottom panel) if log time is plotted against the probit of the proportion dead. A curve which is not straight, but instead has a break in its slope, is called a split probit. Traditionally, a split probit is assumed to reflect either two distinct mechanisms of toxicity at the beginning and later in the exposure, or two distinct subpopulations of individuals in the exposed group that differ in tolerance to the toxicant. A third possibility exists, the log normal model is an inappropriate model for the data, but is rarely assessed.

$$f_{LT50} = S^{\frac{1.96}{\sqrt{N}}} \tag{9.13}$$

where N = the total number of individuals exposed during the test if all individuals died during the test. If there were survivors, the N in this equation must be modified slightly as described originally by Litchfield (1949) or conveniently as described by Newman (1995).

$$Upper\ Limit = LT50 \cdot f_{LT50} \tag{9.14}$$

$$Lower\ Limit = LC50/f_{LT50} \tag{9.15}$$

TABLE 9.2.   Transformations of Times-to-Death Data Used to Select from Among Candidate Models

Candidate Model	Transformation of Mortality[a]	Transformation of Time
Exponential	ln S(t)	t
Weibull	ln [-ln S(t)]	ln t
Normal	Probit [F(t)]	t
Log normal	Probit [F(t)]	ln t
Log logistic	ln [S(t)/F(t)]	ln t

[a]t = duration of exposure; S(t) = the cumulative survival to time t expressed as the proportion of exposed individuals still alive at time, t; and F(t) = the cumulative mortality to time t which is 1-S(t).

Although ecotoxicology has traditionally considered only this log normal model, other models for time-to-death data are commonly assumed or explored in other fields. The most commonly employed include exponential, Weibull, normal, log normal, and log logistic models. Fit to these models may be crudely assessed with a series of linear transformations (Table 9.2). The model with a set of transformations which result in a straight line (e.g., Figure 9.7 for the log normal model) is selected as the most appropriate.

Like dose-response data, time-to-death data may be analyzed with a wide range of methods, ranging from the simple Litchfield method to more involved nonparametric, semiparametric, and fully parametric methods. The nonparametric **product-limit (Kaplan-Meier) methods** do not require a specific model for the survival curve. Product-limit methods allow estimation of survival during exposure and can be used to test for significant differences among treatments. At the other extreme, fully parametric techniques assume a specific model for the survival curve (e.g., log normal or Weibull) and a specific function relating survival to covariates (e.g., exposure concentration, animal size, etc.). The underlying distribution may be selected from a series of candidate models such as those in Table 9.2. These models are usually fit using a maximum likelihood method as described earlier for dose-response data.

Depending on which model is selected to describe the shape of the survival curve, the fully parametric model takes one of two general forms: proportional hazard or accelerated failure time. Selection of an exponential or Weibull model produces a proportional hazard model. A **proportional hazard model** relates the hazard (proneness to or risk of dying at any time, t) of one group (e.g., smokers) quantitatively to that of a reference group (e.g., nonsmokers). Results of proportional hazard models for human mortality are often expressed as easily understood, **relative risks**, risks of one group expressed as a multiple of that of another. For example, the risk of dying from lung cancer may be X times higher for smokers relative to nonsmokers or the risk of surviving a heart attack is Y times higher if one exercises relative to that for someone who does not exercise regularly. Proportional hazard models have the general form,

$$h(t,x_i) = e^{f(x_i)} \, h_0(t) \qquad (9.16)$$

where $h(t, x_i)$ = the hazard at time, t as modified by the value (x_i) of covariate x, $h_0(t)$ = the hazard of some reference group or type which is described by a specific model such as a Weibull model, and $f(x_i)$ = some function of the covariate (x) making the hazards proportional. For example, $f(x)$ may be a simple function making hazards proportional among different animal sizes, e.g., $f(x) = a + b(\log$ animal weight).

Use of the log normal, log logistic, normal, or gamma model results in an **accelerated failure time model**, a model in which the time-to-death (ln TTD_i) of a particular type of individual (e.g., smoker) is changed ("accelerated") as a function of some covariate (e.g., classification relative to smoking habit). For example, any drop in animal size may decrease the expected time until death for an individual exposed to a toxicant. An increase in toxicant concentration may accelerate the expected time until death of an individual. The simple form of the accelerated failure time model is the following,

$$\ln TTD_i = f(x_i) + \varepsilon_i \tag{9.17}$$

The error term (ε) is fit to the assumed model, e.g., log normal model. The function of the covariate modifies the ln TTD and can be any function of the covariate such as that given above for the effect of animal weight (e.g., $f(x) = a + b(\log$ animal weight)). Most applications fitting parametric survival time models allow parameter estimation so that predictions can be made relative to **median times-to-death** (MTTD) or some proportion other than 50% mortality. They also allow one to test for significant effects of different covariates on lethal effect.

Often, especially in clinical studies comparing various treatments, it is impossible or unnecessary to model the underlying survival curve. The exact underlying model is much less important than estimating the influence (hazard or relative risk) of some covariate on survival. For example, the focus of a study may be on determining if a post-surgical treatment improves patient survival relative to the standard treatment (reference treatment). In such a study, the exact form of the survival curve is not required. A semi-parametric method (**Cox proportional hazard model**) is designed to allow examination of proportional hazards without taking on the assumption of any specific model for the baseline or reference hazard.

Incipiency

The LT50 or MTTD may be used to estimate lethal incipiency, the concentration at which 50% of exposed individuals will live indefinitely relative to the toxicant effects. The point at which the line for toxicant A begins to run parallel to the log MTTD axis is an estimate of this lethal threshold in Figure 9.8. Note that there may be no apparent threshold for some toxicants (e.g., Toxicant B). Also, there may be a minimum time required before an effect can be expressed. Toxicant B illustrates such a **minimal time to response**.

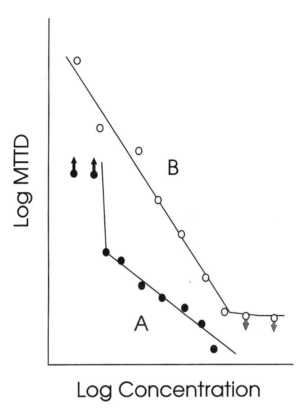

FIGURE 9.8. Incipiency for the time-to-death approach. Chemical A (closed circles) has a distinct incipient MTTD. Below a certain concentration, 50% of exposed individuals will live indefinitely relative to the effect of toxicant A. Note that the two lowest concentrations resulted in less than 50% mortality regardless of how long the individuals were exposed. Some toxicants (B) may show no evidence of incipiency. There may also be a minimum time to get a response, e.g., individuals can only die so fast regardless of the toxicant concentration. Toxicant B in the diagram has such a threshold. (Arrows attached to a point signify that the "true" value is probably in the indicated direction from the point, e.g., the true log MTTD for the last point on curve B is less than the value of the last point.) (Modified from Figure 5 in Sprague [1969] and Figure 4 in Newman [1995].)

Mixture Models

Mixture effects can also be quantified for the time-response approach as done with the dose-response approach. For example, the lethal threshold estimated with the MTTD or some other estimate is used instead of the incipient LC50. However, more involved treatments become possible because more data can be extracted from a toxicity test with these methods. Roy and Campbell (1995) took advantage of this enhanced power of survival time models to quantify the combined effect of aluminum and zinc to young Atlantic salmon (*Salmo salar*).

FACTORS INFLUENCING LETHALITY

Biotic Qualities

A variety of biological qualities can influence effects of toxicants. Some have already been discussed, e.g., developmental stage, lipid pools within the individual, feeding behavior, and induction of detoxification mechanisms. Life stage (e.g., Campbell, 1926) is also important. Two general biological qualities that can influence lethality, acclimation and allometry, will be discussed briefly here.

Acclimation[8] is described here as the modification of biological functions, especially physiological functions, or structures to maintain or minimize deviations from homeostasis despite change in some environmental quality such as temperature, salinity, light, radiation, or toxicant concentration. It is an expression of phenotypic plasticity of individuals in response to a sublethal change in some environmental factor. Often the distinction is made in the literature between acclimation (shifts taking place in a controlled or laboratory setting) and **acclimatization** (shifts taking place under natural conditions). It is important to understand that acclimation is not the shift in population qualities, i.e., not a change in genetic composition or demographic structure, in response to a stressor. Such adaptation or genetic change in a population in response to toxicant stress will be discussed in the next chapter.

Acclimation can occur via pre-exposure to sublethal concentrations of toxicants, enhancing survival of individuals during a subsequent, intense exposure. However, some toxicants cause damage regardless of their concentrations, e.g., cyanide pre-exposure does not enhance survival of rainbow trout because of kidney damage during the first exposure (Dixon and Sprague, 1981). In contrast, pre-exposure of human lymphocytes to low levels of radiation leads to reduced chromatid aberrations during a subsequent exposure to x-rays (Olivieri et al., 1984). Dixon and Sprague (1981) found that, above a certain concentration, pre-exposure of rainbow trout to copper enhanced their tolerance to otherwise lethal concentrations of copper. It is important to note that the enhanced tolerance increased with copper concentration only to a certain point; beyond that point, pre-exposure caused damage and consequent diminished tolerance in pre-exposed fish. Enhanced tolerance was a complex function of both acclimation concentration and time (Dixon and Sprague, 1981). For example, both strongly influenced lethal consequences to salmon exposed to high temperatures after varying intensities and durations of pre-exposure (Elliott, 1991).

Allometry can also influence toxicant effect to individuals (Newman and Heagler, 1991). The classic work of Bliss (1936) used arsenic intoxication of silkworm larvae to estimate the influence of size on toxic impact. First, he took time-to-death data and transformed it to rate of toxic action (rate of toxic action = 1000/TTD). Next, he performed a multiple regression to generate the model,

---

[8]Acclimation may also be used to identify the time allowed for an organism, population, community or ecosystem to stabilize to a set of conditions prior to testing. For example, Kennedy et al. (1995) use the term to describe the establishment of mesocosms for ecotoxicological testing. Acclimation will not be used in this context (relative to an experimental protocol) here.

$$Rate = a + b_1 (\log dose) + b_2(\log weight) \qquad (9.18)$$

where a, b_1, and b_2 are constants derived during model fitting. This relationship was transformed to generate an adjustment factor (weight^h) for the influence of animal weight.

$$Rate = a + b_1 \left[\log \frac{dose}{weight^h} \right] \qquad (9.19)$$

where $h = b_2/b_1$. This approach has been adopted during the last 60 years as the primary approach to incorporating size effects into time-to-death data. The approach (Bliss, 1936) was later modified to dose- or concentration-effect models (Anderson and Weber, 1975).

$$\log LC50 = \log a + b \log weight \qquad (9.20)$$

or

$$LC50 = a \, weight^b \qquad (9.21)$$

Newman et al. (1994) and Newman (1995) demonstrated that a more general approach exploring the various models described for survival time models can lead to a better fit to the data.

Abiotic Qualities

Numerous physical and chemical factors may modify the toxic action of contaminants. Ambient and acclimation temperatures modify impact of some toxicants. For example, toxic impact (LC50) was modified by acclimation and ambient temperature during exposures of a snail (*Potamopyrgus antipodarum*) to cadmium (Møller et al., 1994). Ambient light can also influence the action of photo-labile chemicals by changing the rate at which they break down to more toxic products. **Photo-induced toxicity**, toxicity of a chemical in the presence of light due to the production of toxic photolysis products, was found for bluegill sunfish (*Lepomis macrochirus*) exposed to the polycyclic aromatic hydrocarbon, anthracene (Bowling et al., 1983). Some chemicals can also enhance the **photosensitivity** (sensitivity of cutaneous tissues to the effects of light evoked by a chemical) of individuals. Channel catfish (*Ictalurus punctatus*) treated with the antibiotic, oxytetracycline, become extremely sensitive to sunlight, resulting in skin (sunburn) and eye (lesions) damage (Stacell and Huffman, 1994).[9]

Organic and inorganic constituents of waters may modify metal toxicity (e.g., Andrew et al., 1977; Bradley and Sprague, 1985; Azenha et al., 1995). Freshwater **hardness** (sum of the concentrations of dissolved calcium and magnesium)

---

[9]This is also the case for humans taking certain antibiotics. Physicians and labels on some prescription antibiotics warn patients not to spend much time in the sun because of increased sensitivity to sunburn.

is often related to toxic effect of metals such as beryllium (EPA, 1986), cadmium (Carroll et al., 1979), copper (Hawarth and Sprague, 1978), lead (Davies et al., 1976), and zinc (EPA, 1986). Several mechanisms have been proposed, including competition of hardness cations with toxic metals for binding sites on biomolecules, modification of biological processes, and modification of metal speciation in the exposure waters (Newman, 1995). These effects on metal toxicity are predicted in EPA water quality criteria documents (e.g., EPA, 1985e) with a simple, empirical model:

$$\log \textit{Toxicity Endpoint (e.g., LC50)} = \log a + b \; (\log \textit{Hardness}) \qquad (9.22)$$

or

$$\textit{Toxicity Endpoint} = a \; \textit{Hardness}^b \qquad (9.23)$$

where log a and b = estimates from linear regression of the intercept and slope. Although Equation 9.23 produces a biased prediction of toxic effect (Newman, 1991; Newman, 1993), it is used extensively in the ecotoxicological literature and water quality regulations.

Sediment and soil qualities also influence toxicity as described in earlier discussions of factors modifying bioavailability. Further, QSARs are readily developed for lethality as well as bioaccumulation. The reader is encouraged to review information covered in Chapter 4 as topics discussed there are relevant to this discussion of abiotic factors influencing toxicity.

SUMMARY

In this chapter, lethality was described under acute and chronic exposure scenarios. The possibility of differences in toxic impact at different life stages of an individual was explored along with associated implications. Basic toxicity test designs were outlined and their relative advantages and disadvantages highlighted. The predominant approach to measuring toxicity (dose- or concentration-response design) was detailed and methods of analyzing dose-response data were contrasted. An alternative approach, the survival time design, and methods for analyzing time-to-death data were compared to the dose-response methods. Toxic incipiency and ways of quantifying effects of toxicant mixtures were provided for the dose-response and survival time designs. Finally, biotic and abiotic factors modifying toxicity were explored only briefly because many had already been discussed in the context of factors modifying bioavailability (Chapter 4).

Selected Readings

Anderson, P.D. and L.J. Weber. Toxic response as a quantitative function of body size. *Toxicol. Appl. Pharmacol.* 33, pp. 471–483, 1975.

Christensen, E.R. Dose-response functions in aquatic toxicity testing and the Weibull model. *Water Res.* 18, pp. 213–221, 1984.

Dixon, D.G. and J.B. Sprague. Acclimation to copper by rainbow trout (*Salmo gairdneri*)—A modifying factor in toxicity. *Can. J. Fish. Aquat. Sci.* 38, pp. 880–888, 1981.

Marking, L.L. Toxicity of Chemical Mixtures, in Rand, G.M., and Petrocelli, S.R., Eds., *Fundamentals of Aquatic Toxicology*. Hemisphere Publishing Corp., Washington, DC, 1985.

Newman, M.C. and M.S. Aplin. Enhancing toxicity data interpretation and prediction of ecological risk with survival time modeling: An illustration using sodium chloride toxicity to mosquitofish (*Gambusia holbrooki*). *Aquat. Toxicol. (AMST)* 23, pp. 85–96, 1992.

Newman, M.C. and J.T. McCloskey. Time-to-event analyses of ecotoxicology data. *Ecotoxicology* 5, pp. 187–196, 1996b.

Sprague, J.B. Measurement of pollutant toxicity to fish. I. Bioassay methods for acute toxicity. *Water Res.* 3, pp. 793–821, 1969.

Stephan, C.E. Methods for Calculating an LC50, in *Aquatic Toxicology and Hazard Evaluation. ASTM STP 634*, Mayer, F.L. and J.L. Hamelink, Eds., American Society for Testing and Materials, Philadelphia, PA, 1977.

Effects on Populations

There is an enormous disparity between the types of data available for assessment and the types of responses of ultimate interest. The toxicological data usually have been obtained from short-term toxicity tests performed using standard protocols and test species. In contrast, the effects of concern to ecologists performing assessments are those of long-term exposures on the persistence, abundance, and/or production of populations.

BARNTHOUSE ET AL. (1987)

OVERVIEW

Sometimes, the focus of our efforts is protection of individuals, particularly when those efforts involve effects to humans or endangered species. More often, effects to individuals are measured to predict consequences to populations because the primary goal of most ecotoxicologists is assuring persistence of populations within ecological communities. Qualities used to assess population-level effects include many already discussed, but they are interpreted in a slightly different context. For example, cancer may now be examined in the context of relative incidences in populations occupying different microhabitats or having different demographic qualities. Age-dependent effects of toxicants may be woven into explanations of demographic change. Differences in individual effective doses may be considered in the context of selection for tolerant genotypes in populations. Knowledge of activation mechanisms for carcinogens may be sought to support inferences about causal structure in studies of disease incidence.

Described in this chapter are the means of determining the status of exposed populations. Also discussed are population responses that could lessen adverse effects. First, approaches are detailed for describing the occurrence of toxicant-related disease in extant populations. Associated epidemiological information helps us to assess the imminence of failure of the afflicted population in addition to estimating the probability of an individual with certain qualities being adversely affected. Second, impacts on population demography are described. The advantages are discussed for generating toxicity test data amenable to demographic analysis. Third, we examine the change in population genetics including the evolution of tolerance which might occur after long periods of exposure to contaminants.

EPIDEMIOLOGY

Epidemiology is the science concerned with the cause, incidence,[1] prevalence, and distribution of infectious and noninfectious diseases in populations. Most often, disease is linked through correlation with risk factors[2] such as qualities of individuals and **etiological agents** (agents responsible for causing, initiating, or promoting the disease [Rench, 1994]). An example of an etiological agent might be the high mercury concentrations in seafood taken from Minamata Bay. Two subdefinitions are relevant to studies of disease resulting from chemical (e.g., pollutants or chemicals in the workplace) and physical (e.g., radiation, UV light, high temperatures, or asbestos fibers) etiological agents. Here, **environmental epidemiology** is defined as that subdiscipline of human epidemiology concerned with diseases caused by chemical or physical agents (Rench, 1994). **Ecological epidemiology**, frequently associated with retrospective ecological risk assessments, is the name often given to epidemiological methods applied to determining the cause, incidence, prevalence, and distribution of adverse effects to nonhuman species inhabiting contaminated sites (Suter, 1993).

A range of straightforward metrics may be generated during epidemiological analyses. Those from life tables, as described shortly, were among the first to be applied to human epidemiology. Disease **incidence rate** (I, expressed in units of individuals or cases per unit of exposure time being considered in the study) for a nonfatal condition can be calculated as the number of individuals with the disease divided by the total time that the population had been exposed, e.g., 10 new cases per 1000 person years.

$$I = \frac{N}{T} \tag{10.1}$$

where N = number of diseased individuals or cases, and T = the total time at risk of contracting the disease (Ahlbom, 1993). The T may be expressed as the total number of time units that individuals were exposed to risk during the study period, e.g., per 1000 person years of exposure. **Prevalence** (P) is simply the incidence rate (I) times the amount of time (t) that individuals were at risk.

$$P = I \times t \tag{10.2}$$

If there were 2 cases per 1000 person years, the prevalence in 10,000 person years (e.g., a population of 1000 people exposed for 10 years) would be (2 cases/1000 person years)(10,000 person years) or 20 cases.

---

[1]Incidence and prevalence have slightly different meanings in epidemiology. Disease **incidence** (cumulative) is the number of new individuals scored as having the disease in a certain time interval, e.g., 10 additional cases in the last week. **Prevalence** is simply the total number or proportion of individuals with the disease at a particular time, e.g., 157 cases in New York City during 1957 (Ahlbom, 1993). Often it is expressed as a ratio also, e.g., 157 cases per 10,000 people in 1957.

[2]A **risk factor** is any quality of an individual (e.g., age or dietary habits) or an etiological factor (e.g., chronic exposure to high levels of a toxicant) that modifies an individual's risk of developing the disease.

Occurrence of disease in a population may also be expressed relative to that in another population. Often one is a control or reference population. The simple difference in incidence rates may be used to compare disease in two populations, e.g., 25 more cases per year in population A than in population B. The difference often is expressed in terms of a standard size (10,000 individuals) because the populations will likely differ in size. Also, the ratio of occurrences of the disease in the two populations **(relative risk or RR)** can be expressed as the ratio of incidence rates **(rate ratio)**.

$$RR = \frac{I_A}{I_0} \qquad (10.3)$$

where I_A = incidence rate in population A, and I_0 = incidence rate in the reference or control population. For example, 23 diseased individuals occurring annually in a standard sample size of 10,000 individuals for a heavily industrialized city may be compared to an annual incidence rate of 0.5 individuals per 10,000 individuals in a small town. The relative risk would be expressed as a rate ratio of 46 in this example. Note that relative risk calculated with survival time models described earlier can be used in such epidemiological analyses too.

Relative risk can also be expressed as an **odds ratio** in case-control studies. The number of disease cases (individuals) that were (a) or were not (b) exposed, and the number of control individuals free of the disease that were (c) or were not (d) exposed to the risk factor are used to estimate the odds ratio (Ahlbom, 1993),

$$Odds\ Ratio = \frac{a/b}{c/d} = \frac{ad}{bc} \qquad (10.4)$$

Say, for example, that 750 cases of a fatal disease were documented with 500 of them associated with individuals who had been previously exposed to an etiological agent (a) and 250 of them (b) were associated with individuals who had never been exposed to this agent. In another sample of 500 control individuals showing no signs of the disease, 60 had been exposed (c) and 440 (d) had not been exposed to the agent. The odds ratio would be (500)(440)/(250)(60) or roughly 14.7. This odds ratio suggests that exposure does influence proneness to the disease. An example study employing odds ratios is provided by Rench (1993) in which a disproportionately high number of soil conservationists fell victim to non-Hodgkin's lymphoma relative to a control group, notionally because of high pesticide exposures associated with their jobs.

There exists a wealth of methods such as proportional odds and logistic regression models (SAS Institute, Inc., 1990) for analyzing odds data from epidemiological studies. Easily assessable textbooks such as those written by Ahlbom (1993), and Marubini and Valsecchi (1995) describe these and other statistical methods applicable to epidemiological data. They detail the calculation of confidence intervals for estimates and statistical tests of significance. Although most focus on human epidemiology and clinical studies, there are no inherent obstacles to their application to ecological epidemiology. Unfortunately, most of these powerful methods remain underexploited in ecotoxicology.

Logical rules have been developed to enhance inferential soundness of epidemiology because most approaches in this field rely heavily on inferentially-weak correlations of disease with risk factors. **Nine aspects of disease association** have been identified by Hill (1965) for assigning linkage of a risk factor and disease in environmental epidemiology: strength of association, consistency of association, specificity of association, temporal association, biological gradient (dose-response), biological plausibility, coherence of the association, experimental support, and analogy. The *strength of association* between some risk factor and disease is important to consider. For example, the 200-fold higher prevalence of scrotal cancer for chimney sweeps relative to men in other occupations added strength to the supposition that the cancer was initiated or promoted by some occupational risk factor (carcinogenic PAHs in soot and tar). *Consistency of association*, the consistent observation of the association under numerous, varying conditions, also strengthens conclusions. For example, soundness of the conclusion associating asbestos fibers with lung cancer is enhanced if the incidence of lung cancer is high for either male and female workers in many, diverse occupations sharing one common factor, high asbestos fiber densities in workplace air. However, consistency may also be generated by a bias in data generation so caution must be exercised in applying this rule.

Identification of association under very specific situations (*specificity of association*) may enhance one's ability to assign causation or linkage. For example, only a very specific type of behavior or occupation may have the associated linkage between the risk factor and disease. Indeed, Hill (1965) observed that it was the lack of specificity in the association between lung cancer and smoking that allowed counter arguments to causal linkage to persist for so long. A *temporal association* may be considered in order to reinforce causation. The cause or promoter of the disease should be present before the disease occurs. There is potential for bias here also. For example, a certain occupation may have a higher number of individuals in it that are predisposed to a particular condition, e.g., an unusually sedentary group within the population who are prone to cardiovascular disease may be overrepresented. In such a case, the disease would be associated incorrectly with a pre-existing, workplace risk factor, e.g., some chemical used in the occupation.

Obviously, a clear *biological gradient (dose-response) in the association* strengthens evidence also for a relationship. Although the knowledge is often not available, *biological plausibility* (a plausible, underlying mechanism for the association) can enhance the probability of making a correct linkage between a disease and risk factor. *Coherence of the association* with what is already known about the disease is also very helpful. For example, findings of liver lesions after chronic exposure of laboratory rats to a particular toxicant can support inference of an association between human exposure to that toxicant in the workplace and an increased incidence of liver cancer. *Experimental support of association* (manipulation of the association with measured change in the disease response) can help if practical. This type of support in human studies is limited for ethical reasons and, consequently, tends to be more applicable for nonhuman species. However, these data are available for humans under special circumstances. Hill (1965) provides one example of lung and nasal cancer incidences in nickel refinery work-

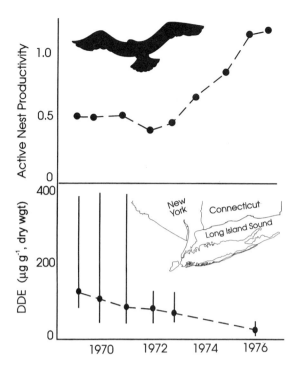

FIGURE 10.1. The gradual increase in active nest productivity (average number of young fledged per active nest) of Long Island Sound osprey. Nesting success was extremely low prior to the widespread banning of pesticides such as DDE. The nest productivity slowly recovered (top panel) as DDE concentrations decreased in eggs from osprey nests (bottom panel), suggesting that DDE was a significant risk factor in nest failure. (Modified from Figure 1 of Spitzer et al. [1978].)

ers before and after workplace exposure routes were controlled. The incidence of cancer dropped significantly after control measures were taken to reduce worker exposure. Figure 10.1 shows an analogous situation with the effect of DDE on nest productivity of osprey (*Pandion haliaetus*). Nestling production gradually increased after a ban on the application of such pesticides. This recovery supports causal linkage between DDE and reproductive damage. As discussed before, the association between DDE and raptor population decline was also supported by biological plausibility, i.e., eggshell thinning.

Analogy is the final quality of associations enhancing the accuracy with which risk factors are identified. Similarity of the association to another well-documented association fosters accuracy. Hill (1965) suggests that the demonstration of thalidomide's effect on fetuses enhanced the credibility of subsequent suggestions of birth defect linkage to other drugs taken by pregnant women. None of these factors alone leads to accurate identification of a real association. Rather, the goal is to identify associations as sufficiently plausible or implausible. Of course, another possible outcome of an epidemiological study is the conclusion that insufficient evidence exists to judge plausibility.

Obviously, these same nine factors can enhance plausibility of association between a risk factor and some adverse consequence in ecological epidemiology. Suter (1993, Table 10.4) outlined Hill's nine factors, providing a slightly more ecological explanation for each. The only change was associated with a greater capacity to perform and to emphasize controlled experiments with nonhuman species. These and similar logical tools will be discussed again in more detail in the chapter describing ecological risk assessment techniques.

POPULATION DYNAMICS AND DEMOGRAPHY

Overview

A **population** is a group of individuals of a species occupying a defined space at a particular time. As recognized early on and studied extensively in ecology and human demography, external factors influence the size, nature, and distribution of nonhuman[3] and human[4] populations. Growth under limitation in artificial systems such as agricultural plots and species tolerance ranges as determined in laboratory trials were used to predict presence and size of populations in the field. Ecotoxicologists have extended this traditional approach to predicting population sizes and probabilities of local extinction of populations exposed to contaminant gradients based on laboratory, small field plot, mesocosm, and enclosure studies.

Often predictions of population viability are made from results of toxicity tests, and ANOVA-derived NOEC and LOEC values. Concentrations that affect such qualities as individual growth, larval and adult survival, and number of young produced per female are pooled to generate these predictions. Decisions may be based on whether or not change was statistically significant. As we discussed, the flaw in this approach is that statistical significance does not dictate biological significance. For this and other reasons already discussed, a movement away from such an approach has begun. (See Chapman et al. [1996] as an example.) A more thoughtful approach might be to decide on a magnitude of change in some relevant quality beyond which the population is assumed to be adversely impacted. The rationale is the following: (1) to move forward in environmental stewardship, consensus must be reached for the level of change required to trigger concern or action by regulators, and (2) consensus opinion should focus on a level of biologically-meaningful change that one can detect with reasonable confidence in most cases (e.g., Hoekstra and Van Ewijk, 1993). Both of these pragmatic approaches (NOEC-LOEC and magnitude of change for some population quality) are compromised as they do not directly answer the question of whether or not the population will remain viable despite the presence of the toxicant. A 20% reduction of a particular quality may be catastrophic to one species population but trivial to another species population: a 20% change in one quality may be trivial but a similar change in another quality may lead to imminent extinction of a population. Fortunately, traditional population and demographic analyses can be used to predict the

---

[3]Perhaps the best known examples are **Liebig's law of the minimum** and **Shelford's law of tolerance**. Liebig's law states that a population's size will be limited by some essential factor in the environment that is scarce relative to the amounts of other essential factors, e.g., phosphorus will limit algal growth in many lakes. Shelford's law states that a species' tolerance along an environmental gradient (or a series of environmental gradients) will determine its realized population distribution and size in the environment. For example, salinity and temperature gradients may define the location and abundance of an oyster species along the east coast of the United States.

[4]The Englishman, Thomas R. Malthus (1766–1834) established a series of assumptions and observations regarding limitations on human populations. **Malthusian theory** was first published in 1798 as the profoundly influential essay, *An Essay on the Principle of Population as it Affects the Future Improvement of Society with Remarks on the Speculation of Mr. Godwin, M. Condorcet and Other Writers*.

possible outcomes of exposure and their probabilities of occurring. Although most toxicity testing methods do not produce information directly amenable to demographic analysis, some ecotoxicologists have begun to design tests and interpret results in this context. Further, current EPA documents describing ecological risk assessment methods clearly recognize and now articulate this inconsistency between traditional toxicity tests and data needs.

> During the past two decades, toxicological endpoints (e.g., acute and chronic toxicity) for individual organisms have been the benchmarks for regulations and assessments of adverse ecological effects. . . . The question most often asked regarding these data and their use in ecological risk assessments is, 'What is the significance of these ecotoxicity data to the integrity of the population?' More important, can we project or predict what happens to a pollutant-stressed population when biotic and abiotic factors are operating simultaneously in the environment?
>
> Protecting populations is an explicitly stated goal of several Congressional and Agency mandates and regulations. Thus it is important that ecological risk assessment guidelines focus upon protection and management at the population, community, and ecosystem levels . . .
>
> (EPA, 1991a)

The focus has begun to shift to population vital rates. This approach, as described here, has much promise for enhancing prediction of population effects of contaminants.

General Population Response

The simplest models of population response treat all individuals identically and predict change in total number or density of individuals over time. Surveys of widespread population trends such as that by Sarokin and Schulkin (1992) may treat populations in this general manner. Changes in total numbers of individuals may be correlated with epidemics of pollutant-linked cancers in the population, or **epizootics** (outbreaks of disease in a large number of individuals) caused by a biological agent to pollution-weakened populations.

Unrestrained, exponential growth of a population can be predicted as a function of the population size (N) and its **intrinsic (or Malthusian) rate of increase** (r) with a simple differential equation.

$$\frac{dN}{dt} = rN \tag{10.5}$$

With knowledge of the initial population size (N_0), its size at any time (t) in the future can be predicted.

$$N_t = N_0 e^{rt} \tag{10.6}$$

Doubling times ($t_d = (\ln 2)/r$) can also be estimated for populations if r is known.

A difference equation can be used instead of Equation 10.5 if population size is measured at discrete intervals such as might be done with a population with nonoverlapping generations. A population of an annual plant or an insect population may have nonoverlapping generations.

$$N_{t+1} = \lambda N_t \tag{10.7}$$

where λ = **finite rate of increase**,[5] and N_{t+1} and N_t = population size at times t+1 and t, respectively. If population size is measured initially (N_0) and at some time in the future (N_t), r can be calculated from λ.

$$\lambda = \frac{N_t}{N_0} = e^r \tag{10.8}$$

All three parameters, λ, r and t_d can and have been used as sensitive metameters for adverse population effects. For example, Marshall (1962) calculated the effects of γ radiation on the intrinsic rate of increase for *Daphnia pulex*. Rago and Dorazio (1984) detailed means to estimate the influence of toxicants on zooplankton population finite rate of increase. These qualities have the advantage over measures such as a simple percentage reduction in reproduction that they are easily fit into predictive ecological models. They also are readily incorporated into results from more complex, life table methods as will be discussed.

Obviously, populations cannot grow exponentially indefinitely. In the simplest form, these equations are modified so that growth rate decreases as the population size approaches some **carrying capacity** of the environment (K, the maximum population size expressed as total number of individuals, biomass, or density that a particular environment is capable of sustaining). For populations with overlapping generations, Equation 10.9 is relevant.

$$\frac{dN}{dt} = rN\left[1 - \frac{N}{K}\right] \tag{10.9}$$

The classic **Ricker model** (Equation 10.10) is relevant to populations with nonoverlapping generations or experimental designs with discrete intervals of population growth.

$$N_{t+1} = N_t\, e^{r\left(1 - \frac{N_t}{K}\right)} \tag{10.10}$$

Even these simplest of models (Equations 10.9 and 10.10) predict very complex behavior for population dynamics under certain conditions (May, 1974; 1976b). According to these models, populations may increase to and remain at K,

---

[5]This term is widely used; however, May (1976a) argued that a better term is the **multiplicative growth factor per generation**. For further discussion of this point, please refer to May (1976a) or Newman (1995).

oscillate around K with gradual convergence to K, oscillate indefinitely around K, or fluctuate chaotically. Consequently, field observations of population densities other than those near K or of population densities that fluctuate widely, do not necessarily reflect an adverse consequence of contamination. Also, recovery of a population after a toxicant-related decrease in population size may not always involve a simple, monotonic increase back to K. Oscillations may occur·and influence the probability of successful recovery or extinction. For example, Simkiss et al. (1993) found that sublethal exposure of blowflies (*Lucilia sericta*) to cadmium modified population dynamics in a food-limited environment.

Populations often unevenly occupy areas with patches of superior and inferior habitat. Subpopulations within a superior habitat may have high rates of increase but those within inferior habitats may have negative growth rates (r). Some subpopulations will act as a source of individuals due to the surplus of offspring produced: others will act as sinks as surplus individuals move in from outside the patch. This source-sink or patch structure results in dynamics distinct from those predicted by the simple models above (Pulliam and Danielson, 1991). For example, a habitat so contaminated that reproduction is impossible may still have a high density of individuals if there is a nearby source of individuals. Maurer and Holt (1996) illustrate the relevance of this concept to ecotoxicology with the example of pesticide impact on wildlife inhabiting a patchy habitat. They found in their models that populations with low rates of increase in uncontaminated patches tend to have a higher risk of extinction than those with high rates of increase. Increased migration rates among patches increased the risk of local extinction.

Demographic Change

> *. . . ten times twelve solar years were the term fixed for the life of man, beyond which the gods themselves had no power to prolong it; that the fates had narrowed the span to thrice thirty years, and that fortune abridged even this period by a variety of chances, . . .*
> Niebuhr's *History of Rome* (as cited in Deevy, 1947)

A more comprehensive analysis can be made by separately examining population **vital rates**, rates at which important processes such as birth, migration, and death occur in populations. Vital rates can be considered for age classes or life stages to obtain a rich understanding of population qualities such as rate of change and stable age structure. Life tables are constructed with these age-specific vital rates to predict population qualities influencing persistence. As suggested by Bezel and Bolshakov (1990), such information is critical for accurate assessment of population effects of pollutants because irreversible shifts in population structure are more often the cause of population extinction than outright death of individuals.

Life tables may include only survival data similar to that modeled in the previous chapter. The conventional notation includes x as the unit of age and l_x as the number of individuals in a cohort that are alive at x. Such l_x **tables or schedules**

TABLE 10.1. Survival (l_x) for *D. pulex* Exposed to 0 and 75.9 R h^{-1} of Radiation. (In this table, the proportion of the original cohort surviving in each treatment is shown instead of raw counts of individuals. These proportions are identical to the cumulative survival or S(t) described for survival data in the last chapter. Derived from Marshall [1962], Table I).

Age Class or x (days old)	Control (0 R h^{-1})	High Dose (75.9 R h^{-1})
0	1.00	1.00
1	1.00	0.98
2	1.00	0.96
3	1.00	0.96
4	1.00	0.96
5	1.00	0.96
6	1.00	0.96
7	0.98	0.96
8	0.98	0.96
9	0.98	0.96
10	0.98	0.96
11	0.98	0.96
12	0.98	0.94
13	0.98	0.94
14	0.98	0.94
15	0.98	0.94
16	0.98	0.94
17	0.98	0.81
18	0.98	0.67
19	0.98	0.29
20	0.98	0.17
21	0.98	0.02
22	0.91	0.00
23	0.81	0.00
24	0.58	0.00
25	0.49	0.00
26	0.35	0.00
27	0.28	0.00
28	0.19	0.00
29	0.14	0.00
30	0.14	0.00
31	0.14	0.00
32	0.07	0.00
33	0.05	0.00
34	0.05	0.00
35	0.00	0.00

summarize information just as the survival models described earlier might; however, no specific underlying distribution need be assumed in the l_x schedule (Table 10.1). The **age-specific number of individuals dying** ($d_x = l_x - l_{x+1}$) and the **age-specific death rate** (probability of dying in interval x or $q_x = d_x/l_x$) may also be derived from l_x.

Some ecologically meaningful statistics can be generated from l_x tables alone as illustrated with Table 10.2. This table is a rendering of the high dose survival

TABLE 10.2. Estimation of Expected Life Span (e_x) from the High Dose in Table 10.1 Assuming 100 Individuals in the Original Cohort. (The l_x in this table is l_x in Table 10.1 multiplied by 100.)

Age (x)	l_x	d_x	q_x	L_x	T_x	e_x (Days)
0	100	2	0.02	99.0	1774.0	17.7
1	98	2	0.02	97.0	1675.0	17.1
2	96	0	0.00	96.0	1578.0	16.4
3	96	0	0.00	96.0	1482.0	15.4
4	96	0	0.00	96.0	1386.0	14.4
5	96	0	0.00	96.0	1290.0	13.4
6	96	0	0.00	96.0	1194.0	12.4
7	96	0	0.00	96.0	1098.0	11.4
8	96	0	0.00	96.0	1002.0	10.4
9	96	0	0.00	96.0	906.0	9.4
10	96	0	0.00	96.0	810.0	8.4
11	96	2	0.02	95.0	714.0	7.4
12	94	0	0.00	94.0	619.0	6.6
13	94	0	0.00	94.0	525.0	5.6
14	94	0	0.00	94.0	431.0	4.6
15	94	0	0.00	94.0	337.0	3.5
16	94	13	0.14	87.5	243.0	2.6
17	81	14	0.17	74.0	155.5	1.9
18	67	38	0.57	48.0	81.5	1.2
19	29	12	0.41	23.0	33.5	1.2
20	17	15	0.88	9.5	10.5	1.1
21	2	2	1.00	1.0	1.0	1.0
22	0	–	–	–	–	–

data in Table 10.1; however, l_x is now expressed as the number of survivors. The average days lived (L_x) is estimated as ($l_x + l_{x+1}$)/2 for each age class. The total days lived (T_x) for the age class is estimated by summing the L_x values from the bottom of the chart (e.g., for x = 21 here) up to the pertinent age class (x). The **expected life span** for individuals of age x can then be estimated as $e_x = T_x/l_x$. Bechmann (1994) measured changes in expected life span of a marine copepod exposed to copper and found an increase in life span at low copper concentrations, suggesting a hormetic or compensatory response by the copepod population.

   Age-specific birth rates (m_x, the mean number of females born to a female of that age class) can be added to produce an $l_x m_x$ **life table**. Much more information can then be extracted from the population. Newman (1995) gave the fictitious example (Table 10.3) of a $l_x m_x$ table for a population living in a contaminated mesocosm. The expected number of females to be produced during the lifetime of a newborn female (R_0 or **net reproductive rate**) can be estimated from this table as the sum of the $l_x m_x$ products in the table. In this case ($R_0 = 0.846$), each female is not replacing herself and the population size will likely decline if conditions do not change. The **mean generation time** (T_c) is estimated as the sum of the $xl_x m_x$ column divided by R_0. (The x used here is the midpoint of the age class, e.g., 0.5 for the 0 to 1 year age class.) The mean generation time for this population is

TABLE 10.3. A $l_x m_x$ Life Table for a Fictitious Population in a Contaminated Meso-cosm. (Modified from Newman [1995].)

Age Class (x)	Midpoint of Class	l_x	m_x	$l_x m_x$	$x l_x m_x$
0 to <1 year	0.5	1.000	0.000	0.000	0.000
1 to <2 year	1.5	0.312	2.238	0.698	1.047
2 to <3 year	2.5	0.095	1.390	0.132	0.330
3 to <4 year	3.5	0.037	0.410	0.015	0.053
4 to <5 year	4.5	0.003	0.400	0.001	0.005
5 to <6 year	5.5	0.000	–	$\Sigma = 0.846$	$\Sigma = 1.435$

1.435/0.846 or 1.7 years. An approximation of the intrinsic rate of increase is ln R_0/T_c or −0.098, indicating a decline in population size through time. The intrinsic rate of increase is more accurately estimated with the **Euler-Lotka equation,**

$$\sum_{x=0}^{\infty} l_x m_x e^{-rx} = 1 \qquad (10.11)$$

The estimate of r generated above may be used initially in this equation. The estimated r is then adjusted up- or downward until the solution to the left side of the equation is sufficiently close to 1. The r resulting in such a condition is deemed adequate for most purposes. Use of the Euler-Lotka equation carries the assumption that the population is stable. If conditions do not change with time, a population with a particular r will eventually establish a stable distribution of individuals among the various age classes. Such a population is called a **stable population**.

For a stable population, the expected contribution of offspring during the life of an individual (**reproductive value** or V_A) is easily calculated for each age class.

$$V_A = \sum_{x=A}^{\infty} \frac{l_x}{l_A} m_x \qquad (10.12)$$

This V_A can be used to reinforce the point discussed previously that it is inappropriate to estimate the consequences to population viability based on reductions in most sensitive life stage survival and NOEC values for lowered reproduction. Reproductive value can be used directly to assess such consequences to populations. Indeed, Kammenga et al. (1996) clearly demonstrated with such an approach that the most sensitive life cycle trait of the nematode *Plectus acuminatus* was not the most critical demographic quality impacted by cadmium exposure. For example, a toxicant that eliminated that age class or set of age classes with the highest reproductive value would have strong, adverse consequences to the population's viability. Conversely, a toxicant may have very minor consequences relative to population viability if it eliminated large numbers of individuals in an age class that had very low reproductive value. Unfortunately, this approach has been largely ignored for assessments of effect although Barnthouse et al. (1987) did effectively include

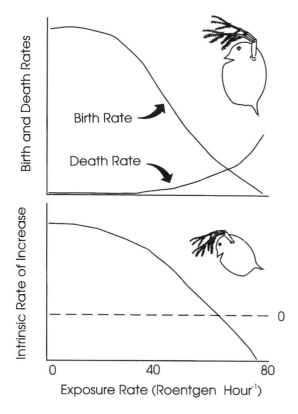

FIGURE 10.2. Birth, death, and intrinsic rate of increase for *D. pulex* exposed continuously to different amounts of ^{60}Cobalt γ radiation. Age-specific survival (l_x) and birth (m_x) rates were measured for females exposed to different levels of radiation, and the intrinsic rate of increase estimated by iterative solution of the Euler-Lotka equation. (Modified from Figures 1 and 2 of Marshall [1962]).

reproductive value in their models of fish population response to contaminants. Martínez-Jerónimo et al. (1993) estimated V_A for *Daphnia* populations exposed to increasing concentrations of Kraft mill effluent.

Marshall (1962) exposed female *D. pulex* to various amounts[6] of ^{60}Cobalt γ radiation (Figure 10.2) and estimated age-specific changes in birth and death rates. The changes in intrinsic rate of increase for the exposure populations were then calculated from these vital rates.

Stable age structure of populations can be estimated if the rate of increase (r or λ) and l_x values are known, and used to suggest contaminant effects. The proportion of all individuals in age class x (C_x) is estimated to be the following,

$$C_x = \frac{\lambda^{-x}\ l_x}{\sum\limits_{i=0}^{\infty} \lambda^{-1}\ l_i}$$

(10.13)

---

[6]Radiation is expressed here as the exposure dose rate. Units of dose are Roentgen per hour. A **Roentgen (R)** is a measure of the amount of energy deposited in some material by a certain amount of radiation. By convention, it is expressed relative to energy dissipation in 1 cc of dry air. Use of R to express dose allows one to normalize for the different amounts of energy that are deposited by different types of radiation. You will remember from Chapter 1 that R was incorporated in the rem, Roentgen equivalent man. The rem is the measure of radiation dose expressed in units of potential effect to humans.

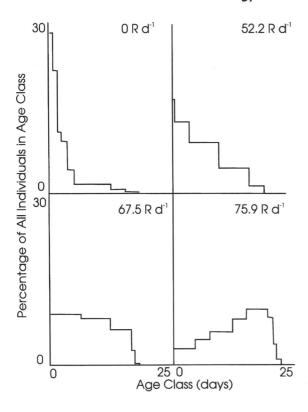

FIGURE 10.3. Stable age structures of *D. pulex* exposed to 0 to 75.9 Roentgens per hour of ^{60}Cobalt γ radiation. Note the gradual trend toward a structure with few neonates as dose rate increases. (Adapted from Figure 3 of Marshall [1962]).

Marshall (1962) applied this approach to predict stable age structures for *D. pulex* exposed to radiation (Figure 10.3). The stable age structure slowly shifted away from a preponderance of neonates at 0 R d^{-1} to a structure composed primarily of older individuals at 75.9 R d^{-1}.

These and similar methods are being used to great advantage in an increasing number of studies of copepods (Daniels and Allan, 1981; Bechmann, 1994; Green and Chandler, 1996) and cladocerans (Winner et al., 1977; Schober and Lampert, 1977; Daniels and Allan, 1981; Hatakeyama and Yasuno, 1981; Van Leeuwen et al., 1985; Day and Kaushik, 1987; Wong and Wong, 1990; Martínez-Jerónimo et al., 1993; Kovisto and Ketola, 1995). Sibly (1996) tabulated a literature search of such studies which included studies primarily of arthropods, but also of algae, gastrotrichs, nematodes, and humans.

Caswell (1996) developed a demographic assay for the effects of pollutants and advocated a shift in attention toward demography. Ferson and Akçakaya (1991) produced software that allows risk assessors to estimate population persistence probabilities based in pollution-induced changes in demographic qualities. A significant literature and theory base are beginning to accumulate on alterations of life history traits associated with contamination (e.g., Adams et al., 1992; Bezel and Bolshakov, 1990; Holloway et al., 1990; McFarlane and Franzin, 1978; Mulvey et al., 1995; Neuhold, 1987; Postma et al., 1995; Schnute and Richards, 1990; Sibly and Calow, 1989; Sibly, 1996).

Energy Allocation by Individuals in Populations

Responses of individuals making up a population have been described relative to energy allocation. These responses can produce significant changes in population demographics. Sibly and co-workers (Sibly and Calow, 1989; Holloway et al., 1990; Sibly, 1996) describe an **optimal stress response** for species exposed to toxicants. The optimal stress response involves a shift in the balance in energy allocation between somatic growth rate and longevity (survival) to optimize Darwinian fitness under stressful conditions. Sibly (1996) gives several examples of toxicants influencing these crucial demographic qualities. Kooijman (Kooijman, 1993; Kooijman and Bedaux, 1996) provides a theory-rich approach utilizing energy budgeting for individuals as the central theme through which survival, growth and reproduction are affected by toxicants. This **Dynamic Energy Budget (DEB) approach** has been formalized into the DEBtox model (Kooijman and Bedaux, 1996).

Toxicant-related effects can be interpreted in the context of the more encompassing **principle of allocation** (there exists a cost or trade-off to every allocation of energy resources). Also discussed as the **concept of strategy**, this principle is used to interpret responses ranging from immediate responses to stress (e.g., early sexual maturity or delayed growth) (Sibly, 1996) to evolutionary responses (e.g., enhanced metal tolerance at the cost of impaired growth rate) (Wilson, 1988) (Figure 10.4). Organisms have a limited amount of energy available that must be optimally allocated among different processes and functions by an individual in order to enhance Darwinian fitness. Energy spent producing defense proteins (e.g., P450) cannot be used for reproduction or growth.

Related to this concept is the **limited lifespan paradigm**, a genetically-defined maximum lifespan is an inherent quality of an individual.[7] Parsons (1995) extends this to the more germane **rate of living theory of aging,** the total metabolic expenditure of a genotype is generally fixed and longevity depends on the rate of energy expenditure. He notes that the immediate response to stress, an increase in metabolic rate, diminishes longevity. He advocates a **stress theory of aging** that selection takes place for resistance to stress, and as an epiphenomena, individuals resistant to stress will predominate in extreme age classes of a population. The diminution of homeostasis under stress with age should be lowest in individuals with highest longevity. He illustrates the concept with populations of *Drosophila*, correlating longevity with reduced metabolic rates under stress and increased antioxidant activity[8]. He refers to the work of Koehn and Bayne (1989) in which high stress resistance was associated with efficient use of metabolic resources. In discussing genetic evidence supporting this stress theory of aging, he notes that different genetically-determined forms of the glycolytic enzyme, glucosephosphate isomerase (GPI or PGI) seem to impart an advantage to individuals in a variety of species under stressful conditions. As we will soon see, this observation is reinforced relative to ecotoxicology by recent genetic studies.

---

[7]Curtsinger et al. (1992) disagree with this concept.

[8]The observation of increased longevity with increased antioxidant activity could also be used to support the **disposable soma theory of aging**. This theory suggests that aging is a consequence of a gradual accumulation of cellular damage via random molecular defects (Parsons, 1995).

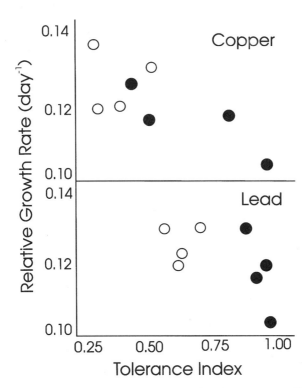

FIGURE 10.4. The cost in terms of growth rate for plant (*Agrostis capillaris*) strains with varying degrees of heavy metal tolerance. Plants were taken from reference locations (open circles) and areas with long histories of heavy metal mining in Wales (closed circles). Growth was measured in near-optimal media with heavy metal concentrations below those causing stress. Tolerance was measured as [2 (root length in heavy metal solution)]/[root length in a heavy metal solution + root length in a control solution]. Plants adapted to high metal concentrations (high tolerance index) had generally slower growth under ideal conditions than nontolerant strains, indicating a trade-off associated with tolerance. (Modified from Figures 1 and 2 of Wilson [1988].)

POPULATION GENETICS

Kettlewell (1955) provided a telling investigation of **industrial melanism** (the gradual increase to predominance of melanic forms in industrialized regions) in peppered moths (*Biston betularia*) of Great Britain. These moths are active at night but remain still on surfaces during the day in order to avoid the notice of visual predators, especially birds. The fitness of rare dark morphs quickly increased relative to the light morphs as surfaces darkened with soot. Although not thought of as such, this premier example of natural selection in the wild is an equally good example to ecotoxicologists of contaminant effects on population genetics. It also reinforces the important point that physical or chemical contaminants do not have to kill or impair individuals outright in order to influence populations. Rapid shifts in population genetics occurred as a consequence of shifting Darwinian fitness relative to a species interaction, predation.

Change in Genetic Qualities

Toxicants can influence population genetics in many ways although ecotoxicologists have a preoccupation with selection-associated changes. By mechanisms already discussed, chemicals and radionuclides can change the genetic qualities of individuals within a population directly. Baker et al. (1996) describe large changes in DNA (the gene for mitochondrial cytochrome b) in voles *(Microtus arvalis* and *M. rossiaemeridionalis)* living near the damaged Chernobyl reactor.

Toxicants can influence population genetics in less direct ways. A toxicant can reduce the **effective population size** (the number of individuals contributing genes to the next generation) and result in a net loss of genetic variation. A **genetic bottleneck** occurs if there are too few individuals available to ensure an allele makes it into future generations. Under less severe conditions, reduction of effective population size may accelerate the rate of loss of a rare allele from a population. **Genetic drift** (random change in allele frequencies in a population) is accelerated at low effective population sizes. The net result of a toxicant's effect on a population, even in the absence of selection, could be the loss of a specific allele or an overall decrease in genetic variability. As genetic variation is the raw material for evolutionary change, this could reduce the ability of a population to adapt to and survive future changes in its environment. Murdoch and Hebert (1994) found reduced variability in mitochondrial DNA of the brown bullhead (*Ameiurus nebulosus*) from the Great Lakes and attributed it to pollutant-induced reductions in effective population size. Kopp et al. (1992) demonstrated a reduction in genetic heterozygosity in stressed populations of the central mudminnow (*Umbra limi*).

Under the presumption that individuals possessing the most genetic variation tend to be most robust,[9] the argument has been made that an increase in genetic variability could occur in populations impacted by toxicants (Mulvey and Diamond, 1991). The individuals most able to survive and reproduce in the polluted environment may be the most heterozygous. For example, Kopp et al. (1992) noted for laboratory assays that central mudminnows (*Umbra limi*) with highest genetic diversity had enhanced abilities to survive stressful conditions (low pH and high aluminum concentrations). Diamond et al. (1989) found that mosquitofish (*Gambusia holbrooki*) with high numbers of heterozygous loci tended to have longer times-to-death during mercury exposure than mosquitofish with fewer heterozygous loci. However, Newman et al. (1989) demonstrated later that the particular heterozygosity effect described by Diamond et al. (1989) was an artifact. Schlueter et al. (1995) found no effect of heterozygosity on time-to-death of fathead minnows (*Pimephales promelas*) exposed to high concentrations of copper. Although plausible mechanisms exist (e.g., multiple heterosis, inbreeding depression or overdominance), presumption cannot be made at this time regarding any

---

[9]Measures of fitness have been correlated with the number of loci found to be heterozygous in individuals (Samallow and Soule, 1983; Koehn and Gaffney, 1984; Danzmann et al., 1986). The mechanism for the enhanced fitness is often assumed to be **multiple heterosis**, a generally higher fitness as a consequence of combined advantages of being heterozygous for each individual locus (heterosis). (**Heterosis** is the general term used to describe the superior performance of heterozygotes.)

TABLE 10.4. Qualities Modifying the Rate of Tolerance Acquisition. (Modified from Table 1 in Mulvey and Diamond [1991].)

Quality	Specific Influence on Tolerance Acquisition
1. Genetic Qualities	
Dominance	Most rapid in early generations if tolerance is controlled by a dominant allele
Single gene versus many genes involved	Most rapid if determined by a single gene
Two or more selection components	Opposing selection components can balance each other or slow tolerance acquisition
Relative differences in fitness	Most rapid if the differences in fitness among tolerant and sensitive individuals is large
2. Reproductive Qualities	
Rate of increase and generation time	Most rapid with high population growth rate and short generation time
Size of population	In general, smaller populations will have less variation than larger populations
3. Ecological Qualities	
Migration	Influx of nontolerant individuals due to immigration could slow tolerance acquisition
Presence of refugia	The presence of refugia such as uncontaminated areas will slow tolerance acquisition
Life stage	Sensitive life stage will have large influence on tolerance acquisition

specific field situation that multiple locus heterozygosity does or does not influence fitness relative to toxicant effects.

Acquisition of Tolerance

Toxicants can also act as selective agents for exposed populations. **Natural selection**, the process by which genes from the most fit individuals are overrepresented in the next generation,[10] can result in enhanced tolerance to toxicants as is amply demonstrated by the adaptation of pests to pesticides. Considerable effort is made to counterbalance tolerance acquisition by insect (e.g., Comins, 1977; Mallet, 1989), rodent (Webb and Horsfall, 1967; Partridge, 1979), and other target species of pesticides. Populations of nontarget species may also adapt and become more tolerant of toxicants. The probability of obtaining, or rate at which a population attains enhanced tolerance is influenced by many factors (Table 10.4).

The likely, but sometimes overlooked, consequence of exposure is local extinction of the exposed population, not enhanced tolerance (Klerks and Weis, 1987; Mulvey and Diamond, 1991). Pollution as a selection force is extreme in its

---

[10]Evolution via natural selection carries several assumptions. It is assumed that surplus numbers of individuals are produced by populations. In a particular environment, individuals vary in their abilities to survive and reproduce, i.e., their fitness. All or a portion of these differences in fitness are heritable. The net result is natural selection.

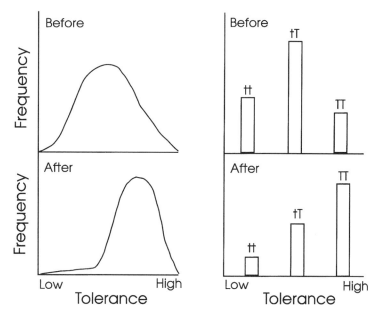

FIGURE 10.5. Shifts in tolerance expected under polygenic (left side of figure) or monogenic (right side of figure) control. With polygenic control, differences in tolerance will appear continuous as shown in the "before" selection panel on the left side of the figure. With selection, the mean tolerance will shift upward and the variation about this mean will narrow. With tolerance determined by a single gene (right side), the distribution of genotypes among homozygous for intolerance (tt), homozygous for tolerance (TT), and heterozygous (Tt) will shift to a predominance of the tolerance allele (T). In this illustration, the T allele is dominant to the t allele. (Modified from Figure 1 of Mulvey and Diamond, 1991.)

rate of change relative to many other environmental factors to which organisms must adapt (Moriarty, 1983). Populations successfully adapting to so rapid a change in environmental conditions are probably exceptional. Therefore, the occasional arguments made to ease regulations based on the premise that "adapted" field populations will be more tolerant than predicted from laboratory assays using "nonadapted" individuals are flawed because they ignore this fact (Klerks and Weis, 1987).

Differences in tolerance[11] of target and nontarget species may be controlled by a single gene (**monogenic control**) or several genes (**polygenic control**) (Figure 10.5). Monogenic control of tolerance to endrin and other cyclodiene pesticides was found for mosquitofish (*Gambusia affinis*) populations inadvertently exposed during agricultural spraying (Yarbrough et al., 1986, Wise et al., 1986). In con-

---

[11]Distinction is made by many authors between the terms, tolerance and resistance. **Tolerance** is often reserved for enhanced ability to cope with a factor due to physiological acclimation. **Resistance** is used if the enhanced abilities are associated with genetic adaptation. Following the lead of Weis and Weis (1989b), tolerance is used here for both acclimation and genetic adaptation.

trast, Posthuma et al. (1993) found polygenic control of heavy metal tolerance in populations of the springtail, *Orchesella cincta*. Enhanced tolerance of this soil insect was associated with differences in metal excretion efficiency among populations.

Tolerance acquisition varies relative to cost (resource allocation) (Figure 10.4). Hickey and McNeilly (1975) found that metal-tolerant plants are at a disadvantage relative to nontolerant plants when grown in a noncontaminated soil. Postma et al. (1995) found that midges (*Chironomus riparius*) from cadmium-tolerant populations had poorer survival, growth, and reproductive success than nontolerant populations when reared in low cadmium conditions. The lowered fitness was attributed to an apparent zinc deficiency in tolerant individuals living under low cadmium conditions.

There may be cross-resistance among toxicants depending on the mechanism underlying enhanced tolerance. **Cross-resistance or co-tolerance** is the condition in which enhanced tolerance to one toxicant also enhances tolerance to another. For example, plants tolerant to one s-triazine herbicide display cross-resistance to other s-triazine herbicides due to elevated levels of a herbicide-binding protein (Erickson et al., 1985). Lead detoxification is higher in isopod populations tolerant to copper (Brown, 1978). Metallothionein gene duplication as an adaptation to elevated levels of one metal (Lange, 1989) could enhance tolerance to other metals. Enhanced rotenone tolerance imparted by elevated mixed function oxidase activity (Fabacher and Chambers, 1972) would likely enhance tolerance to other pesticides detoxified by this mechanism.

Measuring and Interpreting Genetic Change

Allozymes (allelic variants of an enzyme coded for by a particular locus) were first introduced in the late 1970s by Nevo and co-workers (Nevo et al., 1977; 1978; 1981; Lavie and Nevo, 1982; 1986; Baker et al., 1985) to reflect changes in population genetics as a consequence of environmental pollution. Now, they are occasionally useful for assessing population response to toxicants (Gillespie, 1996) although recent DNA techniques may supplant them in the near future. Allozymes have the distinct advantage that they can be rapidly scored in large numbers of individuals using starch gel or some other type of electrophoresis. The different forms of an enzyme notionally coded for at a locus are separated from tissue homogenates of individuals in an electrical field during electrophoresis and biochemical staining is then used to visualize enzyme activity on the electrophoretic medium. Genotype for each sampled individual is then implied from the pattern of staining activity on that medium.

Allele and genotype frequencies in field populations have suggested pollutant-related loss of genetic diversity (Kopp et al., 1992) or selection (Battaglia et al., 1980; Gillespie and Guttman, 1989; Heagler et al., 1993). Interpretation of field results is frequently supported by laboratory studies suggesting differential mortality among allozyme genotypes (Battaglia et al., 1980; Diamond et al., 1989; Newman et al, 1989; Heagler et al., 1993; Keklak et al., 1994; Schlueter et al., 1995). For example, Heagler et al. (1993) interpreted changes in the frequency of alleles associated with a glucosephosphate isomerase locus (GPI-2) of a field population of mosquitofish (*G. holbrooki*) using results of survival analysis of GPI-2 genotypes exposed to high concentrations of mercury in the laboratory.

Too often, the enzyme itself is assumed to be responsible for the observed differences in fitness among genotypes without due consideration of alternative explanations. For example, the different allozymes are thought to have different availabilities of sites to bind with metals, and consequently, susceptibilities to inactivation by the metals. Although this is a reasonable explanation, it is seldom tested rigorously. It was not true in one case in which it was tested. Differences in GPI-2 genotype sensitivity under acute mercury exposure of mosquitofish (*G. holbrooki*) were not a consequence of differential inactivation of allozymes by mercury (Kramer et al., 1992; Kramer and Newman, 1994). Results suggested that differences among genotypes were more readily interpreted in the context of optimal energy resource allocation under general stress. Further, a scored enzyme locus may only be acting as a marker for a closely-linked gene that is actually responsible for the difference in tolerance among genotypes. Such **genetic hitchhiking**[12] is very often given inadequate consideration as a mechanism underlying the observations.

Alleles can be unevenly distributed throughout a population's substructure, e.g., lineages. If differences in tolerance exist within the substructure, this would result in correlations between tolerance and allozyme genotypes that falsely suggest that an allozyme itself is directly linked to tolerance. Lee et al. (1992) reinforced this point by demonstrating a strong family effect relative to mosquitofish (*G. holbrooki*) tolerance to mercury. To my knowledge, no other pollution-related study of allozymes has carefully tested this alternative and equally reasonable explanation for the correlation between allozyme genotype and tolerance. Consequently, conclusions tend to remain ambiguous regarding direct linkage of differential inactivation of allozymes by contaminants to differential tolerance of genotypes to that contaminant.

Population substructuring may also produce an apparent deficit of heterozygotes relative to Hardy-Weinberg expectations.[13] These deficits may be mistakenly attributed to selection against heterozygotes. However, a deficit of heterozygotes can arise if two genetically-distinct groups of individuals are mixed as in the case of unknowingly sampling a highly structured population under the assumption of uniformity in the sample. It can also occur if significant amounts of migration have occurred as in the case of an influx of individuals into a population recently decimated by a pollution event. The **Wahlund effect** predicts that there will be a

---

[12]Endler (1986) defines genetic hitchhiking as "a situation in which a given allele changes in frequency as a result of linkage or gametic disequilibrium with another selected locus . . . [it can] give a false impression of selection at a particular locus . . . Similarly, if there is genotypic correlation among quantitative traits, then selection will appear to affect a trait directly, although it is actually only affected through its correlation with another selected character."

[13]At **Hardy-Weinberg equilibrium**, the frequency of genotypes will remain stable through time. For a two allele locus (i.e., T and t), the frequencies of the genotypes will be q^2 for TT, 2pq for Tt, and p^2 for tt, where q and p are the allele frequencies for T and t, respectively. The following conditions are assumed: 1) the population is large (effectively infinite) and composed of randomly mating, diploid organisms with overlapping generations, 2) no selection is occurring, and 3) mutation and migration are negligible.

net deficit of heterozygotes if two populations, each in Hardy-Weinberg equilibrium but possessing different allele frequencies, are mixed and their combined genotype frequencies quantified. This effect has been ignored in most pollution-related studies to date. Woodward et al. (1996) recently demonstrated such a Wahlund effect during a population study of midges (*Chironomus plumosus*) inhabiting a mercury contaminated lake.

Genotype-related differences in pollutant tolerance are almost exclusively examined relative to survival. However, selection can and does occur at several stages (or components) in the life cycle of an individual (Figure 10.6). There are several such **selection components**: viability selection, sexual selection, meiotic drive, gametic selection, and fecundity selection. **Viability selection** or selection based on differential survival begins at the zygote and continues throughout the life of the individual. This component may be further broken down to viability at different ages or stages of development. **Sexual selection** involves differential mating success of individuals. It may be associated with females or males, i.e., female or male sexual selection. **Meiotic drive** is the differential production of gametes by different heterozygous genotypes. One allele may be underrepresented in the gametes produced by a heterozygous individual. **Gametic selection** involves differential success of gametes produced by heterozygotes. The last component, **fecundity selection**, is the production of more offspring by matings of certain genotype pairs than produced by other genotype pairs. Selection components may act in opposite and balancing directions (Endler, 1986); therefore, measurement of only one, such as viability selection, may result in inaccurate predictions of changes in allele

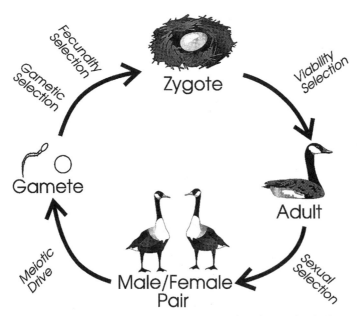

FIGURE 10.6. Components of in the life cycle of an individual in which natural selection (selection components) can occur. Although rarely considered, selection components can be acted upon by contaminants. Please see the text for a detailed explanation.

frequencies under selection pressures from pollutants. For example, Mulvey et al. (1995) found this to be the case with mercury-exposed mosquitofish (*G. holbrooki*). The GPI-2 genotype that was at a disadvantage during acute exposure (viability selection) was not the same as that at a disadvantage relative to female, sexual selection. Unfortunately, selection component analysis is ignored in most studies.

SUMMARY

In this chapter, the importance of assessing effects to populations is emphasized. A brief sketch of epidemiological metrics and logic is provided as applicable to ecotoxicology. Demographic approaches are described which greatly improve our ability to predict the population consequences of toxicant exposure. Examples of their increasing use by ecotoxicologists are provided. The potential influence of toxicants on population genetics is outlined in addition to the possible consequences of such changes. Acquisition of tolerance, factors influencing the rate of tolerance acquisition, and other related processes are described.

Suggested Readings

Barnthouse, L.W. Population-Level Effects, in *Ecological Risk Assessment*, Suter, G.W., II, Lewis Publisher, Chelsea, MI, 1993.

Daniels, R.E. and J.D. Allan. Life table evaluation of chronic exposure to a pesticide. *Can. J. Fish. Aquat. Sci.* 38, pp. 485–494, 1981.

Deevey, Jr., E.S. Life tables for natural populations of animals. *Q. Rev. Biol.* 22, pp. 283–314, 1947.

Klerks, P.L. and J.S. Weis. Genetic adaptation to heavy metals in aquatic organisms: A review. *Environ. Pollut.* 45, pp. 173–205, 1987.

Mulvey, M. and S.E. Diamond. Genetic Factors and Tolerance Acquisition in Populations Exposed to Metals and Metalloids, in *Metal Ecotoxicology. Concepts & Applications*, Newman, M.C. and A.W. McIntosh, Lewis Publishers, Inc., Chelsea, MI, 1991.

Newman, M.C. Effects at the Population Level, in *Quantitative Methods in Aquatic Ecotoxicology*, Newman, M.C., CRC Press, Inc., Boca Raton, FL, 1995.

Rench, J.D. Environmental Epidemiology, in *Basic Environmental Toxicology*, Cockerham, L.G. and Shane, B.S., CRC Press, Inc., Boca Raton, FL, 1994.

Sibly, R.M. and P. Calow. A life-cycle theory of responses to stress. *Biol. J. Linn. Soc.* 37, pp. 101–116, 1989.

Effects on Communities and Ecosystems

> *The accumulation of persistent toxic substances in the ecological cycles of the earth is a problem to which mankind will have to pay increasing attention . . . What has been learned about the dangers in polluting ecological cycles is ample proof that there is no longer safety in the vastness of the earth.*
>
> WOODWELL (1967)

OVERVIEW

Definitions and Qualifications

An ecological **community** is "an assemblage of populations living in a prescribed area or physical habitat; it is an organized unit to the extent that it has characteristics additional to its individual and population components . . . [it is] the living part of the ecosystem" (Odum, 1971). The community is made up of species that interact to form an organized unit (Magurran, 1988) although some species may interact only loosely. Much of the following material is structured around this abstraction.

The impossibility of studying all species, or even all important species, in any community results in studies that focus on some taxonomic or functional subset of the community such as the furbearers of a woodland or fish in a lake. Pielou (1974) suggests that the term **taxocene** should be used to distinguish these taxonomically-defined subsets from true communities. Magurran (1988) suggests the term **species assemblage** for any operationally-defined grouping. Many models and indices are framed in the community context but applied to species assemblages out of necessity. Although such an approach remains valuable and necessary, interpretation of associated results should be tempered with this understanding.

Similarly, the ecosystem concept is also an abstraction or simplification that should not be confused with reality (Newman, 1995). The **ecosystem** concept combines the biota (community) and abiotic environment into an organized system. (See Golley [1993] for a detailed discussion of the ecosystem concept.) Species interact with each other and loosely interact with their physical environment. Biotic and abiotic components act together to direct the flow of energy and cycling of

materials. Obviously, application of this concept to real situations is highly dependent on the scale (time and space), distinctiveness of system boundaries (e.g.,
a distinct, spring-fed lake versus a diffuse bottomland hardwood ecosystem along
a river), and the particular qualities under study (e.g., cation flux from a watershed
ecosystem versus oxygen dynamics of a dimictic lake). This concept must be
applied intelligently to avoid illogical conclusions regarding qualities of an
operationally-defined "ecosystem" using ideal characteristics of the ecosystem
abstraction.

Context

Community and ecosystem qualities are affected by abiotic factors including pollutants (Dunson and Travis, 1991). Despite this, effects at these levels are often addressed in less detail than warranted (Taub, 1989). This neglect probably reflects a
historical bias in the field toward mammalian toxicology which emphasizes effects
to individuals. In illustration of this fact, Clements and Kiffney (1994) noted that
only 12% of 699 environmental toxicology articles published from 1980 to 1982
dealt with populations, communities, or ecosystems. They further noted a disap-
pointingly low percentage (18% of all papers) in a more recent (1992) survey of
the journal, *Environmental Toxicology and Chemistry*. Clearly, a better balance is
needed in ecotoxicology.

   Causal mechanisms for community change are often to be found at the next
lower level of organization, i.e., at the population level. For example, change may
occur because a particular species population's viability was lowered sufficiently by
a toxicant's effect on growth, survival, or reproduction. This scenario is consistent
with the approach advocated in Chapter 1 for maintaining conceptual coherency
in any hierarchical science. However, **emergent properties** must also be considered
carefully at higher levels of organization. Properties emerge in hierarchical systems that cannot be predicted solely from our limited understanding of a system's
parts or components.[1] The counter example to that just given is the indirect loss
of several species because an important keystone species was killed directly by the
toxicant. (A **keystone species** is one that influences the community by its activity
or role, not its numerical dominance.) A species resistant to the direct action of a
toxicant would disappear because another species performing a crucial role in the
community was eliminated. Another example is industrial melanism in which
community processes (i.e., predator-prey interactions) influence population genetics (i.e., predominance of melanism). Causal structure is reversed with interactions among species populations in the community (higher level) producing an
impact to a population (lower level).

   This chapter deals primarily with communities, but processes occurring in
whole ecosystems are discussed toward the end of the chapter. Some properties as-

---

[1]This concept is central to the tedious holistic-reductionistic debate in ecology. Time wasted debating
this obvious point distracts ecotoxicologists from the real challenge, enhancing the inferential strength of
their science regardless of the conceptual vantage taken.

sociated with ecosystems will be addressed again in the next chapter. This chapter begins by describing simple species interactions relative to the influence of toxicant action. Then community qualities, including structure and function, are described in the context of laboratory, mesocosm, and field research. Although much of the ecotoxicological work done with communities has been descriptive, the emphasis in this chapter will be explanatory principles derived primarily from nondescriptive efforts. Field studies and methods are detailed toward the end of this chapter. Most field methods focus on structural changes observed in species assemblages such as soil arthropods or stream macroinvertebrates. Conventional community indices (e.g., species richness) or more specialized indices (e.g., the index of biological integrity) are applied to species assemblages from and around contaminated areas. Less often, community functions are assessed. These functions are discussed briefly toward the end of the chapter.

General Assessment of Effect

A wide range of experimental approaches has been taken to determine the concentration of toxicants below which the community is protected. The **most sensitive species approach** takes the results for the most sensitive of *all tested species* as an indicator of that concentration most likely to protect *all species in the community*. Despite the great advantage of its simplicity, several difficulties arise with this notionally cost-effective approach (Cairns, 1986). One must make the dubious assumption that the tested species and measured effects truly reflect the most sensitive within the community. The most sensitive species approach may not be cost-effective if one considers the high costs of bad management decisions based on flawed approaches (Cairns, 1986). Remember also that the biological significance of the effect remains ambiguous if effect metameters such as an NOEC are used, i.e., the maulstick incongruity.

Recently, a statistical permutation of this most sensitive species approach has appeared for determining concentrations protective of a community. A collection of available NOEC values or other measures of effect are pooled for a sample of species to estimate the concentration below which a predetermined percentage of all species (e.g., 95%) are protected (Figure 11.1) (Van Straalen and Denneman, 1989; Wagner and Løkke, 1991). This concentration is thought to be generally protective of the community. But Hopkin (1993) questioned the assumption that a 5% loss of all species is always acceptable. Arguing in the context of soil species, he noted that elimination of one keystone species such as the earthworm will dramatically influence the soil community even if 95% of all species were protected. It is difficult to justify the assumption that the NOEC values used in any such analysis accurately reflect the effect concentrations within a community or species assemblage (Jagoe and Newman, 1997). Often, values are derived from those of standard test species and are biased toward certain taxa. Beyond this bias, it is difficult to know how many NOEC values are needed to effectively capture the differences among species in an entire community or even a species assemblage.

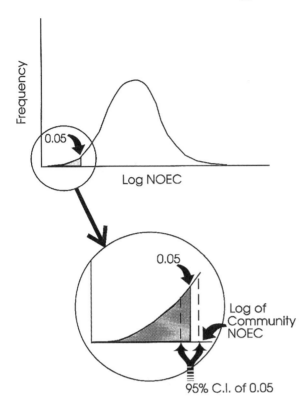

FIGURE 11.1. One method of estimating a community-level NOEC. The 5% quantile is used to estimate the log concentration at which all but 5% of species would be protected. The upper limit of the 95% confidence interval around this estimate is used as the community-level NOEC (Van Straalen and Denneman, 1989; Wagner and Løkke, 1991).

INTERACTIONS INVOLVING TWO OR A FEW SPECIES

Predation and Grazing

An adverse effect on predator-prey interactions can lead to local extinction of a species population even if toxicant concentrations are below those causing diminished growth, reproduction, or survival of individuals in the population. This premise prompted laboratory experiments quantifying such effects. A simple predator-prey arena (Figure 11.2) was used to demonstrate the influence of γ irradiation on the ability of mosquitofish (*Gambusia holbrooki*, formerly *G. affinis holbrooki*) to avoid predation by largemouth bass (*Micropterus salmoides*) (Goodyear, 1972). Mosquitofish were provided with a shallow refuge to simulate normal mosquitofish behavior of avoiding bass predation by staying in the shallows close to the water's edge. The influence of irradiation on predator avoidance over 10 days was dose-dependent with more mosquitofish failing to stay in the refuge as radiation dose increased. This approach was applied again by Kania and O'Hara (1974) to demonstrate that sublethal concentrations of inorganic mercury increase predation in a concentration-dependent fashion. Mosquitofish previously exposed to low concentrations of mercury were incapable of maintaining the most effective orientation relative to predator location in the test chamber. In a slightly more elaborate arena including artificial plants as both prey refugia and predator

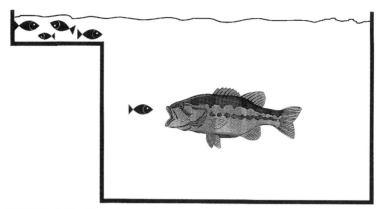

FIGURE 11.2. Experimental arena used to measure the effect of radiation (Goodyear, 1972) or inorganic mercury (Kania and O'Hara, 1974) on mosquitofish avoidance of predation by largemouth bass. The stressors diminished the ability of the mosquitofish to avoid predation by remaining in a shallow refuge.

cover, fathead minnows (*Pimephales promelas*) exposed to cadmium were more vulnerable to largemouth bass predation than were unexposed minnows (Sullivan et al., 1978). Concentrations producing a significant increase in vulnerability were lower than those measured for any other sublethal effect. Increased predation was discussed in the context of the abnormal schooling behavior of the exposed minnows. Similar studies examined the influence of fire ant bait (mirex) on pinfish (*Lagodon rhomboides*) predation of grass shrimp (*Palaemonetes vulgaris*) (Tagatz, 1976). More recently, turbellarian predation of isopods as influenced by cadmium (Ham et al., 1995), and *Hydra* predation on *Daphnia* after lindane (γ-hexachlorocyclohexane) exposure (Taylor et al., 1995) were quantified, but in contrast to the above assays, exposure involved both predator and prey.

Information gleaned from such simple studies can be enriched greatly by applying the principle of allocation (see Chapter 10). Organisms must effectively allocate energy among many activities in order to optimize Darwinian fitness. For example, a predator may change its prey consumption rate as prey densities increase. Such a **functional response** (a change in some predator function, such as prey consumption rate, as a response to changes in prey density) was studied for a largemouth bass-mosquitofish system. Both predator and prey were exposed to ammonia (Woltering et al., 1978). Changes in prey consumption and bass weight were monitored at different ammonia concentrations. Increases in prey consumption rate with an increase in prey density were slowed at high ammonia concentrations, as was the increase in bass weight. In fact, because the mosquitofish were more tolerant of ammonia than bass, the mosquitofish harassed the bass in high ammonia and prey density treatments, resulting in a weight loss of the predator. Clearly, the influence of toxicants on predator-prey interactions is important and complicated.

Atchison and coworkers (Sandheinrich and Atchison, 1990; Henry and Atchison, 1991; Atchison et al., 1996) provide the richest description of predator-prey

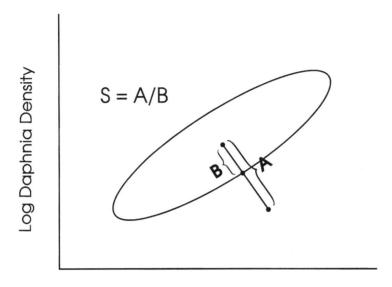

FIGURE 11.3. Calculation of normalized ecosystem strain as described by Kersting (1984). Strain is equal to A/B in this diagram. Ecosystem strain is measured relative to the normal behavior of the system as reflected by the 95% tolerance ellipse. These particular dimensions may be replaced by variables other than algal and grazing *Daphnia* densities. (Modified from Figure 4 of Kersting (1984).)

interactions in the context of energy allocation. They argue from the extensive literature on **optimal foraging theory** (the ideal forager reaches a maximum net rate of energy gain by optimally allocating its time and energy to the various components of foraging) that many important components of foraging are influenced by toxicants. For example, a predator must optimize time and energy spent in prey searching, identification, choice, pursuit and capture, handling, and ingestion. Any or all of these might be altered by the presence of toxicants. Atchison and coworkers cite numerous studies in which predator foraging activities are modified by toxicant exposure.

Kersting (1984) demonstrated that grazing is influenced by toxicants in a study in which he examined the deviation from normal dynamics for a herbivore-plant (*Daphnia magna - Chlorella vulgaris*) micro-ecosystem resulting from pesticide (Dichlobenil) exposure (Figure 11.3). He allowed the *Daphnia-Chlorella* system to come to steady state and plotted grazer density versus algal density through time. A lag function was incorporated in the plot because there was a delay of approximately seven days before the grazer density[2] could respond fully to any change in algal density. A 95% tolerance ellipse was drawn to define the limits of

---

[2]In contrast to a functional response, change in predator or grazer number through increased reproductive output, decreased mortality, or increased immigration in response to changes in prey or food densities is called a **numerical response**.

the system's behavior in the absence of the pesticide. In theory, one would get a point outside of this ellipse only 1 in 20 times as a consequence of random chance alone: points outside of this ellipse would be judged to be outside of normal dynamics. The pesticide was introduced and grazer density versus algal density points plotted. As these points fell outside the tolerance ellipse, the grazer-plant system was judged to have changed as a consequence of the pesticide exposure. A normalized ecosystem strain index (S) was used to quantify the degree to which the system had changed relative to its normal state.

Competition

Interspecies competition, the interference with or inhibition of one species by another, may be influenced by toxicants and also contribute to changes in community structure. This may involve **interference competition** in which one species interferes with another as might occur with territoriality. A toxicant-induced change in aggressive behavior (e.g., bluegill exposed to cadmium [Henry and Atchison, 1979]) could shift the balance of interference competition. **Exploitation competition**, where species compete for some limiting resource such as food, may also be affected by toxicants. For various freshwater zooplankton under toxicant exposure, Atchison et al. (1996) noted differences in filtration rates and suggested that toxicants can produce shifts in exploitation competition among these potentially competing filter feeders.

Atchison and coworkers (Atchison et al., 1996) lament the paucity of studies in behavioral ecotoxicology and attribute this deficiency to the indirect, and relatively complicated, means by which interspecies competition is measured, i.e., as changes in population dynamics of competing species. In the absence of direct information, indirect information was used as evidence to suggest that exploitation competition may be affected significantly by contaminants. Newman (1995) details methods for quantifying the effects of toxicants on interspecies competition.

To digress for a moment, it is implied in this and the previous discussion that changes in species' niches result from toxicant exposure. Here, we define niche with the **Hutchinsonian niche** concept, ". . . the certain biological activity space in which an organism exists in a particular habitat. This space is influenced by the physiological and behavioral limits of a species and by effects of environmental parameters (physical and biotic, such as temperature and predation) acting on it." (Wetzel, 1982). A species in a particular habitat has a **fundamental niche** in which it could exist based on its physiological and other limitations, and a **realized niche** which is that portion of its fundamental niche which it actually occupied. With shifts in competition and foraging behavior, the realized niche of a species is modified by the toxicant. If balanced competition among species in the community is assumed to enhance species packing by fostering optimal niche separation among species, competitive dysfunction may decrease species diversity. As we will see shortly, shifts associated with species abundance curves support this speculation. Interestingly, Chattopadhyay (1996) suggests that toxicants can also stabilize fluctuations in competing species populations under some conditions.

COMMUNITY QUALITIES

General

Several community or species assemblage qualities are measured routinely to assess toxicant effect. The number of species inhabiting a toxicant-impacted site may be compared to the number at an unimpacted site. The presence or absence of indicator species may also be noted. For example, a species that is extremely sensitive to a pollutant may be used much as the proverbial canary in the coal mine. The absence of a particularly sensitive species suggests effect. The decline in osprey populations on Long Island has already been mentioned as an example of such change for a sensitive species. Another example is the disappearance of pH-sensitive species from lakes undergoing acidification. The mysid shrimp, *Mysis relicta*, disappeared from an experimentally-acidified lake when pH was lowered to 6 (Schindler, 1996). Not only did this sensitive species provide an early warning of deteriorating conditions, it was functioning as a keystone species in this Canadian lake. A final example includes mayflies (e.g., *Baetis* sp.), which tend to be sensitive indicators for a variety of pollutants in freshwater systems (Ford, 1989).

Alternatively, a rise to predominance of pollution tolerant species may suggest a deteriorating community. Benthic communities found below high BOD (biochemical oxygen demand) outfalls from sewage treatment plants are typically dominated by heterotrophs tolerant of low dissolved oxygen concentrations. The oligochaete (*Tubifex tubifex*) is a common benthic species at such polluted sites. The *Sphaerotilus* bacterium also forms extensive, filamentous mats below sewage discharges and is used as an obvious indicator species. Indeed, based on this concept, Kolkwitz and Marsson (1908) described a **saprobien spectrum**, a characteristic change in community composition at different distances below a discharge of putrescible organic waste into a river or stream. Characteristic species define zones (e.g., polysaprobic, mesosaprobic, and oligosaprobic zones) below a sewage discharge relative to the oxygen concentrations, amounts of putrescible organic material, and stage of stream recovery.

Several qualities apparently influence community (or "ecosystem") **vulnerability** (susceptibility to irreversible damage) to toxicants (Cairns, 1976). Low **elasticity** (the ability to return to a pre-stressed condition), **inertia** (ability to resist change) and **resilience** (the number of times a community can return to its normal state after perturbation) all contribute to vulnerability. Elasticity is enhanced by the ease with which new individuals can move back into the affected area. Inertia may be influenced by the structural redundancy in the community and previous adaptation of the community to environmental variability. Resilience is influenced by the elasticity of the community and the frequency of perturbation. Too frequent perturbation of a community with low elasticity gradually ratchets the community downward toward a degraded state.

Implicitly, all of this discussion assumes that a community is in some kind of balance and can be expected to return to that balance after a perturbation.

Pratt and Cairns (1996) point out that this concept of community steady-state is pervasive in ecotoxicology. It extends into regulations such as those setting environmental criteria which seek to protect "balanced biological communities."

In contrast to this concept of a community deviating from and then returning to a steady state condition after the stressor is removed, Matthews et al. (1996) suggest that disturbed communities will not return to their original states. This suggestion is consistent with the increasingly expressed view of ecologists that communities are not steady state systems (Pratt and Cairns, 1996). Matthews et al. (1996) argue that communities retain information about occurrences in their past dubbed this argument the **community conditioning hypothesis**. Any dynamics back toward some norm will also reflect the history of the community: one cannot assume that the community will return to its original state. The implication here is that pollution effects to communities will be present long after the toxicant is removed and that any assumed return to an original state is presumptuous. It is this author's opinion that both views (steady state and community conditioning hypotheses) are useful if applied appropriately. One might expect a general recovery of some community qualities to a near "normal," but unique, state after community disruption by toxicants.

Structure

Community Indices

The most commonly used indices of community change are species richness, evenness, and diversity (heterogeneity). **Species richness** is the number of species present in a community. Because the tally of species in a community will increase as more and more individuals are sampled and it is impractical to sample all individuals, species richness is often expressed relative to that of a sample with a standard number of individuals in it. A **rarefaction estimate of richness** produces a number such as 25 expected species in a standard sample of 250 individuals. **Species evenness** is the degree to which the individuals in the community are evenly or uniformly distributed among species. For example, let three species be present in two communities composed of 500 individuals each. In the first community, 450, 41, and 9 individuals are from species A, B, and C, respectively. The numbers of individuals in species A, B, and C in the second community are 134, 138, and 228, respectively. The individuals are more evenly distributed among the species in the second community.

Both species richness and evenness contribute to **species diversity** (= heterogeneity) and are reflected in species diversity indices. Species diversity may be quantified with several formulations; however, the **Shannon** (Equation 11.1) and **Brillouin** (Equation 11.2) **indices** are the most common.

$$H' = \sum_{i=1}^{S} p_i \ln p_i \qquad (11.1)$$

$$H = \frac{1}{N} \ln \frac{N\,!}{\prod_{i=1}^{S} n_i\,!} \tag{11.2}$$

where S = the total number of species, and
    p_i = the proportion of all individuals that are species i as estimated by the number of individuals of species i (n_i) divided by the total number of individuals (N) in the S species.

Both give similar estimates, but Equation 11.2 gives estimates lower than Equation 11.1. This difference arises because the Shannon index is a diversity estimate *for the community* from which the sample was taken but the Brillouin index is a diversity estimate *for the sample*. The diversity in the sample will be lower than the diversity predicted for the entire community from which the sample was taken.

Similarly, species evenness can be estimated for the community (Equation 11.3, **Pielou's J′**) or for the sample (Equation 11.4, **Pielou's J**). The ln S and H_{MAX} in these equations are the maxima for Shannon (H′) and Brillouin (H) indices, respectively. Consequently, these evenness indices are the estimated species diversity divided by the maximum possible species diversity for that community or sample.

$$J' = \frac{H'}{\ln S} \tag{11.3}$$

$$J = \frac{H}{H_{MAX}} \tag{11.4}$$

where

$$H_{MAX} = \frac{1}{N} \ln \left[\frac{N\,!}{([N/S]\,!)^{S-r}(([N/S]+1)\,!)^{r}} \right]$$

with [N/S] equals the integer part of the quotient, N/S, and r is N − S[N/S].

Often, but not always, values for species indices decline as a consequence of pollution. Species diversity dropped in periphyton communities below heavy metal mine discharges (Austin and Deniseger, 1985), in periphyton communities in the presence of high zinc concentrations (Williams and Mount, 1965), and in stream macroinvertebrate communities below coal mine drainage (Chadwick and Canton, 1983). Both diversity and richness dropped for lake algal communities exposed to mining wastes with a few, tolerant species becoming very abundant (Austin et al., 1985). Ford (1989) indicates that richness is a relatively good measure of effect to plankton and benthic communities, although richness may increase slightly at low levels of pollution.

Often a statistically significant difference in species richness or species diversity is used to suggest an adverse impact of toxicants on communities. Aside from the problem of equating statistical and biological significance (i.e., maulstick incongruity), this approach suffers from our lack of knowledge regarding functional redundancy within communities. **Functional redundancy** involves an apparently unaltered maintenance of community functioning despite changes in structure. Species may drop out or be replaced yet the community will still appear to function normally.

There are two unresolved views involving functional redundancy that are germane to the impact of toxicants: the rivet popper and redundancy hypotheses. The **rivet popper hypothesis** suggests that species in a community are like rivets that hold an airplane together and contribute to its proper functioning (Erhlich and Erhlich, 1981). Each loss of a rivet weakens the structure by a small but noticeable amount. The loss of too many rivets eventually leads to a catastrophic failure in function. In contrast, the **redundancy hypothesis** holds that many species are redundant and the loss of some species will not influence the community function as long as crucial (keystone and dominant) species are maintained (Walker, 1991). There are guilds[3] of similarly functioning species to provide consistency of function if one or a few member species are lost. Pratt and Cairns (1996) emphasize the importance to ecotoxicology of determining which of these hypotheses best describes real biological communities. The answer is needed to decide how much toxicant-induced change in a community is required to degrade its functioning. Currently, there is some evidence to support the rivet popper hypothesis (Baskin, 1994). Given our present lack of understanding, it seems prudent to assume that the conservative rivet popper hypothesis should be the working model.

Species abundance curves are also used to describe community shifts as a consequence of toxicant exposure (Figure 11.4). These curves are based on the **Law of Frequencies** (there exists a relationship between the numbers of species and the number of individuals in a community) (Fisher et al., 1943). The numbers of species falling into different abundance classes are plotted against abundance class. In the classic approach of Preston (1948), abundance classes are defined as doublings in abundances, e.g., 2, 4, 8, 16, 32, etc. individuals present for a species. These $\log_2$ classes (e.g., 1 to <2, 2 to <3, 4 to <7, 8 to <15, 16 to <31, . . . individuals) are called **octaves**. A plot similar to that in Figure 11.4 is produced and describes a log normal distribution. This **log normal model** for species abundance is thought to reflect a community structure in which several factors influence species interactions and subsequent allocation of resources.

Patrick (1973) noted that this log normal curve shifts in a predictable way for diatom communities exposed to organic pollution. The mode drops down and the right tail extends out to include more octaves with high numbers of individuals. There is a shift toward more very dominant species and fewer species of intermediate or rare abundances. Herricks and Cairns (1982) suggested that this shift results from a rise to dominance of opportunistic species and a disruption of equilibrium. To May (1976a) and Odum (1985), this shift suggested reversion to an earlier successional stage as a consequence of the disordering effects of pollution. The di-

---

[3]An ecological **guild** is a "group of functionally similar species whose members interact strongly with one another but weakly with the remainder of the community" (Smith, 1986).

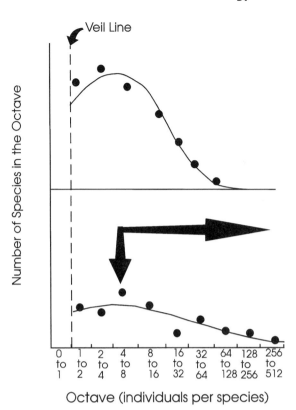

FIGURE 11.4. Species abundance curves (log normal model) before (top panel) and after (bottom panel) toxicant exposure. Note the transition from a common log normal distribution to a curve with a lowered mode and extended right tail. May (1976a) suggests that such a transition may reflect a reversion to an earlier successional stage when interspecies interactions were less important in shaping community species structure (i.e., away from the importance of species interactions toward an r-selection strategy).

verse processes allowing better species packing in a mature community (k-strategy) become less important in shaping community structure than those associated with earlier successional stages, i.e., the classic r-strategy.[4] Based on this premise, Bongers (1990) proposed a community **maturity index** for pollution based on the proportions of species in a soil nematode community that fell into various categories ranging from colonizers (r-strategists) to persisters (K-strategists). Regardless, this shift in the log normal species abundance curve is useful as an indicator of pollutant effect (Gray, 1979; Gerhart et al., 1977).

Finally, more specialized indices may be used to detect changes in community structure. Currently, the most widely applied for aquatic systems is the **index**

---

[4]Although inadequate to fully explain individual success, these two strategies do provide sufficient information here. An **r-strategy** (r = intrinsic rate of increase) or opportunistic strategy may be taken by species coming into an uninhabited/unexploited habitat such as a newly plowed field. Selection favors species that establish themselves quickly, grow quickly to exploit as many resources as possible, and produce many offspring quickly. A **K-strategy** (K = carrying capacity) or equilibrium strategy involves important interactions among species that allow coexistence of many species in the community. Equilibrium species are more effective competitors than opportunistic species. Many factors including interactions among species determine the structure of such a mature community whereas the early successional community structure may be determined more by **niche preemption**, a rapid use and preemption of resources by any species that exploits them before another can. In actuality, the r- and K-strategies are extremes in a spectrum of possible strategies.

TABLE 11.1. Qualities (Metrics) Included in the Original Index of Biological Integrity (IBI). (Modified from Newman [1995].)

Category	Specific Quality
Species richness and composition	1. Total number of species
	2. Number of darter species
	3. Number of sunfish species
	4. Number of sucker species
	5. Number of intolerant species
	6. Proportion of all individuals that were green sunfish, a pollution-tolerant species
Trophic composition	7. Proportion of all individuals that are omnivores
	8. Proportion of all individuals that are insectivorous cyrinids
	9. Proportion of all individuals that are piscivores
Abundance and condition	10. Total number of individuals in the sample
	11. Proportion of total that are hybrids
	12. Total number of individuals with signs of disease or some abnormality

of biological integrity (IBI). Originally, this index combined 12 qualities of fish communities of warm-water, low-gradient streams to determine the degree of stream degradation. These qualities included information on species richness and composition, trophic characteristics, and abundance and condition (Table 11.1). Numerical scores are generated for each quality and summed to produce the IBI for a site. These IBI scores are compared to those expected in the particular area for an undisturbed system. This specialized index has been successfully modified for a variety of aquatic habitats (e.g., Steedman, 1988) and enjoys widespread application today.

Approaches to Measuring Community Structure

Laboratory, microcosm, mesocosm, and field approaches are applied to measuring changes in community structure. These range from straightforward experiments such as those described above for simple species interactions to whole ecosystem exposures such as that shown in Figure 11.5. Laboratory experiments have many advantages, including the ability to randomly assign treatments, the ability to achieve adequate treatment replication, more control of potentially confounding factors, and more control over exposure dosing. Laboratory studies may include two or more species. Laboratory systems designed to simulate some component of an ecosystem such as multiple species assemblages are called **microcosms**. Pontasch et al. (1989) examined changes in stream macroinvertebrate assemblages in response to an industrial discharge by exposing assemblages in laboratory "stream" microcosms. Niederlehner et al. (1985) examined cadmium's influence on species richness of protozoan communities by exposing naturally-colonized substrates to cadmium in laboratory microcosms. In contrast to these aquatic mi-

FIGURE 11.5. The Biology Gamma Forest at the Brookhaven National Laboratory (Long Island, New York, USA) as it appeared in 1964. This eastern deciduous forest which was dominated by white oak, scarlet oak, and pitch pine was exposed to 9500 curies of ^{137}Cs for approximately six months, beginning in 1961 (Woodwell, 1962, 1963). The radiation source was drawn up remotely from inside an underground pipe to expose the woodland. Exposure was many thousand roentgens at this γ source (center of barren spot) and decreased inversely with distance from the source. Zones composed of species of different tolerances ringed the source. Pitch pine (*Pinus rigida*) was the most sensitive with death occurring at 20 r per day. At the other extreme, sedge (*Carex pensylvanica*) was the most tolerant, surviving 350 r per day. (Courtesy of Brookhaven National Laboratory.)

crocosms, terrestrial microcosms may involve plant growth chambers or soil columns (Gillett, 1989).

Between field and laboratory studies are those involving **mesocosms**, relatively large experimental systems also designed to simulate some component of an ecosystem. Mesocosms are delimited and enclosed to a lesser extent than are microcosms. They are normally used outdoors or, in some manner, incorporated intimately with the ecosystem that they are designed to reflect. They differ from microcosms by being larger, being located outdoors as a rule, and as having a lower degree of control by the researcher (Gillett, 1989). Although mesocosms vary considerably in their design (e.g., Figure 11.6), mesocosm studies all have the common goal of obtaining more realism than obtainable with microcosms and more tractability than afforded by field surveys. Liber et al. (1992) conducted a mesocosm-based study to examine natural zooplankton community response to 2,3,4,6-tetrachlorophenol with *in situ* plastic bags extending upward from the sediments

FIGURE 11.6. Two types of mesocosms used to study fate and effects of contaminants. The top panel shows the indoor mesocosms of the Procter & Gamble Company's experimental streams facility (ESF). This system has the great advantage of more control over conditions (e.g., light and temperature) than normally afforded by outside mesocosms. Eight 12-m-long channels allow replication of treatments and production of exposure concentration gradients. The top (head) section of each stream is paved with clay tiles for colonization by algae and microorganisms. Trays of gravel and sand are placed downstream of the tiles and afford substrate for invertebrates. (Courtesy of Mr. John Bowling of Procter & Gamble Co.) The bottom panel shows several outdoor, pond mesocosms used in similar fashion for examining pollutant effects. (Courtesy of Dr. Thomas La Point, Clemson University.)

to the surface of a freshwater body. The bags allowed treatments of different concentrations of toxicant and replication within treatments. Goldsborough and Robinson (1986) used similar *in situ* marsh enclosures to study periphyton assemblage response to the triazine herbicides, simazine and terbutryn. Flowing systems may also be studied with mesocosms as evidenced by the work of McCormick et al. (1991) who studied diatom and protozoan assemblages in experimental stream channels dosed with the surfactant, dodecyl trimethyl ammonium chloride.

Field studies may also vary from surveys of contaminated systems to whole or partial ecosystem manipulations. As experimental manipulations of natural systems afford stronger inference than surveys, a variety of studies have attempted such large-scale and expensive manipulations (e.g., Figure 11.7). To examine ef-

FIGURE 11.7. Whole lake or split lake studies such as shown here afford *in situ* but expensive information on responses of aquatic communities to anthropogenic materials. Although not associated with a conventional "toxicant," this particular split lake study of nutrient enrichment (Canadian Experimental Lake 226: P, N and C added to the bottom section and only N and C added to the top section of the lake) clearly shows the phytoplankton community response. (Courtesy of Dr. Ken Mills.)

fects of radiation exposure, terrestrial systems in several geographical regions were irradiated and changes in associated communities studied. The communities included old fields in South Carolina (Monk, 1966), woodlands in Georgia (Schnell, 1964), and forests and old fields in New York (Woodwell, 1962; 1963). The influence of acidification was examined by Schindler and coworkers (Schindler 1996) by adding acid to an entire experimental lake in Canada. Sensitive species were identified as they disappeared from the lake. Functional redundancy was demonstrated by a shift in lake trout (*Salvelinus namaycush*) predation. As the pH-sensitive fathead minnow (*Pimephales promelas*) population declined, lake trout shifted predation effort to the more pH-tolerant pearl dace (*Semotilus margarita*). Community shifts were also examined as a consequence of copper spiking of streams in Ohio (Winner et al., 1980) and California (Leland and Carter, 1984). High concentrations of copper shifted the insect assemblage away from caddisflies toward more tolerant midges. It also reduced diatom species richness.

More often used than field manipulations are field surveys that provide less structured yet much less expensive observation of the consequences of toxicant introduction to communities. **Biomonitoring**,[5] the widely-applied monitoring with selected sampling protocols of a subset of an entire community with the goal of assessing community condition (Herricks and Cairns, 1982), can involve a simple species listing or much more complex analysis of data. Herricks and Cairns (1982) suggest three general types of biomonitoring efforts. The first simply describes the biota, perhaps summarizing results as a species-abundance list. The second may involve the formulation of a hypothesis which is then tested with field observations. The last type combines these two approaches in order to formally test conclusions (hypotheses) derived from the descriptive phase of the biomonitoring effort. Clearly inferential strength is highest for this last type of biomonitoring.

Function

Changes in community functioning are used less often by ecotoxicologists than structural changes because it is generally believed that feedback loops and functional redundancies make community functions less sensitive to toxicants than community structure (Odum, 1985; Forbes and Forbes, 1994). Regardless, some important community functions can be modified by toxicants. Certainly, modified functioning is implied by any change in functional groups or guilds such as the shift in macroinvertebrate shredder and collector groups measured around coal

---

[5]Qualifiers are frequently made for the term "biomonitoring." As an example, Hopkin (1993) defines the monitoring of community changes along a gradient or among sites differing in levels of pollution as **Type 1 biomonitoring**. **Type 2 biomonitoring** involves the measurement of bioaccumulation in organisms among sites notionally varying in the level of contamination. **Type 3 biomonitoring** attempts to define the effects on organisms using tools such as biochemical markers in sentinel species or some measure of diminished fitness of individuals. **Type 4 biomonitoring** involves the detection of genetically-based resistance in populations of contaminated areas.

mine drainage (Chadwick and Canton, 1983). Blanck (1985) also suggested that natural periphyton photosynthetic activity, measured as $^{14}CO_2$ incorporation, could be used as an ecotoxicological test. Giesy (1978) measured a significant drop in leaf litter decomposition rates at elevated cadmium concentrations, suggesting another important function influenced by toxicants. Cairns and coworkers (e.g., Niederlehner et al., 1985; Cairns et al., 1986; McCormick et al., 1991) demonstrated a clear concentration-response relationship for colonization by protozoa of artificial substrates. They fit colonization data under various toxicant concentrations using the **MacArthur-Wilson model** of island colonization,

$$S_t = S_{EQ} \left(1 - e^{-Gt}\right) \tag{11.5}$$

where S_t = the number of species present at time t, S_{EQ} = the equilibrium number of species for the island, and G = the rate constant for colonization of the island. A sensitive assay was developed and demonstrated with a series of toxicants (Figure 11.8).

Change in community tolerance has also been proposed as a measure of function change with pollutant exposure. It is measured as **pollution-induced community tolerance (PICT)**, an increase in tolerance to pollution resulting from species composition shifts in the community, acclimation of individuals, and genetic changes in populations in the community. Procedurally, a previously-exposed community and an unexposed community may be challenged with a toxicant and

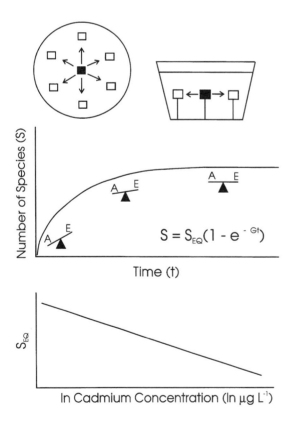

FIGURE 11.8. The protozoan community colonization assay developed by Cairns and coworkers (e.g., Niederlehner et al., 1985; Cairns et al., 1986). A polyurethane foam substrate (filled square in top diagram) is allowed to accumulate species in a natural stream and then brought to the laboratory to serve as a source or epicenter for colonization of other, uncolonized foam substrates (open squares in top diagram). The dynamics of species colonization on foam substrates is measured using the MacArthur-Wilson model for island colonization (center plot) (A = arrives and E = extinctions). This is done with substrates submersed in containers filled with different concentrations of toxicant (bottom plot) to determine the effect of toxicants on the process. (Modified from Figures 1, 2, and 4 in Cairns et al. [1986].)

the difference in responses of the two communities used to reflect adaptation. Kaufman (1982) reported that a periphyton community adapted to copper displayed a smaller decrease in ATP and chlorophyll upon repeated exposure to high copper concentrations relative to a periphyton community with no previous exposure. Similar results were obtained for 4,5,6-trichloroguaiacol-adapted periphyton communities exposed for a second time to this toxicant (Molander et al., 1990).

ECOSYSTEM QUALITIES

Many of the changes predicted by Odum (1985) to occur in ecosystems impacted by toxicants (Table 11.2) have already been discussed in the context of communities. However, a few remain to be discussed more fully. Particularly relevant are those related directly to nutrient cycling and energy flow.

Energy flow in agricultural ecosystems is changed by pesticide application as is obvious from the increased crop biomass. Species dominance and structure of soil arthropod communities can also shift dramatically in ecosystems to which pesticides (e.g., DDT, parathion, or aldrin [Pimental and Edwards, 1982]) are applied. Implied in such changes is a change in energy flow via trophic exchange. Perhaps less obvious is the diminished decomposition rates in soils and recycling

TABLE 11.2. Odum's Predicted Changes in Ecosystems Experiencing Toxicant Stress (Modified from Odum (1985)).

Component/Quality	Predicted Change
Energetics	Increased community respiration
	Imbalance production: respiration, i.e., P/R<1 or P/R>1
	Increased maintenance: biomass, i.e., increase in production/biomass and respiration/biomass
	Increased importance of energy from outside the ecosystem
	Increased export of primary production
Nutrients	Increased turnover of nutrients
	Decreased cycling of nutrients
	Increased loss of nutrients as a result of the two above changes
Community Structure	Increased proportion of species that are r-strategists
	Decreased size of organisms
	Decreased life span of organisms
	Shortened foodchains
	Decreased species diversity with increased species dominance
Ecosystem	Decreased internal cycling and increased importance of input and output from outside sources
	Regression to an earlier successional stage
	Functions (e.g., community metabolism) changed less than structural components (e.g., species richness)
	Decreased positive (e.g., mutualism) and increased negative (e.g., disease or parasitism) interactions

of minerals by soil organisms in agricultural systems and adjacent natural ecosystems. As evidence, application of organochlorine pesticides can shift the elemental composition of crops such as corn and beans (Pimental and Edwards, 1982).

Changes in nonagricultural system's energy flow and material cycling have been demonstrated over a wide range of scales. Acidification of streams in the Great Smoky Mountains diminished leaf decomposition, bacterial production, and microbial respiration rate (Mulholland et al., 1987). Odum (1985) indicates that enhanced losses of calcium from forested watersheds is a good indicator of functional damage to the watershed. At the other extreme of scale, microcosm and mesocosms studies also have been used to examine energy flow and material cycling under the influence of toxicants. For example, copper addition to microcosms reduced primary production and dissolved organic carbon production (Hedtke, 1984). In contrast to the results of Giesy (1978) with cadmium, spiking of experimental streams with triphenyl phosphate (Fairchild et al., 1987) did not lower leaf decomposition rates relative to control streams. Also, rooted flora increased and net nutrient retention increased with treatment. Clearly, these important functions of ecosystems are influenced by toxicants. However, there will be exceptions (e.g., Fairchild et al. [1987]) to general predictions such as Odum's, due to complex changes in the community structure.

SUMMARY

Beginning with a brief discussion of the ecological community, species assemblages, and niche, this chapter outlines general laboratory, mesocosm, and field approaches to determining the influence of toxicants on communities or species assemblages. General methods of estimating community effects include the use of indicator species (sensitive and tolerant species), community level NOEC estimation, and biomonitoring. Simple species interactions, e.g., predator-prey and interspecies competition, were shown to be susceptible to toxicants. The reduction in fitness of individuals participating in such simple interactions was placed into the context of the principle of allocation. Assuming an equilibrium model for communities, qualities contributing to community vulnerability to toxicant effect were detailed, including community elasticity, inertia, and resilience. The question of whether a community can be described accurately with an equilibrium concept was brought up and contrasted with the community conditioning hypothesis. Measures of community structure and function were then discussed relative to toxicant effects. Such changes were placed into the context of community successional regression, functional redundancy theory, and the law of frequencies. Changes in ecosystem energy flow and material cycling, although also implied in discussions of community shifts, were then described briefly. Some of these ecosystem changes will be discussed again in a wider geographical context in the next chapter.

Suggested Readings

Atchison, G.J., M.B. Sandheinrich, and M.D. Bryan. Effects of Environmental Stressors on Inter-specific Interactions of Aquatic Animals, in *Ecotoxicology. A Hierarchical Treatment*, Newman, M.C. and CH. Jagoe, Eds., CRC Press, Inc., Boca Raton, FL, 1996.

Gillett, J.W. The Role of Terrestrial Microcosms and Nesocosms in Ecotoxicological Research, in *Ecotoxicology: Problems and Approaches*, Levin, S.A., M.A. Harwell, J.R. Kelly, and K.D. Kimball, Eds., Springer-Verlag, New York, 1989.

Graney, R.L., J.P. Giesy, and J.R. Clark. Field Studies, in *Fundamentals of Aquatic Toxicology*, 2nd ed., Rand, G.M., Ed., Taylor & Francis, Washington, DC, 1995.

Hopkin, S.P. *In Situ* Biological Monitoring of Pollution in Terrestrial and Aquatic Ecosystems, in *Handbook of Ecotoxicology*, Calow, P., Ed., Blackwell Scientific Publications, London, 1993.

Odum, E.P. Trends expected in stressed ecosystems. *Bioscience* 35, pp. 419–422, 1985.

Woodwell, G.M. The ecological effects of radiation. *Sci. Am.* 208, pp. 2–11, 1963.

Landscape to Global Effects

Even though the pattern of our relationship to the environment has undergone
a profound transformation, most people still do not see the new pattern . . .
The sights and sounds of this change are spread over an area too large
for us to hold in our field of awareness.
GORE (1992)

GENERAL

"Is it bigger than a bread box?" This is the conventional opening to a familiar guessing game in which an object is eventually identified from answers to a series of questions. It reflects our tendency to categorize things by size or scale. This tendency even extends to topics traditionally classified as within or outside the purview of ecotoxicology. Customarily, but not always correctly, ecotoxicology focuses on scales up to the traditional ecosystem, e.g., the fate and effects of pollutants in a lake, stream, field, or forest. Some studies do extend beyond this framework, but they are uncommon.

Divergent answers would result if one asked established ecotoxicologists to decide whether a topic such as global warming, widespread forest decline in central Europe, or global distillation were in the purview of ecotoxicology. Some would feel that, if the context of the problem were bigger than a traditional ecosystem, that it would be better handled in biogeochemistry, landscape ecology, soil sciences, or atmospheric chemistry. There would be a contrastingly uniform affirmation if the question involved PCB bioaccumulation in trout of a lake or a pollution-induced decrease in arthropod species diversity in forest litter. One obvious reason for this bias is that much of ecotoxicology was derived from the science of ecology. Until a few decades ago, the dominating context of ecology was the ecosystem or lower levels of biological organization.

In Chapters 9 and 10, I suggested that the single species bias in much of ecotoxicology grew out of the early transplanting of ideas and approaches from mammalian toxicology. Although still present in ecotoxicology, this single species focus is generally accepted as inadequate to addressing many important topics. Higher level effects can be equally or more important than those at the level of the individual. Similarly, the conventional eco-

FIGURE 12.1. Landscape modification by smelting and mining activities in Copperhill, Tennessee. Copperhill is situated in the Blue Ridge Mountains at the convergence of northern Georgia, western North Carolina, and southern Tennessee. The Ducktown Mining District began smelting circa 1854, and rapidly developed during the next four decades. Sulfuric acid and sulfur dioxide releases were greatly reduced after 1910. Tree growth, as measured from growth rings, was slowed from 1863 to 1912 in the nearby Great Smoky Mountains National Park (88 km upwind) due to the emissions from smelting (Baes and McLaughlin, 1984). This photograph was taken more than 70 years after emission reductions occurred (1982) and shows a desert-like landscape instead of the typical, forested landscape.

system[1] bias is opined here to be inadequate in many cases too. This opinion is reinforced by Cairns, (1993), Catallo (1993), and Holl and Cairns (1995) who argued that a landscape context for ecotoxicology is also needed. This traditional, but now too confining, bias toward the ecosystem or lower levels is designated the **ecosystem incongruity** here.

Supporting examples are easy to find. A landscape example involves copper mining and smelting in Copperhill, Tennessee. By killing vegetation and stripping nutrients from the soil, acidic fumes from smelting transformed a lush forested landscape to the desert-like surroundings shown in Figure 12.1. A larger scale ex-

---

[1]Note that "biosphere" was used instead of the usual "ecosystem" in the definition of ecotoxicology (Chapter 1). The intent in doing so was to untether discussion from the conventional ecosystem context and allow free consideration of landscape, regional, continental, and global scales.

FIGURE 12.2. The TransAlaska Oil Pipeline as it passes across the taiga, a transitional community between the tundra and boreal forest communities.

ample is pollution from the Kuwait oil fires (Figure 1.3), an event influencing significant land (desert and urban) and marine components of a country. The final and most encompassing example is the TransAlaska Oil Pipeline (Figure 12.2). It extends South from Prudhoe Bay (Arctic Ocean) up the North Slope over the Brooks Range to cross the Arctic Circle, Yukon River, and the Alaska Range to end at the Valdez marine terminal on Prince William Sound (Pacific Ocean). For this one project, risk of damage exists for tundra, taiga, boreal forest, river, mountain, lake, fjord, and intertidal "ecosystems." One accident associated with only one segment of this project, the 1989 *Exxon Valdez* spill, spread oil out into parts of Cook Inlet and Alaska Sound, and covered 30,000 km^2 of Alaskan waters. A large hypothetical spill onto the tundra could conceivably have an impact beyond that "ecosystem" because many bird species spend part of their time there and migrate to Asia (e.g., the wheatear, *Oenanthe oenanthe* nesting in rocky fields of the tundra), North America (e.g., sandhill crane, *Grus canadensis* breeding in tundra marshes) and South America (e.g., golden plover, *Pluvialis dominica* nesting on tundra hillsides). Bird populations on several continents could be impacted by an oil spill in Alaska. Clearly, any preoccupation with an ecosystem, rather than a landscape or larger, context would result in an insufficient description of potential consequences of the TransAlaska Oil Pipeline.

There is a second and equally important reason why the ecosystem bias is no longer acceptable. The ecosystem focus draws attention away from important qualities of landscapes that are a heterogeneous matrix of "ecosystems." Unique properties emerge in this landscape[2] context. The source-sink framework for population dynamics discussed in Chapter 10 is an obvious example. Maurer and Holt (1996) modeled effects of pesticides on mobile wildlife in a complex landscape and inclusion of source-sink dynamics emerged as crucial in predicting impact on populations. Another classic example involves **ecotones**, areas of transition between two or more community types (Odum, 1971). Ecotones often have species assemblages with high species richness and high abundance of individuals relative to those of the adjacent communities. There are several reasons underlying this **edge effect**. Species from contiguous habitats are present in the ecotone, increasing species richness. Some species can exploit both habitats in different ways, increasing their abundances at the ecotone. A species may nest in the forest but forage on grains in an adjacent field. Finally, unique species adapted to the ecotone, e.g., estuarine species, add to species richness. Consequently, the application of pesticides to agricultural fields may not have the same predicted ecotoxicological consequences for areas with an extensive network of hedgerows or patches of woods among the fields compared to those without. Such differences due to ecotones become important as the trend toward large agroindustrial farming and away from small farms continues in many parts of the world. Another important class of ecotones, estuaries at the mouths of rivers, are extremely vulnerable to contaminants from upriver sources and from port cities along their shores. Any unwarranted preoccupation with the traditional "ecosystem" context tends to draw attention away from the unique qualities of ecotones and other important features.

The third and final reason why we should extend our spatial context for ecotoxicology is simple: we now have the tools and data to do so. Affordable computer costs and increase in computational power allow diverse data sets, including inexpensive high altitude and satellite data, to be integrated into a coherent and informative form by researchers and managers. Computerized **geographic information systems** (GIS) have emerged to handle these data at a reasonable cost. Most allow one to archive, organize, integrate, statistically analyze, and display many kinds of spatial information using a common coordinate system (Avery and Berlin, 1985). Data of different types such as land use, vegetation, rates of pesticide application, soil type, weather, and air or water quality can be merged and compared statistically to provide invaluable insights for effective stewardship of resources and environmental regulation. Books, such as that by Michener et al. (1994), detail methods for doing so, and a wide range of affordable imagery and maps are available.

Some imagery is produced by remote sensing. **Remote sensing** technologies allow the acquisition and analysis of data without requiring physical contact with the land or water surface being studied. Most determine qualities or characteristics of areas of interest based on measurements of visible light, infrared radiation, or radio energy coming from them (Sabins, 1987). For example, infrared spectral

---

[2]**Landscape** is an ambiguous term used in many contexts. Here, it is used to denote the sum total aspect of any geographical area (Monkhouse, 1965).

characteristics may be used to define vegetation community types over a wide area. Data from sensitive radiation sensors mounted in an airplane are used to map γ irradiation over large areas of U.S. Department of Energy nuclear facilities where releases occurred. Oil slicks on sea surfaces are detected and tracked by their higher radiance of ultraviolet and blue light (Sabins, 1987).

This type of spatial information is quickly becoming incorporated into environmental regulation and management activities. The U.S. EPA now has placed U.S. vegetation types, a Toxic Release Inventory (TRI), air pollution, areas of air quality nonattainment, and Superfund sites into a GIS format (Reichhardt, 1996).

In a departure from the approach used in previous chapters, this chapter will be largely based on examples. Each will be selected to represent an ecotoxicological topic at a particular spatial scale, i.e., landscape, regional, continental, hemispheric or global scale. From the examples, the general trend will become obvious that a contaminant's potential for dispersal and its spatial scale for concern increases with the degree to which it is associated with the more mobile components of the environment (atmosphere mobility > hydrosphere mobility > pedosphere[3] mobility > lithosphere mobility). For example, contaminants associated primarily with the atmosphere, such as those giving rise to acid precipitation, will have effects over wide expanses. Some exceptions to this trend occur if large amounts of a material (e.g., a pesticide) are applied to a wide region[4] committed predominantly to one human activity (e.g., a large agricultural region of North America), or if the human activity giving rise to the contamination is occurring over extensive areas (e.g., lead contamination of North American soils resulting from widespread use of leaded gasoline). An unfortunate corollary is that cause and effect relationships are clearest locally, but become increasingly difficult to assign with distance from a source (Cairns and Pratt, 1990). Consequently, some of the most widespread problems of global concern such as ozone depletion or global warming are quite difficult to assign a cause and effectively remedy.

LANDSCAPES AND REGIONS

Often landscape studies are based on some physical feature such as a watershed. Richards et al. (1993) used GIS methods to categorize land use in a Michigan catchment and linked land use with macroinvertebrate community composition of associated water bodies. There was a direct linkage between agricultural activity and stream substrate quality. In turn, substrate quality influenced the abundances of Ephemeropteran, Plecopteran, and Trichopteran insect taxa in benthic communities. From this study, recommendations for modifying land use and predictions

---

[3]The **pedosphere** is that part of the earth made up of soils and where important soil processes are occurring (Ugolini and Spaltenstein, 1992).

[4]As used here, a geographic **region** is an "area of the earth's surface differentiated by its specific characteristics" (Monkhouse, 1965).

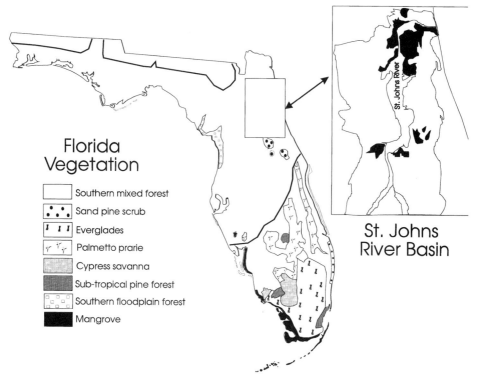

FIGURE 12.3. Three scales (river basin, vegetation type, and ecoregion) of consideration for Florida. (Modified by combining spatial data from Figure 5 of Adamus and Bergman [1995], an ecoregion map from Omernik [1987], and a U. S. Geological Survey vegetation map (Sheet 90).)

of change under various restoration scenarios were generated. Richards and Host (1994) successfully applied this method again to Minnesota catchments along the shores of Lake Superior. A similar approach was taken to categorize and then project future problem areas for nonpoint pollution in the St. Johns River Basin in Florida (Adamus and Bergman, 1995). Analysis integrated information on contaminant amounts and concentrations in surface runoff, sites of storm water treatment and efficiency of that treatment, projected changes in land use, soil types, rainfall, hydrology, and current water quality. The inset of Figure 12.3 shows the predicted sites of significant pollutant generation along the St. Johns River.

Also illustrated in Figure 12.3 are features important in predicting contaminant impact at a larger scale. The dominant vegetation changes considerably in the lower half of the state and determines the specific communities at risk and the *milieu* in which the contaminant effect may or may not be expressed. The scale of an entire state may also be important. Because laws and regulations are applied by states, the arena for dealing with contaminants may be defined by state boundaries. For example, state fish consumption advisories and bans are determined by concentrations in game species. States establish their own, occasionally divergent, criteria using Food and Drug Administration (FDA) action levels or EPA risk-based methods (Cunningham et al., 1994).

Often transcending state borders are **ecoregions** ("mapped classification[s] of ecosystem regions of the U.S. . . . generally considered to be regions of relative homogeneity in ecological systems or in relationships between organisms and their environment"[5] (Omernik, 1987). Inherent in their use is the working principle that contaminant effects vary significantly among ecoregions, e.g., ecoregions with high carbonate soils will be less sensitive to acid precipitation effects than those with mineral chemistries reflecting underlying granite or sandstone mineralogy (Glass et al., 1982). They are used to manage aquatic and terrestrial resources of the United States based on land use, land surface features, vegetation, and soil types. In Figure 12.3, the dark lines crossing central Florida and those dipping down along the northern border of the panhandle define the edges of the three ecoregions in Florida. The Southeastern Coastal Plain ecoregion is the northernmost. The Southern Coastal Plain (approximately the upper half of Florida) and Southern Florida Coastal Plain (approximately the lower half of Florida) ecoregions occupy most of the state.

Hughes and Larsen (1988) applied the ecoregion classification to the formulation of a surface water protection strategy for the contiguous United States. Their aim was to develop a more accurate and appropriate framework for water quality criteria based on ecoregions, rather than the entire United States. Assuming correctly that water bodies within ecoregions were more similar to each other than to water bodies of other ecoregions, such diverse qualities as the index of biological integrity for fish assemblages, phosphorus concentrations, and dissolved oxygen concentrations were successfully classified for various ecoregions.

CONTINENTS AND HEMISPHERES

Problems such as acid rain[6] fit somewhere between the spatial scales of ecoregions and continents (Figure 12.4). Continental networks of precipitation monitoring have documented the spatial scale of the acid precipitation problem and concordance of precipitation pH with sources of acid-generating contaminants (Barrie and Hales, 1984). Industrialized areas emit sulfur and nitrogen oxides that combine with atmospheric water to form H^+, SO_4^{-2}, NO_3^-. Much of the sulfur dioxide produced by North American sources involves the burning of coal that can contain 1.5 to 5% S, and the roasting of Ni, Zn, and Pb sulfides to produce metals. Another source is the burning of oil that may contain 2.5 to 3.5% S (Bridgman, 1994). In

---

[5]It is apparent from Figure 12.3 that some ecoregions such as the Southern Florida Coastal Plain ecoregion are more heterogeneous than others. It contains the Everglades, palmetto prairie, subtropical pine forest, sand pine scrub, cypress savanna, and mangrove swamps.

[6]In equilibrium with gaseous carbon dioxide (Equation 12.1), liquid water in the atmosphere is predicted to have a pH of 5.7. **Acid precipitation**, including rain, fog (Hileman, 1983), snow, or other forms of precipitation, is defined as that with a pH below 5.7.

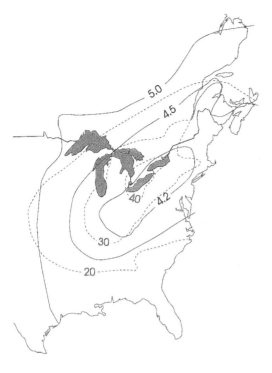

FIGURE 12.4. The pH (solid lines and numbers) and sulfate ion concentration (μm L^{-1}) (dashed lines) in precipitation measured in 1980. In general, pH and sulfate contours coincide with the spatial pattern of sulfur dioxide emissions reported by a joint US/Canadian working group. (Modified from Figures 2 and 3 of Barrie and Hales [1984].)

contrast to sulfur dioxide sources that tend to be industrial or commercial sources with tall smoke stacks, nitrogen oxide sources are primarily near-ground sources such as automobiles (Bridgman, 1994). This, combined with the fact that the acid-producing reactions for sulfur dioxide are slower than those for nitrogen oxides, gives explanation for the general observation that N-related pH problems are more localized than S-related pH problems.

The equations below summarize the general reactions leading to precipitation with high H^+ concentrations (Equations 12.2 to 12.3 for sulfur dioxide and Equations 12.4 to 12.6 for nitrogen oxides [Bunce, 1991]). An oxidation occurs in the first step of Equation 12.3 and in Equation 12.4. Although not explicitly indicated as such, Equation 12.6 is a catalyzed reaction.

$$CO_{2\,(g)} + H_2O_{(l)} \rightleftarrows H_2CO_{3\,(aq)} \rightleftarrows H^+_{(aq)} + HCO^-_{3\,(aq)} \qquad (12.1)$$

$$SO_2 + H_2O \rightarrow H_2SO_3 \qquad (12.2)$$

$$SO_2 \rightarrow SO_3 \rightarrow H_2SO_4 \qquad (12.3)$$

$$NO \rightarrow NO_2 \qquad (12.4)$$

$$2NO_2 + H_2O \rightarrow HNO_2 + HNO_3 \qquad (12.5)$$

$$NO_2 + OH \rightarrow HNO_3 \qquad (12.6)$$

These gases can disperse hundreds to thousands of kilometers from their sources (Cowling and Linthurst, 1981), causing widespread problems in parts of North America, northern Europe (Likens, 1976), and China (Bridgman, 1994).[7]

The impact of low pH precipitation is not solely a function of proximity to and magnitude of a source (Ravera, 1986). Different regions are inherently more sensitive than others. Soil type and underlying mineralogy influence the capacity to buffer pH changes, and consequently, influence sensitivity to acid precipitation effects. An area with an underlying geology of granite, granitic gneisses, or quartz sandstones will have very poor buffering capacity and be sensitive to low pH precipitation. Those areas with sandstone or shale mineralogies are poorly to moderately buffered, and those with limestone or dolomitic geologies will have high buffering capacity and be insensitive to acid precipitation (Glass et al., 1982). Bedrock geology maps can be combined with maps of the distribution of acid precipitation to predict areas of high or low concern. Glass et al. (1982) related bedrock geology, sources of acid precipitation, and stream alkalinity[8] to define pH-sensitivity classes of surface water bodies for New York State. Schindler (1988) examined the acid neutralizing capacity of North American lakes and documented a gradual decrease recently in the northeastern United States (New England and New York), northeastern Canada, and areas of Canada above the Great Lakes.

Effects of acid precipitation to aquatic biota may be sudden or gradual. Releases of pollutants accumulated in snowpack during seasonal thaws may cause high mortality to or diminished spawning success of downstream fish (Cowling and Linthurst, 1981; Bridgman, 1994). Schindler (1988) suggests that, in general, autumn-spawning fishes will be more sensitive than spring-spawning fishes to such releases because their pH-sensitive hatchlings tend to be in shallow, near-shore waters when the spring thaw brings pulses of low pH and high aluminum water. Slow deterioration of aquatic systems involves lowered buffering capacity as acid in precipitation "titrates" the entire system downward toward damagingly low pH conditions. Slow deterioration can involve the shift in equilibria for various biogeochemical processes until a dysfunctional condition emerges. For example, acidic conditions can increase dissolved aluminum flux into overlying waters from solid forms in sediments until toxic concentrations are reached. Low pH conditions can also increase leaching of aluminum and other metals from soils and minerals of the watershed, and have toxic consequences to aquatic biota. Aluminum, calcium, and magnesium leaching can increase in a watershed as a consequence of acid precipitation (Smith, 1981; Schindler, 1988). Cronan and Schofield (1979) showed that atmospheric inputs of sulfuric and nitric acid to the pH-sensitive aquatic systems of

---

[7]Although the discussion here revolves around wet precipitation, fluxes of both dry and wet material can contribute pollutants to sites of effect. **Wet deposition** includes pollutants formed in the liquid media of the precipitation and that incorporated into the precipitation during rain out. **Dry deposition** is the flux of particles and gases such as SO_2, HNO_3 and NH_3 to surfaces (Stumm et al., 1987).

[8]**Alkalinity** is the capacity of a natural water to neutralize acid and is measured by titration of a water sample with a dilute acid to a specific pH endpoint. Most often, it is a function of carbonate ($CO_3{}^{2-}$), bicarbonate ($HCO_3{}^-$), and hydroxide (OH^-) concentrations, i.e., the carbon dioxide-bicarbonate-carbonate buffering of the water. However, dissolved organic compounds, borates, phosphates, and silicates can also contribute to alkalinity.

the Adirondack Mountain region of New York resulted in high dissolved Al concentrations in surface waters. The geology of this sensitive area is dominated by granitic gneisses, resulting in poorly buffered waters. They (Cronan and Schofield, 1979) expressed concerns regarding aluminum toxicity there and in other areas of the United States and Europe that have silicate bedrock.

Regardless of the exact mechanism of demise during the decline in aquatic systems, it is clear that aquatic systems spread over wide areas of continents are being damaged by acid precipitation. Baker et al. (1991) estimated that the atmospheric input of acid anions represents the dominant anion flux into 75% of 1180 acid-sensitive lakes and 47% of 4670 acid-sensitive streams surveyed in the United States. In a survey of 5000 lakes of southern Norway, 1750 had lost fish species and another 900 were seriously impacted due to acid precipitation (Bridgman, 1994). In southern Ontario, 56% of surveyed lakes had reduced fish populations and an extraordinary 24% had no fish at all due to acid precipitation (Bridgman, 1994).

Effects of acid precipitation on terrestrial components of the biosphere are also significant and widespread. At low levels, the nitrogen and sulfur added to a forest in acid precipitation can enhance growth via the **fertilization effect** (Bridgman, 1994). However, acid precipitation can also increase nutrient leaching from foliage and forest soils, and accelerate the weathering of minerals (Smith, 1981). Acid leaching of calcium, magnesium, potassium, and sodium from decomposing forest litter, and calcium, potassium, and magnesium from soils have been measured (Smith, 1981). Leaching of magnesium and potassium from soils may produce a deficiency of these essential plant nutrients. Release of aluminum from solid phases in soils can result in direct toxicity to vegetation (Cowling and Linthurst, 1981; Smith, 1981). Acid precipitation may cause necrotic lesions on foliage, increased plant susceptibility to disease, increased rate of wax erosion from foliage surfaces, and lower nitrogen fixation by legumes via the inhibition of root nodule formation (Cowling and Linthurst, 1981). The composite of all of these effects of acid precipitation on forests is a major explanation forwarded for the widespread forest damage in large tracts of North America (Nihlgård, 1985) and the **waldsterben**, "the widespread and substantial decline in growth and the change in behavior of many softwood and hardwood forest ecosystems in central Europe" (Schütt and Cowling,1985).

Atmospheric dispersal of pollutants in aerosols or particulates can also encompass wide areas. Certainly, dry deposition plays an important role in the acid deposition-related problems just discussed. Much of the widespread metal deposition associated with the Copperhill smelting shown in Figure 12.1 was associated with particulate transport. Elevated iron concentrations were measured in tree rings 88 km from the Copperhill source (Baes and McLaughlin, 1984): particulate-associated iron moved long distances from their source at ore processing. Generally, widespread deposition of particulate-associated cadmium, manganese, lead, and zinc in forests of Tennessee has been documented: one third or more of the annual flux of cadmium, lead, and zinc to a Tennessee Valley forest was from atmospheric deposition (Lindberg et al., 1982). Hirao and Patterson (1974) found that most of the lead in sedge (*Carex scopulorum*) and voles (*Microtus montanus*) inhabiting a remote High Sierra valley in California came long distances from automotive and industrial sources in Los Angeles and San Francisco. Much of the lead (12 kg versus 1 kg from

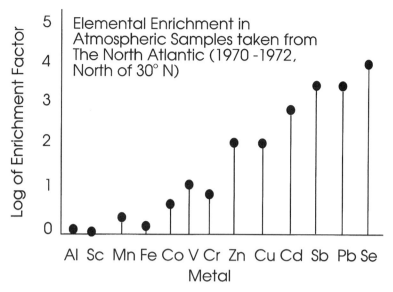

FIGURE 12.5. Metal particulate dispersal over the remote north Atlantic Ocean as evidenced by increased enrichment factors. The enrichment factor (EF_{crust}) for an element is its concentration (X) measured in air samples divided by that expected in the earth's crust: $EF_{crust} = [X/Al]_{air}/[X/Al]_{crust}$. Both air and crustal concentrations are normalized to Al concentrations. Aluminum is a ubiquitous element comprising about 8% of crustal material. Increases in EF_{crust} above 10^0 (=1) imply enrichment from anthropogenic sources. (Modified from Figure 1 of Duce and Zoller [1975])

other sources) entered the valley associated with snow. Still wider transport of Pb has been documented. Analysis of Antarctic ice cores shows an increase in Pb flux to the Antarctic after the worldwide onset of industrialization (Boutron and Patterson, 1983). At the earth's North pole, the flux of particulate-associated contaminants is significant from Eurasia to the Arctic (Pacyna, 1995). In air above remote parts of the northern Atlantic Ocean, levels of many metals are elevated far above background concentrations (Figure 12.5) (Duce and Zoller, 1975). Particulate-associated radionuclides released into the atmosphere during the Chernobyl reactor meltdown were distributed over most of the Northern Hemisphere and are predicted to result in elevated cancer deaths throughout Europe and portions of the former Soviet Union (Barnaby, 1986; Anspaugh et al., 1988).

Another phenomenon with a scale encompassing an entire continent is ozone depletion by chlorofluorocarbons (Figure 12.6). The introduction of chlorofluorocarbons (CFCs)[9] such as CFC12 (dichlorofluoromethane or CF_2Cl_2) and

---

[9]The CFC structures can be derived easily from their names. Ninety is added to the number in the CFC's name, e.g., CFC11 produces 90+11 or 101. This number codes for 1 carbon ("hundreds" digit), 0 hydrogens ("tens" digit), and 1 fluorine ("units" digit). Because the carbon atom forms four covalent bonds and only one is occupied (by a fluorine atom), the other three bonds must be with three chlorine atoms. So, CFC11 is $CFCl_3$.

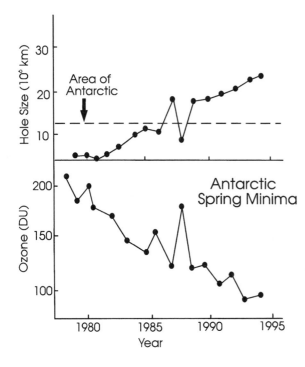

FIGURE 12.6. The increase in size of the Antarctic ozone hole (top panel) and decrease in the spring minima ozone concentration. Ozone concentrations are expressed in Dobson units (DU). One DU is the equivalent of 0.001 mm thickness of pure ozone at 1 atmosphere. To give some scale to this unit, if all of the ozone in the atmosphere were to be brought down to form a pure layer of ozone at sea level, it would be only 3 mm thick (Bunce, 1991). (Data taken from NASA web site jwocky.gstc.nasa.gov, files O3holes2.gif and O3holemin.gif.)

CFC11 (trichlorofluoromethane or $CFCl_3$) has shifted the balance among reactions taking place in the stratosphere to disfavor the maintenance of normal levels of ozone (O_3).

Ozone is formed by the **Chapman mechanism** (Reactions 12.7 to 12.9)(Bunce, 1991). Energy from sunlight of wavelengths < 240 nm and <325 nm is required for Reactions 12.7 and the second step of 12.8, respectively. A catalyst is required for Reaction 12.9.

$$O_2 \rightarrow 2O \tag{12.7}$$

$$O + O_2 \rightarrow O_3 \rightarrow O_2 + O \tag{12.8}$$

$$O + O_3 \rightarrow 2O_2 \tag{12.9}$$

Chloride also becomes involved in the ozone generating and depleting reactions in the stratosphere (Reactions 12.10 to 12.14) (Zurer, 1988). Reactions 12.12 and 12.14 require a catalyst and 12.13 requires light energy.

$$Cl + O_3 \rightarrow ClO + O_2 \tag{12.10}$$

$$ClO + ClO \rightarrow Cl_2O_2 \tag{12.11}$$

$$ClO + O \rightarrow Cl + O_2 \tag{12.12}$$

$$Cl_2O_2 \rightarrow Cl + ClOO \tag{12.13}$$

$$ClOO \rightarrow Cl + O_2 \tag{12.14}$$

Nitrogen species can shift these reactions such that some of the Cl is bound up in nitrogen compounds and unavailable to react with ozone, e.g., $ClO + NO_2 \rightarrow ClONO_2$ (Zurer, 1987; Kerr, 1988a). However, excess Cl from the breakdown of CFCs can overwhelm this sequestering process with a net effect of decreasing ozone concentrations. Molecular chlorine (Cl_2) and hypochlorous acid (HOCl) generated by Reactions 12.15 and 12.16 are readily converted to free radicals which destroy ozone (Zurer, 1987).

$$H_2O + ClONO_2 \rightarrow HNO_3 + HOCl \tag{12.15}$$

$$HCl + ClONO_2 \rightarrow HNO_3 + Cl_2 \tag{12.16}$$

Ozone destruction as a consequence of CFC accumulation in the stratosphere and circulation patterns above the Antarctic have produced an alarming **ozone hole** recently. Although meteorological conditions are less favorable for the formation of a similar hole above the Arctic, i.e., the polar vortex does not last as long, elevated chlorine monoxide (ClO) concentrations and ozone thinning have been reported there too (Zurer, 1989; Kerr, 1992).

The concern for the destruction of vast parts of the ozone layer is heightened by the realization that the expected lifetime of CFCs in the atmosphere is extremely long, i.e., 70 years for CFC11 and 110 years for CFC12 (Thompson, 1992). Any remedial action now will take considerable time to reverse the damage caused by CFC releases. Fortunately, the **Montreal Protocol**, an international treaty to limit and eventually eliminate the use of CFCs, was endorsed by 70 countries and then signed into law by 1987 (Crawford, 1987; Bunce, 1991). Although it will take a long time to reduce CFCs in the stratosphere, there are already encouraging indications that chlorine levels have leveled off in the stratosphere (Kerr, 1996).

Predictions of ozone depletion effects to humans and ecological systems remain vague. Ozone absorbs UV light with wavelengths of 290 to 330 nm or roughly the UV-B range (280-320 nm) as it enters the earth's atmosphere. This UV-B can cause skin cancer and speculations are that the incidence of skin cancers could increase as stratospheric ozone levels drop. However, Bunce (1991) provides the tempering comparison that "for people living in the middle latitudes, the increased risk of skin cancer due to each 1% decrease in ozone levels is equivalent to that posed by moving 20 km closer to the equator." Effects to ecological entities also remain equivocal at this time. Some speculation exists linking ozone depletion to the global decline in amphibian species. Increased UV-B penetration into the surface waters of the oceans is also speculated to decrease marine phytoplankton photosynthesis (Baird, 1995).

BIOSPHERE

General

Some human activities can stretch out to involve hemispheres or the entire planet. As depicted in Figure 12.6, the Antarctic ozone hole extended beyond that continent in the late 1980s. Two general phenomena that are even more encompassing will be discussed in this section: global distillation of persistent organic pollutants and global warming.

Global Movement of Persistent Organic Pollutants

Many persistent organic pollutants (POPs)[10] are subject to extensive movement and redistribution on a global scale (Table 12.1). A POP will vaporize and move in the atmosphere until it reaches a temperature at which it condenses. It then becomes associated with a less mobile solid or liquid phase. The extent to such movement of POPs from their sources of release to cooler latitudes depends on each POP's rate of degradation, vapor pressure, and lipophilicity (Simonich and Hites, 1995).

According to the **cold condensation theory**, POPs in the air will condense onto soil, water, and biota at cool temperatures. Consequently, the ratios for POP concentrations in the air and on condensed phases decreases from warmer to cooler climates (Wania and Mackay, 1995). This leads to the phenomena of **global distillation** in which POPs migrate from warm regions of release to cold regions of condensation (Figure 12.7). This distillation can involve seasonal cycling of temperatures such that movement toward the higher latitudes occurs in annual pulses or jumps (the **grasshopper effect**) (Wania and Mackay, 1996). Because POPs differ in their individual rates of degradation, vapor pressures, and lipophilicities, a **global fractionation** occurs in which some POPs move more rapidly than others toward the polar regions. Some, because of their temperatures of condensation, may be unable to move beyond a certain point toward cooler latitudes. The net result is a redistribution of the different POPs from the equator or site of origin toward the cold polar regions of the earth. The K_{OW}s of POPs also influence their movement via the **retention effect**. Those POPs with high lipophility tend to be held more firmly than less lipophilic POPs in solid phases such as soil and vegetation. Consequently, they spend less time in the atmosphere and are less available for transport in that medium (Wania and Mackay, 1995). All else being equal, the more lipophilic POPs move slower toward higher latitudes than less lipophilic POPs.

Actual global distributions of POPs conform to the global distillation and

---

[10]**Persistent organic pollutants (POPs)** are those organic pollutants that are long-lived in the environment and tend to increase in concentration as they move through foodchains (Wania and Mackay, 1996). According to Wania and Mackay (1996), they are also called **bioccumulative chemicals of concern (BCCs)** and **persistent toxicants that bioaccumulate (PTBs)**.

TABLE 12.1. Persistent Organic Pollutant Mobility in a Global Context (modified from Wania and Mackay [1996]).

Pollutant Classes (Subclassified by number of Cl atoms, rings, or by pesticide type)	Relative Mobility Class			
	Rapidly deposited and retained near source	Preferential deposition and accumulation in mid-latitudes	Preferential deposition and accumulation in polar latitudes	Worldwide dispersion and deposition
Chlorobenzenes			5 to 6 Cl	0 to 4 Cl
PCBs[a]	8 to 9 Cl	4 to 8 Cl	1 to 4 Cl	0 to 1 Cl
PCDDs[a] and PCDFs[a]	4 to 8 Cl	2 to 4 Cl	0 to 1 Cl	
PAHs[a]	> 4 rings	4 rings	3 rings	2 rings
Organochlorine Pesticides	mirex	polychlorinated camphenes, DDT, DDE, chlorodanes	HCB[a], HCCHs[a], dieldrin	
Pollutant Quality				
Log K_{OA} [b]	> 10	8 to 10	6 to 8	< 6
P_L [c]	< –4	–4 to –2	–2 to 0	> 0
T_C [d]	> +30	–10 to +30	–50 to –10	< –50

[a]PCB = polychlorinated biphenyls; PCDD and PCDF = polychlorinated di-benzo-*p*-dioxins and -furans; PAH = polycyclic aromatic hydrocarbons; HCB = hexa-chlorobenzene; HCCHs = hexachlorocyclohexanes.

[b]K_{OA} is the partition coefficient between octanol and air. Like the K_{OW}, it is a measure of lipophilicity.

[c]P_L is the **subcooled liquid vapor pressure** (Pa), a measure of a compound's volatility. Specifically, it is the liquid vapor pressure corrected or adjusted for the heat of fusion, the energy needed to convert a mole of a compound from a solid to a liquid phase. Its use allows the expression of *liquid* vapor pressures at a specific temperature for organic compounds with widely varying melting temperatures.

[d]T_C is the **temperature of condensation**, the temperature (°C) at which the compound condenses or partitions from the gaseous to the nongaseous phase.

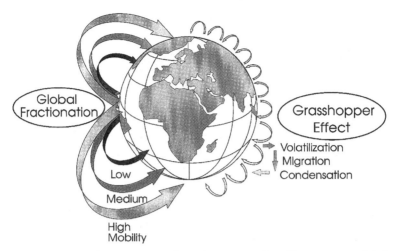

FIGURE 12.7. The movement of persistent organic pollutants on a global scale. (Modified from Figure 1 of Wania and Mackay [1996].)

fractionation scheme outlined above. In a survey by Simonich and Hites (1995) of tree bark from sites around the world, global distillation was apparent for the volatile hexachlorobenzene but not apparent for the less volatile endosulfan and DDT. Clear movement of volatile organic compounds into Arctic systems has also been documented (Chernyak et al., 1995; Muir et al., 1995). Mackay and Wania (1995) produced a model that accurately predicted POP global distributions based on these processes.

Global Warming

Global warming is thought to result primarily from the increased atmospheric carbon dioxide (CO_2) concentrations produced by fossil fuel burning and the worldwide destruction of forests (Woodwell, 1978, Khalil and Rasmussen, 1984). Much of the concern centers around carbon dioxide, methane, and nitrous oxide which have rapidly increased in the atmosphere since preindustrial times (Hileman, 1989). These gases, along with water vapor, are relatively transparent to light entering the earth's atmosphere but absorb long-wave, infrared radiation radiating back from the earth's surface toward space. Because the net energy balance of sunlight influx and infrared radiation eflux from the earth's surface determines the steady-state temperature of the earth (the **greenhouse effect**), increases in these greenhouse gases[11] are believed to produce an increase in global temperatures. Al-

---

[11]**Greenhouse gases** include water vapor, carbon dioxide, methane, nitrous oxide (N_2O), CFCs, methylchloroform, carbon tetrachloride and the fire retardant, halon. Ozone *in the troposphere* may also act as a greenhouse gas (Hileman, 1989). All greenhouse gases are not equal relative to infrared radiation absorption. For example, one molecule of CFC11 absorbs 10,000 more infrared radiation than one molecule of carbon dioxide (Anonymous, 1985).

though there is still some disagreement about whether this is occurring (e.g., Hileman, 1984; White, 1990; Thompson, 1992; Lindzen, 1994; Maswood, 1995), global warming could have widespread effects to human and nonhuman species. Abrupt changes in economic and agricultural activities might occur, sea level may change, and tropical diseases could extend into other regions of the world (Kerr, 1988b). Relative to changes occurring between ice ages, species ranges would have to change very rapidly, with some species becoming extinct and global biodiversity decreasing (Roberts, 1988). Species including important plant species that normally migrate during the ice age-interglacial cycles would be severely challenged by the extremely rapid rate of temperature change. In the current landscape that is highly fragmented by human activity, some would be unable to migrate successfully. Such species in fragmented habitats would be "man-locked" (Roberts, 1988), unable to migrate, and at high risk of extinction.

SUMMARY

In this chapter, an argument was made that a context larger than the ecosystem is required to fully grasp many ecotoxicological problems facing us today. The argument is made using examples of important problems at increasingly wider spatial scales. Unfortunately, as the scale of problems becomes wider, the potential for widespread harm increases and the ability to assign a cause-effect relationship decreases. This makes environmental assessment and management of such problems exceedingly difficult and prone to divergent conclusions.

Suggested Readings

Anspaugh, L.R., R.J. Catlin, and M. Goldman. The global impact of the Chernobyl reactor accident. *Science (WASH.)* 242, pp. 1513–1519, 1988.

Cowling, E.B. Acid precipitation in historical perspective. *Environ. Sci. Technol.* 16, pp. 111A–123A, 1982.

Cowling, E.B. and R.A. Linthurst. The acid precipitation phenomenon and its ecological consequences. *Bioscience* 31, pp. 649–654, 1981.

Hileman, B. Global warming. *Chem. Eng. News* March 13, pp. 25–44, 1989.

Houghton, R.A. The global effects of tropical deforestation. *Environ. Sci. Technol.* 34, pp. 416–422, 1990.

Omernik, J.M. Ecoregions of the conterminous United States. *Ann. Assoc. Am. Geogr.* 77, pp. 118–125, 1987.

Sabins, F.F., Jr. *Remote Sensing. Principles and Interpretation.* W.H. Freeman and Co., New York, 1987, p. 449.

Stumm, W., L. Sigg, and J.L. Schnoor. Aquatic chemistry of acid deposition. *Environ. Sci. Technol.* 21, pp. 8–13, 1987.

Wania, F. and D. Mackay. Tracking the distribution of persistent organic pollutants. *Environ. Sci. Technol.* 30, pp. 390A–396A, 1996.

Zurer, P.S. Arctic ozone loss: Fact-finding mission concludes outlook is bleak. *Chem. Eng. News* March 6: pp. 29–31, 1989.

Risk Assessment of Contaminants

Inferences are movements of thought within the sphere of belief.
JOSEPHSON AND JOSEPHSON (1996)

OVERVIEW

Logic of Risk Assessment

This chapter brings our discussion squarely into the realm of technology as applied to environmental problem solving. Techniques estimating both human and nonhuman risk are described together using the reasonable approach developed recently for the CERCLA assessment process. However, many germane particulars of the approach have been discussed already. Chief among them are the logic of scientific enquiry (Chapter 2), bioaccumulation (Chapters 3 and 4), trophic transfer of contaminants (Chapter 5), biomarkers (Chapter 6), indicators of effects to individuals (Chapters 6 to 8), the NOEL/LOEL approach to sublethal and chronic lethal effects (Chapter 8), models of toxic response including survival time models (Chapter 9), life tables (Chapter 10), and Hill's aspects of disease association (Chapter 10).

The logic of scientific enquiry—indeed, the logic of any effort to enhance belief about physical phenomena—is more complicated than presented in Chapter 2. For example, the straightforward "reject or accept" context for testing the mettle of a working hypothesis is a logical luxury afforded infrequently to the ecotoxicologist. More often, an ecotoxicologist assessing the risk from contamination must bolster belief using tools such as Hill's aspects of disease association. Information is gathered until a balanced judgment can be made based on a **weight of evidence** (preponderance of evidence) approach. Note that this vague term, weight of evidence, might mean that a *reasonable person* reviewing the available information *could* agree that the conclusion was plausible (Apple et al., 1986). At the other extreme, the statistical use of the term implies to Kotz and Johnson (1988) a quantitative or semiquantitative estimate of the degree to which the evidence supports or undermines the conclusion. The first definition seems the most accurate for the majority of risk assessment activities. Ideally, evidence describing clear, consistent, and plausible toxic effects would be judged as having considerable weight. Regardless of its shortcomings, the weight of evidence approach allows movement toward belief using incomplete information. Although this approach is often labeled as less "scientific" (i.e., lacking logical rigor) than that described in Chapter 2, the distinction is one of degree. The more precise the logic and more quantitative its expression, the more the weight of evidence approach resembles the "scientific" approach.

The weight of evidence approach as applied in risk assessment has major elements of abductive inference and probabilistic (Bayesian) induction. (Both abductive inference and probabilistic induction permeate traditional and modern scientific methods too.) **Abductive inference** is simply inference to the best explanation. It uses information gathered about a phenomenon or situation to produce the hypothesis that best explains the data. Josephson and Josephson (1996) give the following example of abductive inference.

1. D is a collection of data.
2. H explains D or, if true, H would explain D.
3. No other explanation (hypothesis) explains D as well as H.
    ∴ H is *probably* true.

This linkage of the weight of evidence approach in risk assessment to abductive logic is not a trivial point. Abductive inference can be formalized in artificial intelligence computer programs (Josephson and Josephson, 1996) that could easily be adopted for risk assessments.

Probabilistic induction uses probabilities associated with competing theories or explanations to decide which is most probably true. Credibilities are assigned to competing explanations based on their associated probabilities (Howson and Urbach, 1989). Instead of the quantal (accept or reject) falsification of a working hypothesis as described in Chapter 2, Bayesian induction considers a hypothesis falsified if it becomes sufficiently improbable.[1] Like the traditional conclusion to designate an accepted explanation as the best current approximation of reality, this conclusion may be reconsidered later if conflicting facts emerge. In reality, if not tradition, these approaches are no less valuable in fostering the growth of knowledge than the classic methods described earlier. Indeed, the enhanced status of a hypothesis surviving repeated and rigorous testing is a straightforward permutation of abductive inference. It would also lead to probabilistic induction if done rigorously and with ample statistical power so as to provide probabilities. Both abductive inference and probabilistic induction can contribute to the weight of evidence approach as applied to risk assessments.

Expressions of Risk

In this chapter, the preoccupation will be on the expression of **risk** as a probability of some adverse consequence occurring to an exposed human or to an exposed

---

[1]Obviously, probabilistic induction is one important logical extension of conventional statistical analyses used in many risk assessments.

ecological entity. For example, one might estimate a less than a 1 in 100,000 chance of dying due to an exposure to a particular toxicant. More precisely, risk is "the product of the probability and frequency of effect [e.g., (probability of an accident) x (the number of expected mortalities)]" (Suter, 1993). The concept, as applied to environmental risk assessment includes the probability of an event occurring that *could* lead to an adverse effect and the probability of an adverse effect given that the event *did occur*. This will be the context for the term in most of this chapter. However, other expressions of risk have already been discussed that are equally valuable. In an epidemiological context (Chapter 10), risk was expressed as a relative risk ratio. In discussions of time-to-event models (Chapter 9), risks were expressed as relative risks and modeled as proportional hazards. Although not used as often as probabilities in environmental risk assessments, relative risks are often used to explain risk to the general public. For example, people living in U.S. states with very low selenium levels in soils and waters have a higher risk (expressed as a percentage above the national average) of dying from heart disease than those living in states with normal levels of selenium (Anonymous, 1976). Survival functions (Chapter 9) and tabulations of age-specific life expectancies (Chapter 10) are also useful expressions of fatal risk that can easily be incorporated into predictive models.

In communicating with the public, the conventional expression of risk as a probability (e.g., 10^{-5} or 1 chance in 100,000 of the consequence occurring) can be less intuitive than its expression as a change in life expectancy. Consequently, expression of risk in terms of life expectancy is worth exploring for a moment before proceeding to the more conventional context of risk. Also this expression of risk is very closely tied to and extends the mathematical foundations laid down in Chapter 9 and Chapter 10. The **loss of life expectancy** (LLE) is estimated as the simple difference between the life expectancy with (E_x) versus without (E) the risk factor being present (Cohen and Lee, 1979).

$$\Delta E = E_x - E \tag{13.1}$$

Loss of life expectancy is expressed in days, months, or years depending on the magnitude of the risk factor's effect. For example, the LLE for the average American (age 0 to 55 years) due to cancer is 0.34 (males) and 0.32 (females) years (Cohen and Lee, 1979). A uranium miner (1970-1972 statistics) has a mortality rate of 232×10^{-5} yr^{-1}, or a LLE of 1160 days relative to that of the average person. Cohen (1981) gives an example of a *gain* in life expectancy (negative LLE) that is a particularly fascinating statistic to the author. There is a gain of 500 days if one's occupation is that of a university teacher. There is an intuitive statistic with which I can live!

Risk Assessment

Assessment of contaminant-associated risk is mandated in key U.S. federal laws, including RCRA and CERCLA as amended by SARA. It is also implied by use of the term, "unreasonable risks" in FIFRA and TSCA (Suter, 1993). For this reason, the

EPA has developed numerous documents and regulations ensuring that the intent of these laws is met relative to protecting human health and the environment. For example, assessment of risk to humans is detailed in guidelines for Superfund sites (EPA, 1989d, 1996) and to ecological entities in several EPA guidance documents (EPA, 1989c, 1996). The remainder of this chapter is a condensed version of these guideline documents.

As defined in such regulations, a **risk assessment**[2] is the process by which one estimates the probability of some adverse effect(s) of a present or planned release to either human or ecological entities. Two general categories of risk assessments exist: retroactive and predictive. A **retroactive risk assessment** deals with an existing condition such as a contaminated seepage basin and the **predictive risk assessment** deals with a planned or proposed condition such as a planned discharge of a waste. In some situations, the predictive risk assessment may also deal with a future consequence of an existing situation. For example, a predictive assessment may estimate the consequences of a contaminated plume of groundwater that will outcrop soon to a stream or reach nearby drinking water wells.

A risk assessment is carried out by a **risk assessor**, a person or group of people "who actually organizes and analyses site data, develops exposure and risk calculations, and prepares a risk assessment report" (EPA, 1989d). The assessment is provided to a **risk manager**, "the individual or group who serves as primary decision-maker for a site" (EPA, 1989d). This distinction between the roles of the risk assessor and manager is important: the assessor does not make any decisions regarding the action to be taken as a consequence of the assessment although potential remedial actions may be detailed in the assessor's report. A decision is the responsibility of the risk manager who must also weigh all costs, benefits, and risks. As a human risk example, the extremely low risk of cancer associated with consuming residual pesticide in food may be deemed acceptable relative to the widespread economic and nutritional risk associated with abandoning pesticide use in agriculture. A common ecological example involves the certain risk of damage via the destruction of habitat during removal of contaminated soil from an area that is predicted to have a small, but measurable, toxic risk to an endemic species. Is it wise and in the spirit of federal law to destroy an invaluable and fragile habitat in order to remove a contaminated soil that presents a very small risk?

Permutations of the **NAS paradigm** (National Research Council, 1983) (Figure 13.1) are used for both human and ecological risk assessments. There are four components to this paradigm: hazard identification, exposure assessment, dose-response assessment, and risk characterization. In the first, relevant data on the situation are gathered and chemicals of potential concern highlighted. **Exposure** (contact with the contaminant) assessment estimates the magnitude of releases, identifies possible pathways of exposure, and estimates potential exposure. Dose-

---

[2]Formally, a risk assessment is different from a hazard assessment. A **hazard assessment** compares the expected environmental concentration (EEC) to some estimated threshold effect (ETT) with the intent of deciding if (1) a situation is safe, (2) a situation is not safe, or (3) there isn't enough information to decide. Often a **hazard quotient** (HQ = EEC/ETT) is used as a crude indicator of hazard. (This use of the EEC divided by some endpoint value for adverse effect is called the **quotient method**.) A risk assessment is like a hazard assessment except that it has as its goal the generation of a quantitative estimate (probability) of some adverse effect occurring.

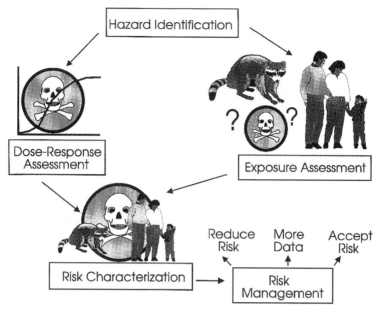

FIGURE 13.1. The NAS paradigm for risk assessment. The risk assessor identifies the hazard, assesses the potential for exposure (or current exposure) and the dose-response relationship for the relevant entities being protected and the relevant toxicants. This information is integrated to characterize the risk. At that point, the assessment is provided to the risk manager who then decides to accept or reduce the associated risk or, if insufficient information is available to make a good decision, to seek out more information.

response assessment gathers together relevant toxicological data relating exposure to relevant effects. Risk characterization then integrates this information to assess the potential or existing risk of adverse effect. Also included in this fourth component is a statement of the uncertainty involved in the risk estimate and the general quality of data used in the assessment. These components will be described to particulars for human and ecological assessments.

HUMAN RISK ASSESSMENT

General

> . . . scientific activity is not the indiscriminate amassing of truths; science is selective and seeks the truths that count for most, either in point of intrinsic interest or as instruments for coping with the world.
>
> Quine (1982)

Obviously, human risk assessment has as its goal the estimation of the probability or likelihood of an adverse effect to humans occurring as a result of a defined exposure. For Superfund sites, it is one part of a **remedial investigation and feasibility study (RI/FS)**[3] which has the larger goal of implementing "remedies that reduce, control, or eliminate risks to human health and the environment" or, more specifically, the accumulation of "information sufficient to support an informed risk management decision regarding which remedy appears to be most appropriate for a given site" (EPA, 1989d). The human risk assessment part of the RI/FS has three components: a baseline risk assessment, the refinement of the initial remediation goals, and the evaluation of the risk associated with candidate remedial activities. The first stage involves the application of the NAS risk assessment paradigm to the specific site with the intention of determining if the contaminants at the waste site present, or could present in the future, a risk if left alone, i.e., the **"no action" alternative** to remediation of the site. In the assessment, consideration should include exposure from one or several routes, and effect to one or several subpopulations of humans. The four sections that follow give the details for the baseline human risk assessment based on the NAS paradigm.

Hazard Identification (Data Collection and Data Evaluation)

From among the many chemicals present at the site, a tentative list of those of most concern (**chemicals of potential concern**) is made in this phase of the risk assessment. The concentrations of these chemicals are compared to background concentrations. Concentrations are measured in all relevant media, e.g., air, surface water, groundwater, soils, sediments, game species, or edible fish. In contrast to ecological risk assessments, contaminant concentrations in biological media most relevant in human risk assessment are edible parts, not the entire organism. Concentrations in all relevant media from the waste site are compared to background levels.

The various contaminant sources are described. Qualities of the environmental setting are also compiled, especially those qualities "that may affect the fate, transport, and persistence of the contaminants" (EPA, 1989d). (As part of the entire RI/FS, a preliminary series of tests may also be done to assess various treatability or remediation alternatives at this stage.) The quality of the data needed for the assessment should also be defined, e.g., types of effects data or required detection limits for the chemicals of potential concern. Also, preliminary exposure scenarios are formulated at this early stage. Usually, risk assessors establish a formal scoping process to identify what data must be collected at this point.

After the data have been collected, they are applied to do the following: (1) determine the analytical data quality and adequacy of sampling, (2) compare site concentrations to background concentrations, (3) identify chemicals of potential

---

[3]The **remedial investigation (RI)** has three parts: characterization of the type and degree of the contamination, human risk assessment, and ecological risk assessment (EPA, 1994). The **feasibility study (FS)** explores the various options for remediation. The chapter will focus on the last two parts of the RI.

concern from the initial list of compounds, and (4) determine if the data are adequate to proceed. A rough screening is done to highlight contaminants for further consideration. According to the EPA (1989d), a chemical concentration may be multiplied by a toxicity value to generate an associated **risk factor**.

$$R_{ij} = C_{ij} \, T_{ij} \tag{13.2}$$

where R_{ij} = the risk factor for the chemical i in association with media j, C_{ij} = the concentration of i in media j, and T_{ij} = a toxicity value for i in j. It is further suggested that the total risk score for a mixture of contaminants in a medium can be estimated as the sum of the individual R_{ij} values: $R_j = R_{1j} + R_{2j} + R_{3j} + \ldots R_{ij}$. The size of the risk factor dictates whether the chemical is retained for further consideration. Obviously, the main virtue of this approach is expediency, not accuracy. It is felt that the crude estimate of risk is adequate, assuming effect additivity and a Selyean stress context for mixture effects.

Determining a **toxicity value** can involve either one of two methods. One method uses the slope of a published effect-dose relationship: R_{ij} = Slope * C_{ij}. The other uses a RfD value in a similar manner. Details of these approaches will be described below in Dose-Response Assessment.

Note that, although the methods below will focus on long exposure periods, short duration exposures such as those associated with consumption of tainted drinking water may be assessed using one-day, ten-day, or other short duration health advisories. The U.S. EPA Office of Drinking Water has developed such health advisories based on an ingestion for individual chemicals. These **health advisory concentrations** identify concentrations below which no impact to human health is expected for the specified duration of exposure. They are usually derived by a process like that described below for RfD values. Often they are applied by public officials in dealing with spills, short-term exposures, or similar situations, although longer-term advisory concentrations are also available.

Exposure Assessment

The goal of human exposure assessment is to determine or estimate the route, magnitude, frequency, and duration of exposure (EPA, 1989d). In such an assessment, exposures are estimated for specific subpopulations (e.g., hypersensitive individuals, elderly, children, remediation workers) by specific routes (e.g., dust inhalation during remediation, drinking water, game consumption). **Exposure pathways** (the avenues by which an individual is exposed to a contaminant including the source and route to contact) may include inhalation, ingestion from various media, and dermal contact and absorption.

The **reasonable maximum exposure (RME)** is calculated for chemicals of potential concern. This conservative estimate of exposure is computed differently depending on the route of exposure. Exposure concentrations are often taken from among the highest measured, e.g., the concentration at the 95% upper confidence limit instead of the mean concentration. The intake is then estimated using an equation such as the general Equation 13.3 (EPA, 1989d).

$$I = C \frac{(CR)(EFD)}{BW} \frac{1}{AT} \tag{13.3}$$

where I = the intake (e.g., [mg of contaminant])[kg of body mass]$^{-1}$[day]$^{-1}$), C = contaminant concentration (e.g., mg L^{-1}), CR = contact rate (e.g., L day^{-1}), EFD = an estimate of frequency and duration of exposure composed of EF (exposure frequency, e.g., days year^{-1}) and ED (exposure duration, e.g., years), BW = body weight or mass (kg), and AT = the time over which exposure is averaged. The AT may be estimated as ED x 365 days per year for a noncarcinogen or 70 years x 365 days per year for a carcinogen. The exact formulation of this equation changes depending on the exposure route, but the general approach remains the same, allowing estimation of the appropriate RMEs. Appendix 5 is a listing of those formulae as supplied in EPA guidelines (EPA, 1989d). EPA documents and other sources are drawn on for specific variable values used in such calculations.

Dose-Response Assessment

The goal of the dose-response assessment is to gather all information useful in establishing a relationship between the extent of contamination and the likelihood and/or magnitude of an adverse effect. A wide range of information is drawn upon for human risk assessments including human epidemiological data, data derived from study of nonhuman animals, and predictive models such as QSARs. General mechanistic information is also sought in making judgments. The most valuable data are from long-term studies involving humans, e.g., Japanese atomic bomb studies discussed in Chapter 14. But such studies are often less structured than nonhuman animal studies. There is an obvious reason for this. Most human information comes from effects observed after accidents, often occupational accidents with very high exposures. From these unstructured experiences, it is difficult to isolate dose effects from covariates (e.g., age, sex, or other risk factors) and to extrapolate downward to the lower, chronic exposure scenarios normally associated with environmental contaminants. The use of nonhuman animal studies carries the uncertainty of extrapolation to humans. Most animal to human extrapolations involve allometric modification of effects, e.g., adjusting effect by Weight$^{2/3}$ (or Weight$^{3/4}$) if mg day^{-1} intakes are used. If intake is expressed as mg kg^{-1}day^{-1}, the allometric adjustment for the difference in weights between the study animal and humans is Weight$^{1-2/3}$, or Weight$^{1/3}$. (See Chapter 9 for further explanation.) Sometimes, the allometric PBPK models discussed in Chapters 3 and 4 are applied to reduce inaccuracies in extrapolation. In general, results with high consistency among animal species provide increased confidence in extrapolating to humans (EPA, 1989d). As you will see, various uncertainty factors are employed to compensate for the uncertainty associated with such problems. For carcinogens, a weight of evidence classification has also been established to aid the risk manager in understanding the degree of uncertainty involved in each risk calculation.

For noncarcinogenic effects, the **reference dose (RfD)** is applied to risk estimation. It is the best estimate of the daily exposure for humans, including the most sensitive subpopulation, that will result in no significant risk of an adverse health effect if not exceeded. It is assumed that it is accurate to only within an order of magnitude, i.e., 10-fold of the true value. There are different types of RfDs. In most assessments, a **chronic RfD** is assumed unless indicated otherwise. It is the RfD associated with exposures spanning an individual's lifetime. There are also **subchronic RfDs (RfD$_s$)** that are derived from short-term exposure data and **developmental RfDs (RfD$_{dt}$)** that focus on developmental consequences of a single, maternal exposure during development. Which RfD is most appropriate depends on the exposure scenario of concern. Calculations may be done for several scenarios if warranted. For example, a chronic scenario may assess the risk of leaving the site as it exists, and an acute scenario may assess risk for workers during remediation.

RfDs incorporate toxicity data (i.e., an NOAEL) and qualitative uncertainty factors (UFs) associated with these data. First, all data sets for the toxicant in the appropriate media are compiled. In the absence of data for the appropriate media or exposure route, data from other forms or routes of exposure may be used after adjustment. Unless there is sound evidence indicating otherwise, the conservative assumption is made that humans are as sensitive as the most sensitive species tested. Under this assumption, the human or nonhuman animal study with the lowest-observed-adverse-effect-level (LOAEL) is taken to be the **critical study**. The associated effect is called the **critical toxic effect**. Notice that the LOAEL comes from a suite of LOAELs derived by methods discussed in Chapter 8, and that there are serious problems associated with the application of LO(A)ELs and NO(A)ELs. Once the LOAEL is found, the no-observed-adverse-effect-level (NOAEL) or highest toxicant level tested that had no adverse effect is identified for the critical study/effect. This NOAEL is the measure of toxicity used in the derivation of the RfD. The LOAEL can be used after adjustment if an NOAEL is not available.

Uncertainty factors (UFs) and a modifying factor (MF) are now applied to the NOAEL to compensate in a conservative direction for uncertainty or unaccounted factors. **Uncertainty factors** are often, but not always, factors of 10 that lower the NOAEL to compensate for various sources of uncertainty including variation in sensitivity within the human population (UF$_H$), extrapolation from other species to humans (UF$_A$), use of subchronic rather than chronic or lifetime exposure data (UF$_S$), and use of a LOAEL instead of the NOAEL (UF$_L$). Uncertainty factors can take other values, e.g., the EPA's database (IRIS) lists those for hexavalent chromium as UF$_H$=10, UF$_A$=10 (data from rats), and UF$_S$ = 5 (data involved a chronic, but less than lifetime, exposure). Expert opinion may be applied to modifying a NOAEL even further. Expert opinion is incorporated through a **modifying factor** that ranges from greater than 1 to 10. The RfD is then generated using the NOAEL (or LOAEL), UFs, and MF. Obviously, the default values for the UFs and MF are 1 if the associated uncertainties or modifying circumstances are insignificant. Commonly, the UF$_H$ is set at 10 to ensure that the most sensitive humans are protected. UF$_L$ is 10 if LOAEL is used instead of NOAEL in Equation 13.4; otherwise, it is 1.

$$RfD = \frac{NOAEL}{(UF_H)\,(UF_A)\,(UF_S)\,(UF_L)\,(MF)} \qquad (13.4)$$

Clearly, the approach just described for noncarcinogenic effects is based on a threshold model of effect. Below a certain level, the human individual is protected from any adverse effect. That protective level is assumed to be above the RfD but below the LOAEL.

In contrast, a nonthreshold model is assumed in dealing with the risk of carcinogenic effects. A **slope factor (SF)** (risk or probability of occurrence per unit of dose or intake) for the risk-dose model is then used to estimate the probability of a cancer under a particular exposure scenario. Although based primarily on intake, models may be expressed in terms of concentration, dose, or intake. The slope factor is defined by the EPA as "a plausible upper-bound estimate of the probability of a response per unit intake of a chemical over a lifetime" (EPA, 1989d). Usually, the upper 95% confidence limit for the estimated slope of the risk-dose or risk-intake curve is used as the slope factor. The exposure dose or intake is multiplied by the slope factor to estimate risk.

$$Risk = (CDI)(SF) \qquad\qquad (13.5)$$

where Risk = the probability of developing cancer, and CDI = the chronic daily intake averaged over a lifetime (70 years) (mg kg^{-1} day^{-1}). Estimation of daily intakes is illustrated generically in Equation 13.3 and specifically for various sources in Appendix 5 at the end of this book. All risk estimations are accompanied by a qualitative **EPA weight of evidence classification** because the strength of evidence for specific chemicals being human carcinogens varies widely. This classification informs the risk manager about the strength of the evidence supporting the risk calculation.

The EPA has compiled RfDs, slope factors, drinking water health advisories (one-day, ten-day, longer-term, and lifetime advisories), and important associated information into a large database, **Integrated Risk Information System or IRIS**. At this time, IRIS is accessible through the U.S. Government's Right-To-Know web site (http://www.rtk.net/T866). As an example of how easily these data may be found, the author just spent less than ten minutes logging onto IRIS to retrieve the following information for cadmium. The cadmium RfD is 0.0005 mg kg^{-1} day^{-1} based on proteinuria[4] (the critical effect) in humans after imbibing cadmium in drinking water. The NOAELs were 0.005 mg kg^{-1} day^{-1} for water and 0.01 mg kg^{-1} day^{-1} for food. The UF$_H$ is 10 and MF is 1 for Equation 13.4. Confidence in these data was judged to be high because of the extensive human and animal data sets, and sound PBPK models available to the assessor.[5]

---

[4]**Proteinuria** is the presence of protein in the urine. The suggestion is kidney damage caused by cadmium that has accumulated in the renal cortex.

[5]Here, again, is another opportunity for confusion if the distinct goals of scientific, technical, and practical ecotoxicologists are not kept in mind and respected. Data from very sound scientific studies may be judged of "low" confidence relative to its use in estimating an RfD. On the other hand, an RfD is meaningless in a scientific sense. What might be good for one purpose is inadequate for another.

Risk Characterization

The data collected in the previous steps are now combined in this last step to generate a statement of risk. To this end, the exposure information including intake rates and dose-response data are used in the calculations below. The final statement of risk may be qualitative or quantitative. Regardless, it must include an explanation, specific details, and qualifiers for the final expression of risk.

A hazard quotient (Equation 13.6) is estimated for each noncarcinogenic effect using the exposure level or intake (E) and associated RfD. If the quotient does not exceed 1, the human population is assumed to be safe. Hazard quotients for chronic, subchronic, and shorter-term exposure may need to be estimated also depending on the exposure scenario of concern.

$$Hazard\ Quotient = \frac{E}{RfD} \qquad (13.6)$$

However, the quotient estimated above is not useful if there is more than one chemical of potential concern. Under the (dubious) assumption of additivity of effects, a hazard quotient can be calculated for situations involving several (x) chemicals of potential concern.

$$Hazard\ Index_{Total} = \sum_{i=1}^{x} \frac{E_i}{RfD_i} \qquad (13.7)$$

where E_i = exposure levels (or intakes), and RfD_i = reference doses for the i chemicals expressed in similar units and covering the same exposure durations as E_i. With temperance, summations could be done for contaminants in different media. Again, a quotient less than 1 suggests protection of the human population. For chronic exposures, a chronic hazard index can be estimated by Equation 13.7 if chronic RfD_is are used and CDI_i is used as E_i. Similarly, a subchronic hazard index is generated with a subchronic RfD_i and subchronic daily intake rates (SDI_is).

Risk of carcinogenic effects are estimated with Equation 13.5 assuming a multistage model of carcinogenicity. This linearized, multistage model is appropriate only at low doses. Consequently, it should be replaced by the first-order or **one-hit risk model** if estimated risk is 0.01 or higher,

$$Risk = 1 - e^{-(CDI)(SF)} \qquad (13.8)$$

Of course, other plausible models such as a gamma multiple hit or Weibull models exist as discussed regarding the dynamics of carcinogenesis in Chapter 7 and 9 and expressed mathematically in Chapter 9 (Equations 9.16 and 9.17).

If several carcinogens are considered together, a total cancer risk is approximated as the simple sum of the individual risks assuming independence of effects. This may not be valid as suggested from our previous discussions of carcinogenesis. Again, risk can also be summed across different media with an understanding of the limits of such a calculation.

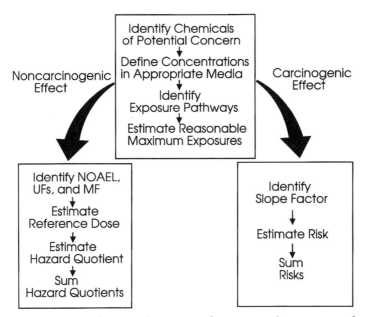

FIGURE 13.2. The general sequence of steps toward assessment of the hazard/risk associated with noncarcinogenic and carcinogenic effects of contaminants. The final step of summing hazard quotients or risks for the individual chemicals or media may not be required or may not be appropriate (see text for more details).

Summary

I prefer the errors of enthusiasm to the indifference of wisdom.
Anatole France (Quoted in Casti [1989])

The techniques described above for assessing risk from noncarcinogenic and carcinogenic contaminants (Figure 13.2) can be criticized easily for inconsistency with scientific knowledge. However, as detailed in Chapter 2, the goal of this crucial process is not scientific and should not be judged from that context alone. The goal is to protect human health. Proponents are acutely aware of many approximations and compromises in the approach. They conditionally accept those errors in order to fulfill their immediate goal. Indeed, they incorporate uncertainty factors and conservative slope factors so as to ensure that the results will be biased toward excessive caution and away from any possible harm to humans. These modifications are not made to more accurately define the threshold levels for the contaminant effects. Unfortunately, this dictates that more time and money than necessary be spent in remediation. It also assures that more valuable habitat than necessary will be damaged or destroyed during unnecessary or excessive remediation activities.

ECOLOGICAL RISK ASSESSMENT

"If seven maids with seven mops
Swept it for half a year,
Do you suppose," the Walrus said,
That they could get it clear?"
"I doubt it," said the Carpenter,
And shed a bitter tear.

Carroll (1872)

General

The first two parts of a remedial investigation (characterization of the contamination and human risk assessment) have been described briefly to this point. Now let's consider the third part of a remedial investigation, the ecological risk assessment. Like human risk assessment, the goal of ecological risk assessment is the estimation of the likelihood[6] of a specified adverse effect or ecological event due to a defined exposure to a stressor.[7] Relevant effects may range from the suborganismal to the landscape scale. Unlike human risk assessment, ecological risk assessments must consider many species with diverse niches and phylogenies. It may even consider ecological entities, e.g., communities, composed of many species occupying a heterogeneous landscape. Also, in contrast to human risk assessment in which extrapolation to one species (humans) is often done from many species (e.g., mouse, rat, or dog toxicity data), ecological risk assessment extrapolates from one or a few species to many.

Ecological assessments may be retroactive or predictive. The predictive assessment generally adheres to the NAS paradigm (Figure 13.1), but retroactive assessment relies less on this paradigm and more on surveys of contamination and ecological impact, models of fate and effects, and epidemiological data.

The ecological risk assessment process is organized slightly differently from the NAS paradigm, although the overall logic remains the same (Figure 13.3). The first step is problem formulation, a process involving both the risk assessor and risk manager. Next is the analysis step. The analysis step has two components similar to the NAS paradigm's exposure and dose-response assessments. It also has parts of the hazard identification component of the NAS paradigm. In the analysis and risk characterization stages, there may be reexamination of various actions or decisions as new information arises. In the last step, risk characterization, the in-

---

[6]Likelihood is interpreted even more loosely in ecological assessments than in human assessments (Norton et al., 1992). "Descriptions of risk may range from qualitative judgements to quantitative probabilities. While risk assessments may include quantitative risk estimates, the present state of the science often may not support such quantitation" (EPA, 1996). The term may not always imply the generation of a probability with an associated statistical statement of confidence.

[7]**Stressor** is defined in ecological risk assessments as "any chemical, physical, or biological entity that can induce adverse effects on *ecological components*, that is, individuals, populations, communities, or ecosystems" (Norton et al., 1992). Obviously, there is extreme latitude in this definition and, as discussed before, many effects at higher levels will more often be ambiguous than clearly adverse.

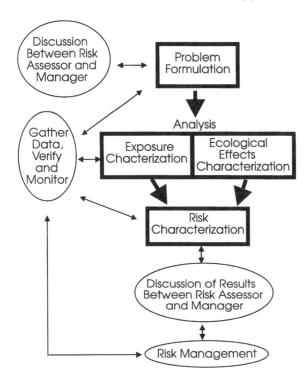

FIGURE 13.3. The general framework of an ecological risk assessment. The boxes and thick arrows reflect the risk assessment proper and are similar to the NAS paradigm (Figure 13.1). The narrow, two-way arrows reflect interactions between the risk assessor and risk manager during the process. They also reflect the continual accrual and evaluation of information during the entire process. (Modified from Figure 2 in EPA [1994].)

formation generated in the analysis step and the context developed in the problem formulation step come together. After the risk characterization step, the risk assessors and managers review the results relative to the original needs set out during problem formulation and any needs that may have emerged during the process.

Problem Formulation

Problem formulation includes the initial planning and scoping that establishes the framework around which the assessment is done (Norton et al., 1992). It includes the selection of assessment endpoints, a conceptual model, and a plan of analysis. The **assessment endpoint** is the valued ecological entity to be protected (e.g., bald eagles nesting by a contaminated lake) and the precise quality to be measured for this entity (e.g., adult survival and nesting success of bald eagles).[8] Including the measured quality in the definition ensures that the assessor specifies clearly an effect or improvement after remediation that can be quantified. For example, if one incorrectly decided that the assessment endpoint was the vague "functional in-

---

[8]In many publications (e.g., EPA, 1989c; Norton et al., 1992), the distinction is made that the **assessment endpoint or receptor** is the ecological entity or value to be protected (e.g., a population of the endangered bald eagle nesting by a contaminated lake) and the **measurement endpoint** is a measurable response to the stressor (e.g., number of fledglings produced per nest each year) that is related to the valued qualities of the assessment endpoint (e.g., reproductive success of the bald eagles). Some logical or quantitative model must link the two endpoints.

tegrity of the stream ecosystem," it would be very difficult to know what exactly should be measured to establish whether an adverse effect is present or to document any improvement after remediation. Sometimes, the specific quality of the assessment endpoint cannot be measured in the valued entity and it is measured in a surrogate instead. For example, risk assessment for an endangered species may require unacceptable destructive sampling to determine egg viability or body burden of a contaminant. Instead, these qualities may be measured for a closely related or ecologically similar species, and associated results applied to predicting risk to the endangered species.

The EPA (1996) suggests three qualities for a good assessment endpoint: (1) it should be ecologically relevant to the ecosystem being assessed, (2) it should be susceptible to the stressor, and (3) it is desirable, but not necessary, that it be valued by society. The last quality increases the value of the assessment to the risk manager. Suter (EPA, 1989c) further suggests that an ideal assessment endpoint should have an unambiguous operational definition and that it should be readily measurable or predictable from measurements. He gives the example of "balanced indigenous populations" as an ambiguous, and therefore compromised, assessment endpoint. Assessment endpoints are often, but not always, legally protected entities (e.g., survival and reproduction of an endangered or threatened species) or economic entities (e.g., successful reproduction of a salmon species). They may also be important ecological qualities, e.g., species diversity or some reflection of biodiversity. Suter (EPA, 1989c) tabulated (Table 13.1) some examples of assessment and measurement endpoints that may be applied to various levels of ecological organization.

The **conceptual model** links the assessment endpoint and the stressor of concern. It evaluates possible exposure pathways, effects, and ecological receptors. Conceptual models include hypotheses of risk and a diagram of the conceptual model. The **risk hypotheses** are clear statements of postulated or predicted effects

TABLE 13.1. Examples of Endpoints for Ecological Risk Assessments (Taken from EPA [1989c].)

Level of Ecological Organization	Assessment Endpoint	Measurement Endpoint
Population	Extinction	Occurrence
	Abundance	Abundance
	Yield or Production	Reproductive Performance
	Age or Size Class Structure	Age/Size Class Structure
	Mass Mortality	Frequency of Mass Mortality
Community	Market Sport Value	Number of Species
	Recreational Value	Species Diversity/Richness
	Change to a Less Useful or Appealing State	IBI
Ecosystem	Productive Capacity	Biomass
		Productivity
		Nutrient Dynamics

FIGURE 13.4. A conceptual model diagram for the pesticide spraying of nonbiting midges as described in Chapter 1 and depicted in Figure 1.2. Here, the assessment and measurement endpoints are separated for clarity. They could have been combined as the assessment endpoint. The valued ecological entity is the western grebe population.

of the stressor on the assessment endpoint. The **conceptual model diagram** (Figure 13.4) shows the pathways of exposure and illustrates areas of uncertainty or concern. It is a visual aid for communicating to the risk manager the model from which the risk hypotheses emerge.

In the final step of problem formulation, the risk hypotheses are examined carefully and a plan of analysis is produced. "Here, risk hypotheses are evaluated to determine how they will be assessed using available and new data" (EPA, 1996). An **analysis plan** defines the format and design of the assessment, explicitly states the required data, and describes the methods and design for data analysis. It describes what will or will not be analyzed. Measurement endpoints, those qualities that will be measured to assess effect to the assessment endpoint, are also stated in the analysis plan. A measurement endpoint may involve measurements derived directly from the valued ecological entity or from its surrogate.

Analysis

Again, the analysis step has two components (exposure characterization and ecological effects characterization) that are very similar to the exposure and dose-response assessments of the NAS paradigm. The exposure and ecological effects characterizations are done in tandem with considerable exchange of information occurring between the two components.

Some aspects of hazard assessment of the NAS paradigm are inserted as part of the analysis step of an ecological risk assessment. The gathering of relevant data and identification of chemicals of potential concern done in the hazard assessment step of the NAS paradigm are also done in the analysis step of the ecological risk assessment. However, some of the initial data gathering also occurred during the problem formulation step of the ecological risk assessment.

Exposure Characterization

Exposure characterization describes the characteristics of any contact between the contaminant and the ecological entity of concern. It summarizes this information in an exposure profile. Temporal and spatial patterns in contaminant distribution are defined in addition to the amount of contaminant present. The source of the contaminant, any potential co-stressors, transport pathways (e.g., outcropping of contaminated groundwater to the stream, or ingestion of prey after biomagnification), and type of contact are defined. In this characterization of exposure, quantitative methods described earlier for human exposure are applied. Concentration, duration, and frequency of exposure must be considered, including consideration of factors such as seasonal cycles and home ranges of species. For example, the contaminated region may be used only 10% of the time by a free ranging species. Or a contaminated food may be ingested only during one season. Estimates of exposure duration and frequency should take such factors into consideration. The final **exposure profile** "quantifies the magnitude, and spatial and temporal pattern of exposure for the scenarios developed during problem formulation" (Norton et al., 1992).

Ecological Effects Characterization

Ecological effects characterization "describes the effects that are elicited by a stressor, links these effects with the assessment endpoints, and evaluates how effects change with varying stressor levels" (EPA, 1996). It also specifies the strength of evidence associated with the effects characterization, and level of confidence in the causal linkage between the contaminant and the effect (Norton et al., 1992). Information generated with many of the methods described previously is brought together to develop a stressor-response profile for the valued ecological entity.[9] Which methods are applied would depend on the nature of the exposure and the ecological entity of concern. Unfortunately, most dose-response information are generated for effects to individuals, yet those of most use here are dose-response information for populations, communities, ecosystems, and landscapes. The unfortunate consequence of this imbalance between available information and need is compromised ecological risk assessments. (See quote at the beginning of Chapter 10.)

---

[9]Databases like IRIS are being developed for ecological effects. For example, Oak Ridge National Laboratory has databases on its web site (http://www.hsrd.ornl.gov/ecorisk/benchome.html) for aquatic biota, wildlife, terrestrial plants, sediments, and soil invertebrates/microbial processes. Details are provided in extensive documentation obtained from this same web site.

Risk Characterization

Risk characterization draws together the information from previous steps to produce a statement of the likelihood of an adverse effect to the assessment endpoint (EPA, 1996). Risk may be expressed in several ways including a simple qualitative judgment or hazard quotient. It could involve a richer interpretation including description of the influences of concentration and temporal variations on estimates of effect. It could employ complex models that also generate some estimate of confidence in the risk predictions.

The final statement of risk must include details about the adequacy of the data going into the judgment, uncertainty involved in the conceptual model or calculations, and weight of evidence for each causal relationship. If a surrogate was used in place of the assessment endpoint, confidence in the associated extrapolation should be addressed. For example, the confidence in the extrapolation would be high if there was very little variation among raptors in the dose-effect relationship and a surrogate raptor with a very similar niche was used for bald eagles. Some statement about the significance of the adverse effect should also be made so that the risk manager may better judge the seriousness of risk consequences.

Summary

The expedient compromises and conservative inaccuracies already discussed for human risk assessments are also present in ecological risk assessments. More are added because the manifestation of effects could involve several species or levels of ecological organization. There is a tendency to address the level at which most information exists, i.e., the individual level. Understandably, implications are most often to the level of population viability.

CONCLUSION

Risk assessment technology is evolving quickly. Indeed, terminology and emphasis have changed from 1989 to 1996 as evidenced in the EPA documents used to frame this chapter. It is anticipated that this treatment will be passé within just a few years. Regardless, the basic risk assessment paradigm is intelligent and insightful; it will probably remain intact for the near future.

What is painfully needed at this time is a sound data set for effects at all levels of organization. Also, basic ecotoxicological testing approaches produce data that are inadequate to the task, e.g., NOAEL approaches for estimating toxic thresholds. As an important example, temporal dynamics are given minimal consideration in most ecotoxicological tests. As we discussed in Chapter 9, only tradition in the field inhibits the construction of more complete models incorporating time effectively. Neglected methods for expressing many toxicant-related effects in terms of risk probabilities are also discussed in Chapter 9. Finally, unjustified conceptual shortcuts are made in ecological risk assessments such as assuming that the most sensitive life stage is the critical life stage for a population or that effects are additive. As we already discussed in earlier chapters, these are dubious assumptions. Hopefully, these and other shortcomings in risk assessments will be resolved soon.

Suggested Readings

EPA. *Risk Assessment Guidance for Superfund. Volume I. Human Health Evaluation Manual (Part A). Interim Final. EPA 540/1-89/002 December 1989.* NTIS, Springfield, VA, 1989.

EPA. *Ecological Assessment of Hazardous Waste Sites: A Field and Laboratory Reference Document. EPA 600/3-89/013 March 1989.* NTIS, Springfield, VA, 1989.

EPA. *Summary Report on Issues in Ecological Risk Assessment, EPA/625/3-91/018 February 1991.* NTIS, Springfield, VA, 1991, p. 46.

EPA. *Proposed Guidelines for Ecological Risk Assessment, EPA/630/R-95/002B August 1996,* Risk Assessment Forum, U.S. Environmental Protection Agency, Washington, DC, p. 247.

Neely, W.B. *Introduction to Chemical Exposure and Risk Assessment.* CRC Press, Inc., Boca Raton, FL, 1994, pp. 190.

Page, N.P. Human Health Risk Assessment, in *Basic Environmental Toxicology*, Cockerham, L.G. and Shane, B.S., CRC Press, Inc., Boca Raton, FL, 1994.

Suter, G.W. II. *Ecological Risk Assessment.* Lewis Publishers, Chelsea, MI, 1993, p. 538.

Risks from Exposure to Radiation

Thomas G. Hinton
Savannah River Ecology Laboratory

> *It is still an unending source of surprise for me to see how a few scribbles on a blackboard or on a sheet of paper could change the course of human affairs.*
> STANISLAW ULAM

INTRODUCTION

This chapter introduces the reader to the fundamentals of radioactive contamination and its associated risks. Emphasis is placed on topics that illustrate differences between radioactive and nonradioactive contaminants. Three key pathways by which organisms are exposed to radioactive contamination and the concepts used to calculate dose are highlighted. The effects of exposure to radiation and the data from which risk factors have been derived are presented. Both human and ecological risks are considered. The chapter ends by addressing the widely held view among radiation protection organizations that if man is adequately protected then so are the aquatic and terrestrial biota.

FUNDAMENTALS OF RADIOACTIVITY

During much of the fifteenth to seventeenth centuries a respectable branch of chemistry was engaged in trying to "transmutate" ordinary base metals into gold; to change the ordinary into the extraordinary. Fortunes were spent, careers wasted, and heads rolled with the finality of the guillotine, but no one could get one element to change into another. Human attempts at alchemy failed.

Radiation, however, is Nature's alchemist. The process of radioactive decay transforms one element into another. Indeed, there are entire chains of transformations that occur naturally within our ecosystems. For example, uranium-238 changes into thorium-234, thorium-234 into protactinium, and eventually—approximately 10^{10} years later and having undergone 14 different transformations—the original U atom changes from radioactive bismuth into stable lead. At each step the resulting product loses all the characteristics of the parent element and acquires the characteristics of the newly formed daughter element. Characteristics such as color, melting point, hardness, even physical state, are changed; for example, radium, a solid, transforms into radon, a gas!

These transformations occur because radionuclides are excited atoms. They have an excess amount of energy and, therefore, are unstable. Radionuclides gain stability by releasing the excess energy in the form of electromagnetic photons (x or gamma rays) and/or particles (alpha or beta). In doing so, changes in the atomic structure occur, resulting in a gain or loss of a proton. The number of protons largely determines the characteristics of an atom. All elements above bismuth (z = 83) in the periodic table are naturally unstable (radioactive), and a few of the lighter elements have one or more radioactive **isotopes.**

If only the seventeenth century alchemists could have duplicated what was occurring naturally all around them. It was not until 1942, when Enrico Fermi and his colleagues engineered the first sustained nuclear chain reaction, that man's attempts at alchemy proved successful. Their goal, however, was not to produce gold; they, instead, sought the tremendous energy released from the fission of U atoms.[1] Their "scribbles on a blackboard" led to the development of nuclear energy, and nuclear weapons, thus profoundly affecting the twentieth century (Rhodes, 1986).

Types of Radiation

Three principal types of radiation are emitted from radionuclides during radioactive decay: electromagnetic photons, beta particles (β), and alpha particles (α). Electromagnetic photons consist of gamma rays (γ) and x-rays that differ in their source. **Gamma rays** are emitted from the nucleus and **x-rays** are emitted from the

---

[1]**Nuclear fission** is the splitting of atomic nuclei with neutrons, resulting in the release of energy, other neutrons and radioactive fragments called **fission products**. Naturally occurring ^{235}U and man-made ^{233}U and ^{239}Pu have high probabilities of undergoing fission when bombarded by neutrons. The fissile nucleus is split into two or more fragments. The most probable mass partitioning results in fragments having mass numbers roughly in the ranges of 90 to 106 and 134 to 144. Fission products contain excess energy and are thus radioactive, most decay by the emission of beta particles. Approximately 200 MeV of energy is released for each fission, as well as 2 or 3 neutrons. The neutrons can cause additional fission and thus a chain reaction can take place, governed by the density and geometry of fissile nuclei and the presence of material which slow or capture the neutrons. A nuclear reactor controls the fission process at a specific rate, thereby producing heat to boil water, turn turbines and generate electricity. In a fission weapon the geometry of the constituents are such that the chain reaction proceeds in an explosive manner. A nuclear explosion is not physically possible in a reactor because of fuel density, geometry and other considerations (Lapp and Andrews, 1972; Whicker and Schultz, 1982; Rhodes, 1986)

shells of electrons that surround the nucleus. Once emitted, differences are not discernible. The essentially massless γ rays are photons similar to those of visible light and radio waves, all with the same velocity, but γ rays have shorter wavelengths and much greater energies.[2] All photons travel at the velocity of light and have energies inversely proportional to their wavelengths. The most energetic x-rays have wavelengths less than 10^{-12} cm, while the wavelengths of weak x-rays approach 10^{-6} cm. The energy of a photon affects its ability to penetrate matter. Energetic photons have sufficient energy to pass entirely through a human, and require a meter thickness of dense material, such as concrete, to be fully absorbed (i.e., to have all of their energy dissipated into the surrounding media).

Gamma rays are emitted at monoenergetic energies. Thus the γ ray emitted by cesium-137 (^{137}Cs) always has an energy of 662 keV, cobalt-60 (^{60}Co) has two gamma rays that always have energies of 1170 and 1330 keV, and naturally occurring potassium-40 (^{40}K) emits a γ photon at 1440 keV. This unfaltering reliability of the emitted photon energies facilitates identification of unknown gamma-emitting radionuclides.

Beta (β) particles are electrons or positrons ejected from the atom during the radioactive decay process. They have the same mass as an electron (5.49×10^{-4} amu),[3] and may be either positively (**positron**) or negatively (electron) charged. Beta particles have a continuous spectrum of energies with a maximum characteristic for that radioisotope. The average β energy of a radioisotope is equal to approximately 1/3 the characteristic maximum energy. Beta particles are less penetrating than γ photons. A sheet of aluminum a few millimeters thick can stop β radiation, as can 1 to 2 cm of flesh. Tritium, one of the commonly emitted radioisotopes from nuclear reactors, emits low energy β particles.

Alpha (α) particles are a chunk of the nucleus ejected from a radioactive atom to reduce excess energy and gain stability. They are relatively huge in mass (7,345 times larger than a β particle), consist of two neutrons and two protons, and carry a +2 charge. Their emissions are monoenergetic. The large mass and double charge of α particles cause them to react strongly with matter. Their energy is quickly deposited within a very short distance of the material they interact with, greatly reducing their penetrating abilities. High energy α particles are stopped by a sheet of paper, or outer skin surfaces. Thus, their hazard is greatest when they are inhaled or ingested. The α emissions from plutonium, for example, are of concern when Pu is ingested and translocated to bone surfaces, or when Pu is inhaled and the sensitive lining of the lung exposed.

More detailed information on the types of nuclear decay and the physics of radiation can be found in Wang et al. (1975), Johns and Cunningham (1980), and Knoll (1989).

---

[2]Energy can be expressed in terms of electron volts (eV). Gamma rays have energies of 10^2 to 10^7 eV compared to 20 to 80 eV for visible light and 10^{-4} to 10^{-8} eV for radio waves. An electron volt is the energy acquired by an electron when it passes through a potential difference of 1 volt. 1 eV = 1.602 X10^{-12} erg. Kiloelectron volts (keV; 10^3 eV) and megaelectron volts (MeV; 10^6 eV) are commonly used extensions of the base unit.

[3]Masses of atomic constituents are measured in atomic mass units (amu), defined as 1/12 the mass of an atom of ^{12}C. There are 1.6605×10^{-24} g per amu.

Concentrations, Decay Constants, and Half-Life

Concentrations of nonradioactive contaminants are generally expressed on a mass basis (e.g., mg lead kg^{-1} tissue). In contrast, concentrations of radioactive contaminants are expressed on an activity basis (Bq ^{137}Cs kg^{-1} tissue). The base unit of activity is the Becquerel (Bq), equal to one disintegration per second.[4] A **disintegration** is the event in which a radioactive element releases photons or particles to gain stability, i.e., radioactive decay. Thus, **activity** is a measure of the rate at which a given quantity of radioactive material is emitting radiation

Each radionuclide has a characteristic half-life ($T_{1/2}$) and decay constant (λ). **Half-life** is the amount of time required for one-half of the number of radioactive atoms to decay. At the end of one half-life 50% of the original activity will remain, after a second half-life 25% of the original activity will still be present, and so on. Half-life is an indication of a radioactive element's instability. Those with very short half-lives are particularly unstable and decay quickly.

The decay constant (λ) indicates the fraction of radioactive atoms (N) that will decay per unit time (t)

$$\frac{dN}{dt} = -\lambda N \tag{14.1}$$

and is related to half-life by:

$$T_{1/2} = \frac{0.693}{\lambda} \tag{14.2}$$

The activity of a radioactive sample can be calculated at any time (A_t) if the original activity (A_0), decay constant, and elapsed time (t) are known, using the formula:

$$A_t = A_0 \, e^{-\lambda t} \tag{14.3}$$

Radionuclide Detection

Samples containing radionuclides have two unique traits that facilitate determining their contaminant levels. If the sample contains γ-emitting radionuclides, some of the γ rays emitted from the contaminant within the tissue of the living plant or animal actually pass through the tissues and can be detected externally to the organism. Very sensitive instrumentation is used to detect the emitted γ rays,[5] without having to destructively process the sample or take an aliquot from it. This provides a powerful way of resampling a living organism, and thus determining

---

[4]The traditional unit of activity (Curie) has been replaced in the International System of units with the Bq; 1 Ci = 3.7 x10^{10} Bq.

[5]See Knoll (1989) and Wang et al. (1975) for information on radioanalytical instruments.

the accumulation or elimination of the radioactive contaminant with time. Such techniques are not applicable to α and β emitters because most of the radiation is absorbed within the organism and not externally emitted. Detection of α and β-emitting contaminants generally requires sample preparation that precludes repetitive sampling on living organisms.

The second interesting aspect of radioactively contaminated samples is that the level at which instruments detect the radiation is improved merely by analyzing the sample for a longer period of time. A longer assay allows activity within a sample to be better distinguished from background activity and electronic noise of the measurement instruments. The longer a radioactive sample is assayed, the more radiation is emitted and thus measured. Nonradioactive samples either have an adequate level of contaminant to be detected or they do not. Assaying a sample containing stable potassium for a longer period of time will do nothing to improve the probability of detecting it. In contrast, if radioactive ^{40}K is not distinguishable in a sample after a 10 minute assay, analyzing the same sample for 10 hours might increase the signal to noise ratio sufficiently to detect it.

These traits, coupled to the extreme sensitivity of today's radioanalytical instruments, make it possible to detect minute quantities of radiation. For example, radioactive contaminant levels on PAR Pond, a 10 km^2 lake on the Department of Energy's Savannah River Site, are such that detection of ^{137}Cs is easily accomplished within all components of the ecosystem. The lake contains a total ^{137}Cs inventory of 1.6×10^{12} Bq (44 Curies). When converted to mass, the entire 10^{12} Bq amount to a total of only 0.5 g of ^{137}Cs, which is distributed throughout the entire lake ecosystem! Samples from the lake ecosystem containing as little as 0.3 Bq g^{-1} of ^{137}Cs can easily be assayed in one hour, an activity equivalent to a mass of 1.0×10^{-13} g of ^{137}Cs within the sample.

Effects

Damage from exposure to radioactive contaminants is initiated by **ionization**,[6] caused by the energetic rays and particles released during radioactive decay. Ionization occurs in biological material exposed to radiation, resulting in some probability of molecular or genetic damage, either directly or through a multistep process. Part of the process often involves the formation of free radicals.[7] Free radicals are a principal cause of damage from exposure to radiation because they can

---

[6]If the radiation has sufficient energy to eject one or more orbital electrons from the atom or molecule that it interacts with, then the process is referred to as ionization. **Ionizing radiation** is characterized by a large release of energy (approximately 33 eV per event), which is more than enough to break strong chemical bonds; for example, only 4.9 eV are required to break a C=C bond (Hall, 1978).

[7]Free radicals are atoms or molecules with an unpaired or odd orbital electron. In addition to orbiting around the nucleus, electrons also spin, either clockwise or counterclockwise, about their own axis. In atoms or molecules with an even number of electrons, spins are matched; for every electron spinning clockwise, another is spinning counterclockwise. This leads to a high degree of chemical stability. Molecules or atoms with an odd number of electrons, free radicals, have an electron that is left with an unmatched spin, creating instability.

easily break chemical bonds within DNA molecules. The OH• free radical, formed when cellular water is ionized, is among the most common.

Ionization results in biological damage. For humans, the biological consequence upon which most risk calculations are based, following exposure to radiation, is cancer. Four steps are thought to occur in cancerous tumor formation (Hall, 1978). The first step, *initiation*, involves damage to the DNA, and most likely damage to both strands of the DNA double helix. Although a portion of the double strand damage is repaired, completely error-free repair is not expected. *Promotional events* in the intra- and extra-cellular environment, brought about by dietary constituents, hormones, or other environmental agents, cause the initiated cells to abnormally proliferate. Further gene mutations within the rapidly dividing cell population cause a *conversion* to full malignancy. The *progression* stage of the disease allows invasion of adjacent normal tissues, eventually resulting in fatality if not successfully treated. It is important to realize that the time between the breakage of chemical bonds and the expression of a biological effect, the latency period, may take from days to dozens of years depending on the circumstances involved.

Cancer and genetic disorders are classified as **stochastic health effects**; meaning that the initiation of effects is probabilistic, and that the risk of incurring cancer or genetic effects is proportional, without threshold, to the dose in the relevant tissue. The severity of a stochastic health effect is independent of the dose. In contrast, **nonstochastic health effects** (acute radiation syndrome, opacification of the eye lens, erythema of the skin, and temporary impairment of fertility) are dependent on the magnitude of the dose in excess of a threshold (ICRP, 1977).

DOSE

Risk can be defined as the probability of a deleterious effect from a specific exposure to a contaminant. The derivation of risk factors, or slope factors if the Environmental Protection Agency terminology is used, requires detailed knowledge about dose-response relationships. For nonradioactive contaminants, dose is often expressed as a concentration (e.g., mg of lead kg^{-1} tissue). For radioactive contaminants, however, dose goes beyond concentration and refers to the energy deposited within biological tissues as a result of radioactive decay (i.e., 1 Joule of energy kg^{-1} tissue = 1 Gray).[8]

If it is assumed that all of the released energy is deposited within the tissue, **absorbed dose rate** (Gy d^{-1}) can be calculated from a given radionuclide concentration using the general equation:

$$Dose\ rate(Gy\ d^{-1}) = (C)(SEE)(1.602 \times 10^{-13}\ Joules\ MeV^{-1})\left(\frac{dis}{s\ Bq}\right)\left(\frac{86,400}{d}\right) \quad (14.4)$$

---

[8]The traditional unit for absorbed dose was the rad, 100 rad = 1 Gy.

where C = concentration (Bq kg^{-1}), SEE = specific effective energy (MeV dis^{-1}), and dis = disintegration. The radionuclide specific effective energy is found in reference tables (ICRP, 1983).

Dose, however, is not entirely adequate to relate the amount of radiation absorbed to its effects. The effectiveness of absorbed energy at causing biological damage depends on the physics of radioactive decay, characteristics of the exposed tissues (some organs and tissues are more sensitive to radiation than others), and characteristics of the radiation. For example, a 1 MeV α particle creates approximately 20 times more damage over a given distance than a 1 MeV γ ray, because the α particle has a higher **linear energy transfer** (LET): more of its energy is transferred into surrounding tissues per micron distance traveled. The relative biological effectiveness of various radiation types has been incorporated into quality factors that, if multiplied by absorbed dose, estimates *dose equivalent*. Dose equivalent was expressed traditionally in rem, but now in the International System of Units (SI) of sievert, where 1 Sv = 100 rem (ICRP, 1977; 1991). Dose equivalent normalizes the different types of radiation to the same propensity for biological damage. Dose equivalent has no analog for chemical carcinogens. There is no common basis, similar to the quality factors used in radiation dosimetry, for relating biological effectiveness of one nonradioactive chemical to another.

A further improvement in radiation dosimetry was made in 1977 (ICRP) when the concept of **effective dose equivalent** was introduced. Recognizing that biological effects from a uniform irradiation of the whole body are different than effects from a similar dose concentrated in specific tissues (as happens if some radionuclides are ingested—e.g., ^{131}I goes directly to the thyroid gland), effective dose equivalent weights the dose to different organs or tissues. Thus, the fractional contribution of organs and tissues to the total risk of stochastic health effects is normalized to when the entire body is uniformly irradiated (ICRP, 1991). Exposures with equal effective dose equivalents are assumed to result in equal risks, regardless of the distribution of the deposited energy among different body tissues.

The last consideration made when calculating dose involves the time period of exposure from internally deposited radioactivity. Dose rates are integrated over a 50 year period for adults and 70 years for children because a one-time intake of radionuclides commits an individual to a future dose, due to the time required for the contaminant to be removed from the body by biological elimination and radioactive decay. The final product of all quality factors, organ weighting factors, and integration of dose over the individual's working life time results in the **committed effective dose (CED)**. The fundamental unit to which risk factors are multiplied in order to estimate the probability of an individual human acquiring a fatal cancer from exposure to radiation is the CED (NCRP, 1993).

The CED is a rigorous attempt to specify not only the amount of energy absorbed by tissues, but to account for numerous factors that influence the biological effectiveness of that energy at causing damage. From this perspective radiation dosimetry is better developed than current methods used to determine dose from nonradioactive contaminants. Table 14.1 compares various parameters used for radioactive and nonradioactive contaminants.

The unit of CED is intended for use in setting human exposure limits or in

TABLE 14.1.   A Comparison of Parameters Used to Describe Various Aspects of Radioactive and Nonradioactive Contamination

Parameter	Contamination Type	Measures	Typical Unit
Activity	Radioactive	Rate at which radiation is emitted	Disintegration per second (Becquerel)
Concentration	Nonradioactive	Mass of contaminant per mass of tissue	mg lead per kg tissue
Concentration	Radioactive	Activity of contaminant per mass of tissue	Bq ^{137}Cs per kg tissue
Dose	Nonradioactive	Concentration	mg lead per kg tissue
Dose	Radioactive	Energy absorbed in biological tissue due to radiation emitted from contaminant	Energy absorbed per kg tissue (1 joule per kg = 1 Gray; traditional unit was rad)
Dose Equivalent[a]	Radioactive	Normalizes the different types of radiation to the same propensity for biological damage	Absorbed dose x radiation specific quality factors results in Sieverts (Sv; traditionally expressed as rem)
Effective Dose Equivalent[a]	Radioactive	Normalizes for different biological effects when a dose is concentrated in specific organs compared to the same dose distributed uniformly over the whole body	Sv
Committed Effective Dose[a]	Radioactive	Integrates effective dose equivalent over a 50 y period to account for physical and biological decay of ingested radioactivity	Sv

[a]No analog for chemical carcinogen

assessing risk in general terms (e.g., for hypothetical exposure situations). The basic framework of radiological protection is designed to provide an appropriate standard of protection against ionizing radiation without unduly limiting the beneficial uses of radiation (Clark, 1995). The 1990 dose limit recommendations (ICRP, 1991) keep doses below the relevant threshold for deterministic effects and demand that all reasonable steps are taken to reduce the incidence of stochastic effects to acceptable levels (Clark, 1995).

ENVIRONMENTAL TRANSPORT

In order to estimate risks from exposure to radioactive contaminants, knowledge of contaminant transport and fate is required. All processes governing the transport of stable contaminants described in Chapters 3 to 5 are applicable to radioactive contaminants as well. Any process by which an organism can acquire a nutrient, or stable contaminant, is also a plausible route for radioactive contaminant uptake. Thus, processes such as ingestion, inhalation, root uptake, and surface absorption are often necessary considerations. It is important to recognize that the movement of an element through the environment is not affected by whether or not it is radioactive. Radioactive cesium (^{137}Cs) moves through the soil and is taken up by plants in the same manner and rates as stable cesium (^{133}Cs). There are, however, characteristics of radioactive contaminants that can cause their transport kinetics to differ from stable analogs—particularly important is the contaminant's half-life.

If a radioactive isotope has a short half-life, it may decay into another element before it has time to participate in all of the environmental pathways that its stable isotope does. For example, ^{131}I has a half-life of eight days. With such a short life it is not possible for aerially deposited ^{131}I, released from a nuclear reactor, for example, to migrate through the soil and into the rooting zone of plants. Such a transport process takes time and the ^{131}I will have decayed long before it reaches the plant roots. Ubiquitous stable iodine, however, is in contact with plant roots and can be taken up by the plant. Thus, in this case, the behavior of the radioactive element does not entirely duplicate the stable one.[9]

The other situation where a radioactive contaminant may not behave like its stable counterpart occurs when a recent introduction of a radionuclide is not in equilibrium with the environment and, therefore, behaves differently than its equilibrated stable isotope. Cesium uptake by plants represents a good example. Cesium's bioavailability is partially dependent on the soil clay content. Cesium has a strong tendency to bind to three different exchange sites on clay: surface sites, frayed edged sites, and nonexchangeable interlayer sites (Cremers, 1988). Each site has different kinetics associated with the attachment of cesium atoms. Cesium adsorbs to the surface sites most readily, but also is least tightly bound there, and can be dislodged by other ions with a greater affinity for those sites (such as K^+ or NH_4^+). If a cesium atom gets all the way into the interlayer sites then the clay tends to collapse, tightly trapping the cesium atom within. Cesium is not easily displaced from these inner sites by other ions, and the exchange rates are very

---

[9]Iodine-131 is a particularly important dose-contributing contaminant following a nuclear accident such as Chernobyl. It was not necessary for the iodine to migrate to the rooting zone, however, for plants to become contaminated. I-131 was deposited on the surface of the grasses and taken up by the leaves. Cattle forged on the contaminated grass, ingested the contaminant and a portion was transferred to their milk. Iodine was deposited in the thyroid of humans who drank the contaminated milk, and increased their risk of thyroid cancer. The Chernobyl accident occurred on 26 April 1986. Such cancers have a latency period of about 10 years, which is why we are just now observing an increase rate of thyroid cancers in children exposed to high levels of ^{131}I following the Chernobyl accident.

slow. The exchange dynamics of cesium on the frayed edged sites are intermediate in tenacity and exchange rates.

The interaction of cesium with the various exchange sites in the soil are time dependent. Different dynamics of these exchange sites cause newly deposited ^{137}Cs to behave differently than stable cesium that has been in the soil sufficiently long to attain equilibrium. ^{137}Cs deposited from the Chernobyl accident was taken up by plants at a faster rate than stable cesium in the same soils (Salbu et al., 1994; Bunzl et al., 1995). Stable cesium (^{133}Cs) is a natural, ubiquitous component of soils, and in essence, has been in the soil since its formation. It is in equilibrium with all of the exchange sites within the clay matrix. In contrast, newly deposited ^{137}Cs has a higher probability of dominating the surface and frayed edged sites, locations where other ions can more readily displace the ^{137}Cs into the soil solution, and thus increase its availability for root uptake. Once ^{137}Cs has remained in the soil long enough, it will reach a similar equilibrium to that of stable cesium and their kinetics will be similar. This concept of radionuclides changing bioavailability with time has been termed **aging**, and has been observed numerous times in field and laboratory studies (Schimmack et al., 1989; Sanzharova et al., 1994; Velasko et al., 1993).

Newly deposited stable cesium would also undergo a similar aging process, but it would be difficult to distinguish its dynamics from the stable cesium already in the soil. The ^{137}Cs isotope serves as a tag, or **tracer**, that provides information about its behavior as a contaminant. By using a radioactive tracer we also learn about clay particles and the behavior of stable cesium within the environment, illustrating that radioactive tracers can be powerful research tools to increase our knowledge of chemical, physical, and biological processes.

Excellent information on the environmental transport and fate of radionuclides can be found in Eisenbud (1973), Till and Meyer (1983), and Whicker and Schultz (1982). Table 14.2 provides general environmental transport properties of some common radionuclides.

Models Using Rate Constants

Predicting the impact of radioactive contamination to humans and the environment generally requires some form of mathematical model to simulate the contaminant's transport through the environment. Models have a wide range of complexities depending on their intended use. Radioactive contaminant transport has been most commonly predicted with computer simulation models that use first-order differential equations to describe the rate at which contaminants move among environmental components (Chapters 3 to 5). Details of such models are beyond the scope of this introductory chapter, but it is important to understand that the *rate* at which a contaminant moves among all the components is required. It is not too difficult to measure the current location and concentration of an environmental contaminant. One can sample the water, soil, various species of plants and animals, and determine the concentration (Bq kg^{-1}) currently in each component. The difficult task is being able to predict what concentrations will be in those same components in 10 weeks or 10 years. The challenge is to identify the ecological processes by which contaminants move from one environmental component to another, and then to determine the corresponding **rate constants**. Simulation models that utilize rate constants to predict the movement and future concentrations of radioactive contaminants in the environment are exemplified by PATHWAY (Whicker and Kirchner, 1987) and ECOSYS-87 (Müller and Pröhl, 1993).

Screening Level Models

Many decisions regarding radionuclide releases and their potential impacts are based on results from much less sophisticated calculations referred to as "screening level models." The complex intricacies of radionuclide movement are replaced by a handful of dominant processes, and then further compensated for by inserting very conservative numbers for those processes. For example, among the parameters governing the uptake of a radionuclide by a plant are (1) characteristics of the radionuclide (chemical form, physical form, particle size, solubility, valence state), (2) local climatic conditions (humidity, temperature, solar irradiation, barometric pressure), (3) soil conditions (pH, moisture content, texture, clay type and abundance, percent organic matter, abundance of competing ions), and (4) plant characteristics (species, growth stage, area of leaves, area of roots, depth of roots, type of leaf surface). Rather than describe each of these individual parameters by separate mathematical formulas, a screening model might aggregate them all into a single parameter, a soil-to-plant transfer factor, and then choose a number for the transfer factor that would conservatively maximize the estimate of contaminant uptake by the plant.

Similar aggregation and insertion of conservatism is done for other processes. Calculations are conducted and the model prediction is then compared to some predetermined action level, such as a maximum allowable concentration or dose limit. If the simplified, conservatively-obtained model prediction is less than the action level then compliance with the legal limit has been demonstrated (NCRP, 1989). The assumption is that the overly-conservative aggregation of parameters has caused the model prediction of contaminant concentration to be much larger than reality. Because the exaggerated model prediction is less than the regulatory limit, the risk manager is confident that the true concentration is also less than the limit.

When applying mathematical models to assess radionuclide contamination, it is often recommended that the simplest model that adequately addresses the problem be used first (NCRP, 1984). Recent trends in risk analyses, however, are moving away from screening level models toward models that produce realistic predictions with associated estimates of uncertainties.

Models Using Equilibrium Conditions and Dose-Conversion Factors

At a level of complexity slightly greater than screening models, the next three subsections introduce the reader to a general approach of deriving the committed ef-

TABLE 14.2 General Ecological Properties of Selected Radionuclides (Adapted from Whicker and Schultz [1982].)

Radionuclide ($T_{1/2}$; emission)	Sources[a]	Nutrient Analogs[b]	Important Exposure Modes	Degree of Food Chain Transport[c]	Successive Trophic Level Concentration[d]	Critical Organs (Vertebrates)	Gastro-Intestinal Assimilation	Biological Retention
^{3}H (12 y; β)	Cosmic Fission Activation	H	Ingestion, Uptake, absorption, inhalation	High	Approaches Unity	Total body	Complete	Low (days)
^{131}I (8 d; β, γ)	Fission	I	Ingestion, absorption, inhalation	High	Up to 10 times	Thyroid	High	Moderate (weeks-months)
^{40}K (1.3×10^9 y; β, γ)	Primordial	K	Ingestion, absorption, uptake, external γ	High	Approaches unity	Total body	High	Moderate (weeks)
^{137}Cs (30 y; β, γ)	Fission	K	Ingestion, absorption, external γ	High	Approaches 3.0	Total body	High	Moderate (weeks-months)
^{90}Sr (28 y; β)	Fission	Ca	Ingestion, absorption, uptake	High	< 1.0	Bone	Moderate	High (years)
^{222}Rn (3.8 d; α, γ)	^{238}U decay series	None	Inhalation of daughters	Negligible	Negligible	Lung (from daughters)	Negligible	Negligible

Radionuclide	Origin		Routes of exposure	Mobility		Target organs		Persistence
^{60}Co (5.2 y; β,γ)	Activation	Co	Ingestion, adsorption, inhalation, external γ	Moderate-high	< 1.0-10^2	GI, total body, lung	Moderate	Low (days)
^{144}Ce (285 d; β,γ)	Fission	None	Ingestion, inhalation, adsorption, external γ	Low-moderate	< 0.1	GI, bone, lung, liver	Very low-negligible	Moderate (1-5 years)
^{232}Th (1.4 x 10^{10} y; α,γ)	Primordial	None	Ingestion, inhalation,	Very low	< 10^{-2}	Bone, lung	Very low-negligible	High (years)
^{238}U (4.5 x 10^9 y; α,γ)	Primordial	S, Se?	Ingestion, inhalation, uptake, external γ	Low-moderate	< 1.0	GI, kidney, lung	Very low	Moderate (months)
^{239}Pu (2.4 x 10^4 y; α,γ)	Activation	None	Ingestion, inhalation, adsorption	Very low	< 10^{-2}	Bone, lung	Very low-negligible	High (years)

[a] Activation—produced in a nuclear reactor or during a nuclear explosion due to interactions with neutrons.
Cosmic—produced in the atmosphere by cosmic ray interactions with matter.
Fission—produced as a by-product from the fission of U or transuranics in nuclear reactors or nuclear explosions.
Primordial—radionuclides that appeared at the earth's formation and are still present today due to their extremely long half-lives.
^{238}U decay series - a daughter product in the primordial ^{238}U decay scheme.
[b] The presence of a nutrient analog is important as it often increases the ecological mobility of the radionuclide.
[c] This is a relative ranking of the radionuclide's bioavailability.
[d] Some contaminants tend to bioaccumulate up successive tropic levels - radionuclides generally do not, although some exceptions are noted.

fective dose from radioactive contaminants by assuming equilibrium conditions and using dose conversion factors. The three subsections cover the principal pathways by which humans come in contact with radioactive contaminants. Recall, that before we can assign a risk, it is first necessary to determine the dose.

Inhalation Pathway

If the source term of the contaminant is atmospheric, then the committed effective dose from inhaling airborne material can be calculated from:

$$CED(Sv) = (C)(IR)(ET)(EF)(EED)(IDF) \qquad (14.5)$$

where C = concentration in air (Bq m^{-3}), IR = inhalation rate (m^3 h^{-1}), ET = exposure time (h d^{-1}), EF = exposure frequency (d y^{-1}), EED = effective exposure duration (y), and IDF = inhalation dose factor (Sv Bq^{-1}). The effective exposure duration accounts for radioactive decay by:

$$\int_0^t e^{-\lambda t} dt$$

where t is the residence time, and λ is the physical or radioactive decay constant. Some parameters in Equation 14.5 depend on the particular scenario being modeled, such as the inhalation rate, exposure time and frequency. In any risk calculation, assumptions about the contaminant and exposure scenario must be made. For example, calculations are often made for a future hypothetical resident of a contaminated site. Standardized data for numerous assumptions related to dietary consumption patterns, breathing rates, skin surface area available for contact, and exposure frequency and duration for various levels of occupancy are located in guidance documents for Superfund sites (EPA, 1991c). Other standardized parameters can be found in NCRP (1984; 1989), Pao et al. (1982), and Yang and Nelson (1984). A brief listing of some useful parameters is provided in Table 14.3.

The inhalation dose conversion factor (Equation 14.5) converts the inhaled activity to a collective dose equivalent per unit intake. Dose conversion factors for each radionuclide are found in Eckerman et al. (1988) and Eckerman and Ryman (1993) for inhalation, ingestion, submersion in a contaminated plume, submersion in contaminated water, and external irradiation from contaminated soils. The authors have taken the primary guides[10] for radiation protection, issued by the International Commission on Radiological Protection (ICRP, 1977) and EPA (1987), and derived dose conversion factors and **annual limits on intake** (ALI).

---

[10]Primary guides are the current recommendations for radiological protection (e.g., the committed effective dose for a radiation worker in a given year should not exceed 50 mSv).

TABLE 14.3. Usage Factors for Calculating Dose to Humans

Parameter	EPA (1989)	NCRP (1991)
Drinking water	2 L d^{-1} (adult, 90th percentile)	800 L y^{-1}
	1.4 L d^{-1} (adult, average)	
Ingestion of surface		
water while swimming	50 mL h^{-1}	
Swimming frequency	2.6 h d^{-1} (national avg.)	
	7 d y^{-1} (national avg.)	300 h y^{-1}
Soil ingestion	200 mg d^{-1} (1 –6 years of age)	
	100 mg d^{-1} (>6 years of age)	
Inhalation	30 m^3 d^{-1} (adult, upper limit)	8000 h y^{-1}
	20 m^3 d^{-1} (adult, average)	
Ingestion rate		
fish	0.284 kg meal^{-1} (95 percentile)[a]	Freshwater:
	0.113 kg meal^{-1} (50th percentile)[a]	10 kg y^{-1}
	6.5 γ d^{-1} (averaged over a year)	Marine:
	48 d y^{-1} (average per capita)	20 kg y^{-1}
milk[b]		300 L y^{-1}
meat[b]		100 kg y^{-1}
vegetables[b]		200 kg y^{-1}
Home grown (fraction)[c]	Fruit: 0.2 avg; 0.3 worse case	
	Vegetables: 0.25 avg; 0.4 worst case	
	Beef: 0.44 avg; 0.75 worst case	
	Dairy prod.: 0.40 avg; 0.75 worst case	
Fruit, vegetables and grain		200 kg y^{-1}
External exposure		
contaminated surface from air deposition		8000 h y^{-1}
shoreline		2000 h y^{-1}
submersion in water		300 h y^{-1}
submersion in air		8000 h y^{-1}

[a]Pao et al. 1982
[b]NCRP (1989) recommended screening values; derived from NCRP (1984), Yang and Nelson (1984), Rupp (1980) and Rupp et al. (1980).
[c]For contaminated backyard gardens, the fraction of ingested food that is contaminated and consumed daily.

Ingestion Pathway

Dose from consuming contaminated food can be calculated as:

$$CED(Sv) = (C)(INR)(EF)(EED)(INDF) \qquad (14.6)$$

where C = concentration in food (Bq kg^{-1}), INR = ingestion rate (kg d^{-1}), EF = exposure frequency (d y^{-1}), EED = effective exposure duration (y), and INDF = ingestion dose factor (Sv Bq^{-1}). Ingestion dose conversion factors are listed in Eckerman et al. (1988) and account for the fractional uptake from the small intestine to blood, and then to specific organs, of the common chemical forms of the radionuclides. They also account for losses from tissues via excretion and radioactive decay.

External Irradiation

In addition to the pathways of contaminant transfer already discussed above and in Chapters 3 to 5, external irradiation has to be considered with radioactive contamination. This pathway is generally not needed for nonradioactive contamination. For example, if a lake is contaminated with stable lead, bound largely to the sediments, swimming over the contaminated sediments poses no health risk to fish. A fish would receive a chemical dose from the lead contamination in its food, water consumed, and lead contamination passed through the gills. However, if the same system was contaminated with radioactive ^{214}Pb, the fish would receive an additional exposure merely by swimming above the contaminated sediments. Some of the emitted γ rays from ^{214}Pb would be projected out from the sediments, upon radioactive decay, and irradiate organisms in the overlying water column.

Once emitted from the sediments, however, the radiation is not capable of traveling indefinitely. Attenuation occurs exponentially and can be expressed as

$$I = I_0 e^{-\mu d} \tag{14.7}$$

where I_0 is the initial γ incident rate (photons cm^{-2} s^{-1}) dependent on the activity of the source, μ is the attenuation coefficient (cm^{-1}), and d is the thickness of the attenuating material (cm; the water column in our example). Attenuation coefficients for most materials can be found in Johns and Cunningham (1980). Attenuation is dependent on the energy and type of radiation (i.e., α, β, or γ), as well as the density of the absorbing medium (e.g., air versus lead shielding). In our example, a one Mev γ ray would have an attenuation coefficient of 0.0706 in water. Calculations using Equation 14.7 reveal that 50% of the ^{214}Pb γ rays would be attenuated by water within 9.8 cm of the sediment surface. Thus, bottom-dwelling catfish would receive a greater external dose from the contaminated sediments than pelagic sunfish because the emitted γ rays are quickly attenuated by the water.

Calculating the external dose to humans exposed to radioactively contaminated soil, water column, or plume of air is possible using external dose coefficients provided in Eckerman and Ryman (1993). Equation 14.8 can be used in situations involving contaminated soil.

$$CED(Sv) = (C)(EF)(EED)(ISF)(SDC) \tag{14.8}$$

where C = the concentration in soil (Bq m^{-3}), EF = exposure frequency (s y^{-1}), EED = effective exposure duration (y), ISF = indoor shielding factor (unitless), and SDC = soil dose coefficient ((Sv m^3)(Bq s)$^{-1}$). An indoor shielding factor accounts for time spent indoors where the irradiation from the contaminated soil would be reduced. Shielding factors are discussed and provided in Eckerman and Ryman (1993). The soil dose coefficient is dependent on the distribution of contaminant within the soil, as well as the energy of radiation emitted.

External dose from a plume of contaminated air can be calculated as

$$CED(Sv) = (C)(EF)(EED)(ADC) \tag{14.9}$$

where C = concentration in air (Bq m^{-3}), EF = exposure frequency (s y^{-1}), EED = effective exposure duration (y), and ADC = air dose coefficient ((Sv m^3)(Bq s)$^{-1}$).

A similar calculation can be performed for immersion in water by replacing the air concentration with water concentration and using the appropriate dose coefficient (Eckerman and Ryman, 1993).

Each of these calculations (Equations 14.5, 14.6, 14.8, and 14.9) result in an estimate of the committed effective dose for a particular pathway of contaminant exposure. Effective dose equivalents for all pathways are summed to yield a total CED. The total CED can then be multiplied by an appropriate risk factor to obtain the probability of harmful effects. An example of this approach related to a DOE Superfund site is provided by Whicker et al. (1993). Derivation of the risk factors is the subject of the next section.

DERIVATION OF RISK FACTORS

Epidemiological Studies

Human epidemiological data form the basis for determining the probability of deleterious effects from exposure to radionuclides. In contrast, risks from exposure to most chemical carcinogens are largely extrapolated from laboratory experiments using nonhuman subjects (EPA, 1989e). The latter involves a greater degree of uncertainty. The major epidemiological studies used to determine risks from radionuclide exposures are presented below.

In the late 1890s, the diagnostic potential of newly discovered x-rays was realized. X-rays provided a unique tool for physicians to peer inside the body of a living human. The harmful effects of overexposure to radiation, however, were immediately observed, often in parallel with the discovery of radiation's beneficial medical uses. Early radiologists placed their hands in the x-ray beam to gauge its intensity from the red discoloration to the skin. For a short while skin erythema was a commonly used metric, but the developed side effects of infections, tumors, and in extreme exposures—a required amputation of fingers, quickly alerted practitioners to the damage of radiation.

The delayed effects of radiation became apparent when some individuals hired to paint luminous dials on clocks and watches subsequently developed bone cancer. The painters would often lick their radium-ladened brushes to sharpen the point, thereby ingesting some radioactive material. Other groups of exposed individuals from which the effects of radiation have been observed include uranium miners exposed to radon gas, and individuals medically treated with acute partial-body x-rays to counter the effects of various illnesses. Some of the miners developed lung cancers, and certain medical patients developed an assortment of solid cancers.

The single most important source of information to understand radiation effects comes from the Japanese atomic-bomb survivors. A very meticulous reconstruction of each individual's physical position relative to ground zero at the time of the blast has been performed, taking into consideration the presence of buildings and other structures that shielded the individuals from the radiation. This has allowed careful dose estimates to be made for individuals that are now being followed for lifetime health consequences.

The average dose to the Japanese cohort of 75,991 individuals was 0.12 Gy.[11] The ongoing study of the Japanese population indicates that 344 of the 5,936 cancer deaths are attributable to radiation. No dose-related increase in genetic effects has been observed (NCRP, 1993). Such epidemiological studies have resulted in today's consensus that induction of cancer is the predominant risk from exposure to radiation rather than genetic effects (NCRP, 1993). The EPA (1989d) concurs that the risk of cancer may be used as the sole basis for assessing the radiation-related human health risks for a site contaminated with radionuclides.

Dose-Response Relationships

When a conflation of the Japanese data occurs with other sources from which radiation effects to humans have been observed (Table 14.4), it becomes evident that the doses at which a significant increase in carcinogenesis have been documented are relatively high (0.1 to 0.2 Gy). Such doses are 100 to 200 times greater than the maximum allowable exposures received by today's public. Most importantly, effects have not been demonstrated at the lower doses.

A problem arises as to how to extrapolate effects observed at high doses to potential effects at low doses. Three fundamental dose-response options are a linear extrapolation, a linear-quadratic relationship, and a threshold-type response (Figure. 14.1). Much debate centers on which is the appropriate dose-response relationship. Choosing the appropriate extrapolation model is critical because the risk factors used to estimate the probability of harmful effects are derived from the dose-response relationships. Some method is required, however, to project the risks beyond those observed to have occurred at high doses to those that might occur at the lower doses encountered in today's radiation environment.

Threshold Option

Threshold models suggest that there is a dose below which there is no risk. Proponents for a threshold response (Curve C, Figure 14.1) cite as supporting evidence that an increased cancer incidence has not been observed in populations exposed to unusually high background radiation,[12] three times the nominal rate (NAS/NRC, 1990).

The international committees responsible for establishing radiation protec-

---

[11]Epidemiological studies use the base unit of absorbed dose (Gy) rather than Sv because of the additional uncertainties associated with the quality and weighting factors.

[12]**Natural radiation background** is composed of cosmic radiation emitted from stars and long-lived terrestrial radionuclides that are ubiquitously present in earth's soils. An NCRP (1987) monograph thoroughly describes background radiation and cites that the estimated average CED from background radiation is 3.0 mSv y[-1] in the United States.

TABLE 14.4. Features of Some Epidemiological Studies from Which Radiation Risk Factors Were Derived. (Adapted from Clark [1995].)

Parameter	Life Span Study (LSS) -Japanese Atomic Bomb Survivors	Ankylosing Spondylitis Study (ASS)	Canadian Tuberculosis Patients Given Chest Fluoroscopies	Children in Israel Irradiated for Ringworm of the Scalp	UK National Registry for Radiation Workers
Population Size	75,991	14,106	31,701	10,834	95,217
Period of follow-up	5 to 40 years after exposure	Up to 38 years (mean of 13 years)	Up to 30 years (mean of 27 years)	Up to 32 years (mean of 26 years)	Up to 34 years (mean of 12.7 years)
Ranges of: (i) ages at exposure	All	Virtually all ≥15 years	At least 10 years	0 to 15 years	18 to 64 years
(ii) sexes	Similar numbers of males and females	83% males	Female	Similar numbers of males and females	92% male
(iii) ethnic groups	Japanese	Western (UK)	Western (N. America)	African, Asian	Western (UK)
Setting in which exposure was received	War	Medical: therapy for nonmalignant disease	Medical: diagnostic	Medical: therapy for nonmalignant disease	Occupational
Range of organs irradiated	All	All but mainly those in proximity to spine	Mainly breast and lung	Mainly brain, bone marrow, thyroid, skin, breast	All
Range of doses	Mainly 0 to 4 Gy	Mainly 0 to 20 Gy	Mainly 0 to 10 Gy	Brain: 0 to 6 Gy (mean: 1.5 Gy) Thyroid: 0 to 0.5 Gy (mean: 0.09 Gy)	Mainly 0 to 0.5 Sv (mean: 0.034 Sv)
Dose rate	High	High	High, but highly fractionated	High	Low

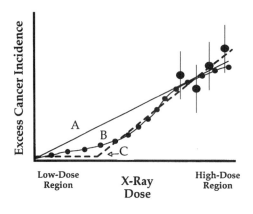

FIGURE 14.1. A hypothetical illustration of how various models can be used to extrapolate the incidence of cancer from data generated at high doses to the low dose region. Curve A represents a linear extrapolation, curve B depicts a linear-quadratic relationship, and curve C illustrates a threshold-type response (adapted from Hall, 1978). All three curves pass through the data, yet each gives quite different results when extrapolated to the low dose region.

tion criteria, however, have assumed that a threshold does *not* exist, partially because of the method by which cancers are formed. It is generally accepted that tumors initiate from damage to single cells, although as described earlier, a complex series of multistage promotional events are required for a neoplasia to progress to full malignancy. A single mutational event in a critical gene in a single target cell can create the potential for neoplastic development (Clark, 1995). This means that there is some probability, albeit very low, that a single radiation track (the lowest dose and dose rate possible) hitting the nucleus of an appropriate target cell could cause damage to the DNA and initiate a tumor (Clark, 1995). Thus, at the DNA level, there is no basis for assuming that a dose exists below which the risk of tumor induction is zero.[13]

Linear vs. Linear-Quadratic

A linear model suggests that there is no dose-rate effect. That is, a dose given over a short period of time (e.g., 50 mSv in one day) is just as effective at causing damage as the same dose protracted over a longer period of time (e.g., 50 mSv over 50 yr). This model is contrary to numerous animal experiments where significant dose rate effects have been observed (Ullrich et al., 1987). Animal experiments over a wide range of doses suggest that the linear-quadratic model (Curve B, Figure 14.1) is more appropriate. Human data, however, are inconclusive. For all cancers, other than leukemia, the Japanese data fit a linear dose-response model (Curve A, Figure 14.1). The same data also fit a linear-quadratic model (Curve B). Neither model fits the data significantly better than the other. Interestingly, the Japanese data also indicate that the dose response curve of solid cancers differs from that of

---

[13]The Health Physics Society, a professional organization responsible for protecting humans from over-exposure to radiation, recently questioned current radiation protection criteria. They issued a position statement stating that doses below 0.1 Sv are either too small to be observed or are nonexistent and, therefore, should not warrant expenditures of conducting full-scale risk analyses. (Mossman et al., 1996).

leukemia. The data for leukemia suggest a linear-quadratic model is significantly better than a linear dose-response relationship (NCRP, 1993).

The United Nations Scientific Committee on Effects of Atomic Radiation (UNSCEAR, 1986) concluded that linear extrapolation from high-dose data to low doses (less than 0.2 Gy) could result in an overestimation of risk by a factor as high as five. The Committee on Biological Effects of Ionizing Radiation (NAS/NRC, 1990) also considered the animal experimental data and found that the linear model overestimated risk from 2 to 10 times for the endpoints of specific locus mutation, reciprocal translocations, tumor formation and longevity.

Currently, a linear dose-response model is used to obtain risk factors. The linear model produces the largest risk factors, and the agencies concerned with health protection have elected to use the most conservative numbers until sufficient data are accumulated to warrant otherwise. However, the ICRP (1991) has recommended that estimates of cancer risks associated with exposures to low doses or low dose rates be reduced by a factor of two when results are based on exposure to high doses at high dose rates.

Currently Accepted Risk Factors

Recommended risk factors, furnished by the International Commission on Radiological Protection, are derived from the dose-response relationships obtained in the epidemiological studies presented above. The **risk factors** give the probability of a deleterious effect for each mSv of dose received. The ICRP (1991) has quantified risks into four categories:

Deleterious Effect	Risk Factor (mSv^{-1})
Fatal cancer	5.0×10^{-5}
Severe genetic effects	1.3×10^{-5}
Nonfatal cancer	1.0×10^{-5}
Total detriment	7.3×10^{-5}

EPA slope factors differ slightly from the ICRP recommendations. The incidence of cancer is the endpoint of interest in the EPA slope factors, rather than the probability of fatality from a cancer used by the ICRP.

RISKS TO HUMANS FROM EXPOSURE TO RADIATION

Risk, in the context of this section, is the lifetime probability of a human experiencing a deleterious effect due to radiation exposure. Risk from exposure to radionuclides is calculated as the committed effective dose (presented earlier in the chapter) times a risk factor:

$$Risk = CED \ (mSv) \times Risk \ Factor \ (mSv^{-1}) \qquad (14.10)$$

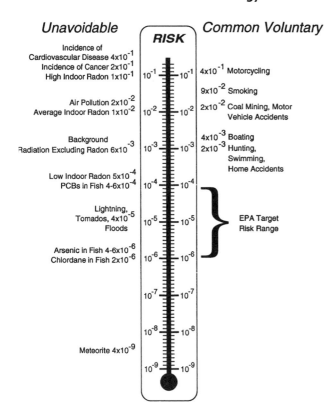

FIGURE 14.2. A comparison of risks from various voluntary and unavoidable sources. Numbers indicate the probability of death from the disease, natural phenomena, contaminant, or activity. (Adapted from Hoffman, 1991).

Using the risk factors presented above, individuals with a CED of 2.0 mSv, an amount twice that recommended by the ICRP for public exposure, would have a probability of acquiring a fatal cancer during their lifetimes of 1.0×10^{-4} (a probability of 1 in 10,000; Equation 14.10).

Declaration of acceptable risk is purely a societal decision. Figure 14.2 compares the EPA (1989d) guideline of acceptable risks (10^{-4} to 10^{-6}) to other risks we are routinely subjected to. Risks associated with radiation, and currently deemed acceptable, are of a similar magnitude to the probability of an individual being struck by lightning (4×10^{-5}). Such risks are orders of magnitude less than the natural cancer incidence (unrelated to radiation exposure) of 2×10^{-1}.

ECOLOGICAL EFFECTS FROM RADIOACTIVE CONTAMINATION

The propensity for nonhuman species to take up radioactive contaminants is demonstrated by an increased ^{131}I burden in the thyroids of herbivores following radioactive releases from atmospheric testing of nuclear weapons (Figure 14.3). Radioactive releases generated in China were quickly transferred to Colorado deer and elk (Whicker and Schultz, 1982), illustrating the mobility and potential impact of some contaminants.

I-131 in Thyroids of Colorado Herbivores

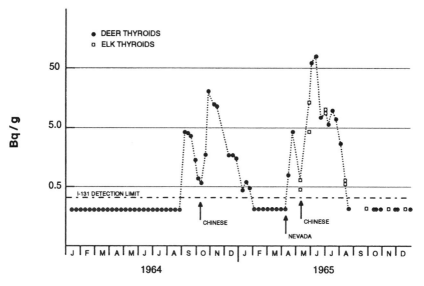

FIGURE 14.3. [131]I concentrations in Colorado mule deer and elk thyroids follow-ing nuclear testing activities in China and the U.S. during 1964 and 1965 (Adapted from Whicker and Schultz, 1982).

Notable reviews on effects of radiation have been prepared on specific groups of organisms: protozoa (Wichterman, 1972), brine shrimp (Metalli and Ballardin, 1972), insects (O'Brien and Wolfe, 1964), amphibians (Brunst, 1965), reptiles (Cos-grove, 1971), birds (Mellinger and Schultz, 1975), plants (Sparrow et al., 1958), terrestrial and aquatic animal populations (Turner, 1975; Blaylock and Trabalka, 1978), and plant communities (Whicker and Fraley, 1974). The National Council on Radiation Protection and Measurements has examined the effects of ionizing radiation on aquatic organisms (NCRP, 1991), the International Atomic Energy Agency has considered whether or not nonhuman species are adequately protected by radiation standards designed for humans (IAEA, 1992), and the lower limits of radiosensitivity in nonhuman species have been reviewed (Rose, 1992).

Research on effects to plants and animals documented over the last 70 years have led to the formulation of some basic paradigms. The effects of radiation on reproduction have been most extensively studied in mammals, and the majority of results suggest that natality is a more radiosensitive parameter than mortality (Carlson and Gassner, 1964). Among the vertebrates, mammals are generally more radiosensitive than birds, fish, amphibians, or reptiles (Casarett, 1968). Rose (1992) reviewed the literature for lower limits of radiosensitivity in nonhumans and found that the lowest dose from an acute exposure with measurable effects was 10 mGy. This dose was delivered to pregnant rats, and ultimately impaired the reflexes of their offspring (Semagin, 1986). Mice are among the organisms whose ovaries are the most sensitive to irradiation, and Gowen and Stadler (1964) found that reproduction was impaired in females at doses of 200 mGy, with permanent sterility occurring at 1000 mGy. Males were less sensitive and had impaired reproduction at 3200 mGy (Rugh and Wolff, 1957).

Studies on radiation effects to nonmammalian organisms have indicated higher radioresistence and, as with mammals, that the early life cycle stages are the most radiosensitive (NCRP, 1991). The lowest dose reported to have an impact on amphibians was 20 mGy, a dose that was lethal to newt eggs (*Triturus alpestris*; Peters, 1960). Anderson and Harrison (1986) found that radiation doses in excess of 10 mGy were necessary to damage the most sensitive stages of fish development.

The National Council on Radiation Protection and Measurements (NCRP, 1991) recently established a maximum dose rate of 10 mGy d^{-1} from chronic exposure for the protection of populations of aquatic organisms. The NCRP recognized that other environmental stresses might act in combination with radiation and cause an impact at the maximum reference level of 10 mGy d^{-1}. Therefore, they conservatively recommended that a comprehensive ecological evaluation of the radiation exposure and environmental stressors be conducted when populations are exposed to 2.4 mGy d^{-1}. The International Atomic Energy Agency (IAEA, 1992) has also addressed the issue of effects of ionizing radiation on plants and animals, and concluded that "There is no convincing evidence from the scientific literature that chronic radiation dose rates below 1 mGy d^{-1} will harm animal or plant populations."

Sufficient data exist on acute lethality to rank groups of organisms according to their sensitivities to radiation (Figure 14.4). Mammals seem to be the most sensitive group, and humans are among the most sensitive mammals. This has caused some organizations to conclude that if man is adequately protected from radiation, then other organisms are also likely to be sufficiently protected (ICRP, 1977). This assumption has been generally accepted and adopted by those establishing radiation protection standards, even though "sufficient protection" has never been quantified.

The International Atomic Energy Agency (1992) set about to determine if the statements made by the ICRP were consistent with current knowledge, and whether radiation protection standards were needed for aquatic and terrestrial biota. The IAEA approached the problem by developing hypothetical contamination scenarios that resulted in humans acquiring a maximum permissible dose (1 mSv y^{-1}). They then calculated the dose rates aquatic and terrestrial biota would receive if living in the same hypothetical environment. Doses received by the biota were then compared to doses known to have caused observable effects to non-humans. They concluded that limiting the dose to the most exposed group of humans to 1 mSv y^{-1} would lead to dose rates to plants and animals in the same area of less than 1 mGy d^{-1} (notice the different dose rates: y^{-1} for humans and d^{-1} for nonhumans). Because convincing evidence does not exist within the scientific literature that chronic radiation doses below 1 mGy d^{-1} will harm plant or animal populations, the IAEA (1992) concluded that specific radiation protection standards for nonhuman biota are not needed. They recognized, however, that under some situations site-specific analyses would be required, such as when endangered species are involved.

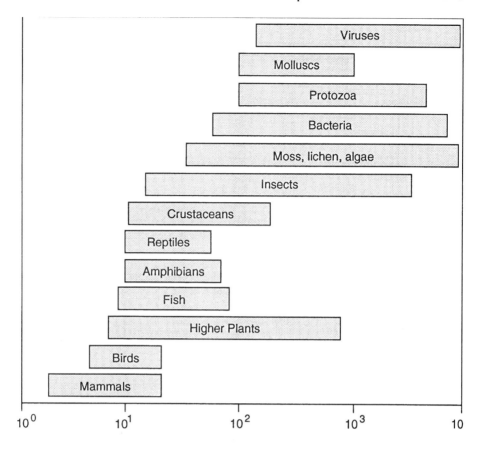

Acute Lethal Dose Range (Gy)

FIGURE 14.4. A comparison of the radiation sensitivity among groups of organisms (Adapted from Whicker and Schultz, 1982).

CONFIDENCE IN RISK ANALYSES

The conclusion of the IAEA should not imply that ecological risk analyses are not needed, or that our understanding of risk to humans and the environment is complete. Numerous uncertainties exist in risk analyses that result in our having a reduced confidence in the risk estimate. A thorough risk analysis should include uncertainty estimates that indicate the level of confidence in the risk prediction (NCRP, 1996). Estimating the uncertainty associated with risk analyses is critical to a meaningful use of the risk estimate. A cumulative uncertainty of a factor of 10 would not be unreasonable in a risk analysis conducted for human exposure to radiation. Some of the dominant sources of uncertainty are (Sinclair, 1993): (1) estimating the dose received by the Japanese cohort; (2) using data from a population that received high dose rates and extrapolating to the low dose range; (3) using epidemiological data derived from a population with strikingly different cultures and life styles, some of which seems to influence cancer incidence. For example, the natural stomach cancer incidence among male Japanese (unrelated to the atomic weapons explosion) is about six times greater than among males from North America, and incidence of breast cancer is six times greater in North American females compared to Japanese (NCRP, 1993). If natural cancer incidences differ among populations, then the question arises as to how pertinent are radiation risk factors to populations other than those from which the factors were derived.

As complex as human risk analyses are, determining risks to other environmental components is even more so, and with much greater uncertainties. Problems exist for the following reasons:

1. Measuring the dose received by most nonhuman species is not a trivial task. Dose measurement instruments can be placed on some species, but recapturing the same animal and recovering the instrument is problematic.
2. Mathematical models that predict dose to free-ranging organisms have not been developed and validated as well as human dosimetry models.
3. Tables of dose conversion factors that transform a unit concentration of contaminant to a dose received by the organism exist for only a few nonhuman species. A recent compilation of dose conversion factors for nonhumans (Amiro, 1997) has many shortcomings when compared to dose conversion factors derived for humans. Amiro (1997) states that the factors for nonhumans are generic and useful only as conservative indicators of radiological dose. They are not intended to give exact dose estimates. Precise doses are not possible because the nonhuman dose conversion factors (1) do not account for the relative biological effectiveness of different radiation types (e.g., γ vs α emissions); (2) conservatively assume that all energies emitted by radionuclides from within the organism are also absorbed by the organism, resulting in an overestimation of dose to smaller organisms; and (3) do not consider the shape of different organisms. Thus, the same dose conversion factors are used regardless of an organism's geometry, which "could range from an ant to an elephant, or from a lichen to a tree" (Amiro, 1997). The dose conversion factors presented by Amiro (1997) are useful, but it is apparent that they are far less rigorous than those calculated for humans.
4. There are no risk factors established for nonhumans. Dose-response data pertinent to nonhumans are inadequate to quantify the probability of deleterious effects for the common exposure conditions (i.e., chronic, low-dose conditions).
5. Agreement has not been reached as to what the proper endpoint should be for an ecological risk analysis. The endpoint for humans is health of the individual. However, risk to individuals is generally not considered for nonhuman species. Seldom are we concerned with an individual fish or tree, but rather that their populations remain viable.[14] The critical endpoint for humans (cancer incidence) is not likely to be relevant for most populations of plants and animals.

---

[14]Concern for individual non-human species does occur when the species is endangered.

Contaminants can exert effects on levels of biological organization ranging from molecules to ecosystems. Often because of the complexity of documenting effects at the levels of population, community, and ecosystem, most
ecotoxicological research concentrates on effects at lower levels of organization (cellular or molecular). Also, many of the mechanisms that connect effects between individuals and populations are poorly understood. However, it is
the population level effects that are of primary concern for nonhuman species.
Cellular damage generally represents a sublethal endpoint that may provide
good early warning signs of potential contaminant impact, but specific relationships between cellular damage and consequences at upper levels of biological organization have not been made (Clements and Kiffney, 1994; Underwood
and Peterson, 1988). Current environmental risk analyses that use
cellular/molecular effects as endpoints are of questionable value, primarily because of large uncertainties about the significance of cellular/molecular effects
to the population. Many cellular and molecular effects are repaired within the
organism, or are removed from the population through natural selection.

6. Most of the dose-effect data for nonhumans are not pertinent to conditions
under which the majority of organisms are now exposed (chronic, low-level
exposures). Most of the data in the literature were generated under conditions
of high dose rates and short-term laboratory experiments.

The findings of a recent Committee on Risk Assessment Methodology, con-
vened by the National Research Council, are particularly pertinent. They identified
four major areas where scientific consensus was lacking in ecological risk analysis
(Barnthouse, 1994):

1. extrapolation across scales of time, space, and ecological organization,
2. quantification of uncertainty,
3. validation of predictive tools, and
4. economic valuation of ecological resources.

Until these areas are adequately addressed, as well as numerous others brought
out in this chapter, ecological risk analyses will have greater uncertainties than
those currently associated with estimating risk to humans.

SUMMARY

Risk, when rigorously defined, should include a probability statement about the
incidence of a deleterious event. The science supporting the determination of hazards to humans from radiation exposures is sufficiently developed to meet that rigorous definition. Documentation of effects from human exposures to radiation
were initiated with the discovery of x-rays in the late 1890s and have since developed into a science that crosses disciplines of nuclear physics, human physiology,
chemistry, biology, and the medical professions. The uncertainties associated with
risks to humans from exposure to radiation are probably less than for many nonra-

dioactive contaminants. The concept of dose seems to be much better developed in the radiological sciences, and human epidemiological data are available to derive the risk factors for radioactive contaminants, whereas most risk factors for nonradioactive contaminants must rely on extrapolation from animal laboratory data.

Quantifying ecological risks from radioactive contamination is more difficult, indeed, the term "risk" may not be appropriate, as the science is not sufficiently developed to attach a probability statement to the analysis. Thus, ecological risk analyses are more qualitative than quantitative. Risk factors for radiological or nonradiological contaminants do not even exist for nonhuman organisms.

Risk factors are a powerful tool for making decisions about contaminated sites. They provide a common metric with which comparisons can be made. Most importantly they are generated from scientific input. Many decisions regarding the management of contaminated sites will be influenced by sociopolitical motivations and the public's perception of the hazard. Risk analysis is the primary mechanism by which science can enter into the decision making process. It is, therefore, critical for risk analyses to be conducted with the utmost professionalism and scientific rigor. Anything less will result in the risk analysis process losing credibility and, thereby, the loss of scientific input into issues concerning public health.

ACKNOWLEDGMENTS

This work was supported in cooperation by the University of Georgia and the Department of Energy Award Number DE-FC09-96SR18546. The critical reviews by C. Bell, J. Graves, J. Joyner, M. Malak, L. Marsh, M. Mulvey, S. Rich, C. Strojan, and F. W. Whicker were much appreciated.

Suggested Readings

Eisenbud, M. *Environmental Radioactivity*. Academic Press, New York, 1973.

Hall, E. J. *Radiobiology for the Radiologist.* Harper & Row Publishers. San Francisco, CA, 1978.

IAEA (International Atomic Energy Agency). *Effects of Ionizing Radiation on Plants and Animals at Levels Implied by Current Radiation Protection Standards.* IAEA, Vienna, Austria, 1992.

Lapp, R. E. and H. L. Andrews. *Nuclear Radiation Physics.* Prentice-Hall, Englewood Cliffs, NJ, 1972.

NAS/NRC (National Academy of Sciences/National Research Council, Committee on the Biological Effects of Ionizing Radiations). *Health Risks of Exposure to Low Levels of Ionizing Radiation, BEIR V.* National Academy Press, Washington, DC, 1990.

UNSCEAR (United Nations Scientific Committee on the Effects of Atomic Radiation). *Sources, Effects and Risks of Radiation.* Publication E.88.IX.7, United Nations, New York, 1988.

Whicker, F.W. and V. Schultz. *Radioecology: Nuclear Energy and the Environment*, CRC Press, Inc., Boca Raton, FL, 1982.

Conclusions

There will come soft rains and the smell of the ground,
And swallows circling with their shimmering sound;

● ● ●

Not one would mind, neither bird nor tree
If mankind perished utterly;

And Spring herself, when she woke at dawn,
Would scarcely know that we were gone.

TEASDALE (1937)

OVERVIEW

So far, we have discussed the diverse goals of ecotoxicology and the many disciplines contributing to our understanding of contaminant fate and effects. We discussed the accumulation of contaminants in components of the biosphere, including transfers to individuals and trophic levels. Then effects from the subindividual to the biosphere levels were discussed in a series of chapters (Chapters 6 to 12) that took up most of the book. Finally, the practical issue of estimating risk to humans and the environment due to contamination was explored. One of the intentions of detailing risk assessment methods was to point out the many shortcomings in the associated technologies and science, and to emphasize the need for more effort in this extremely important aspect of ecotoxicology. Now all that is left to do is remind the reader of the importance and context of ecotoxicology. That is the goal of this brief chapter.

PRACTICAL IMPORTANCE OF ECOTOXICOLOGY

In looking back at the enormous human suffering of World War I, Sarah Teasdale wrote the poem from which the above excerpt was taken. She correctly observes that, although the suffering was of such enormity that it will stand out in history for centuries, it was trivial in the larger context of Nature. The biosphere can do just fine without our help. Indeed, Stephen Jay Gould (quoted in Wheeler, 1996) states: "Humans are simply a dot on one side of the curve of biological complexity ... a dot that could easily disappear."

Does this mean that we are free to contaminate the biosphere with abandon? In the context of ensuring that life continues on the earth, this is certainly the case. Our influence on the permanence of life on earth is minuscule. One meteorite striking the earth at the end of the Cretaceous period had more influence on the trajectory of life on earth than we will ever have. In a few billion years, life will end when the Sun expands to engulf our planet: our efforts cannot change the physics of star aging. However, this is certainly not true in the context of ensuring that the biosphere remains compatible with and appealing to human life. It would be a catastrophic mistake to continue to contaminate the biosphere. Presently, we are having enormous influence on the biosphere relative to its ability to provide an appealing home for us and to valued species. For this reason, ecotoxicology has now become an essential part of human knowledge and activities—not because life on earth needs us to protect it, i.e., the Lorax incongruity. It is extremely important to understand that ecotoxicological knowledge and activities are no more or less important than our industrial knowledge and activities. Few people would abandon the right to clean water and an aesthetically pleasing environment. Similarly, few people would deny themselves or their children the fruits of our industry—an industry that produces waste. The key is striking an informed and insightful balance between these two facets of our lives. "The ultimate goal of development and of our concern for the environment is to increase the quality of life" (Gallopin and Öberg, 1992).

SCIENTIFIC IMPORTANCE OF ECOTOXICOLOGY

It may not seem to be true if you are reading this chapter as part of a college course that is now drawing to an end, but there is also intrinsic value to the material described so far. There is value in simply learning more about the world. With ecotoxicology, one has the added value of the knowledge having obvious and immediate utility.

Our world is filled with fascinating concepts and more appear daily. In the two years that passed between writing the book preface and this final chapter, many new discoveries have been made about the world. The conclusion from the Viking lander missions that life did not exist on Mars has recently been thrown open to question based on evidence from a Martian meteorite found on the Antarctic ice. In a recent issue of *Science* magazine (February 28, 1997), there was an announcement of a new subatomic particle. The first mammalian clone, a sheep named Molly, was produced from a mature cell of another individual. What were once the props of science fiction writers are now the realities of our world. New facts enhance our appreciation of the world and facilitate wise decisions. Although I admit to more than a small bias here, those associated with ecotoxicology are certainly among the most interesting and useful.

References

Adams, S.M. Biological indicators of stress in fish. *Am. Fish. Soc. Symp.* 8, pp. 1–8, 1990.

Adams, S.M., W.D. Cumby, M.S. Greeley, Jr., M.G. Ryon, and E.M. Schilling. Relationships between physiological and fish population responses in a contaminated stream. *Environ. Toxicol. Chem.* 11, pp. 1549–1557, 1992.

Adamus, C.L. and M.J. Bergman. Estimating nonpoint source pollution loads with a GIS screening model. *Water Resour. Bull.* 31, pp. 647–655, 1995.

Adolph, E.F. Quantitative relations in the physiological constitutions of mammals. *Science (WASH DC)* 109, pp. 579–585, 1949.

Agard, D.A. To fold or not to fold . . . *Science (WASH DC)* 260, pp. 1903–1904, 1993.

Aguilar, A. and A. Borrell. Reproductive transfer and variation of body load of organochlorine pollutants with age in fin whales (*Balaenoptera physalus*). *Arch. Environ. Contam. Toxicol.* 27, pp. 546–554, 1994.

Ahlbom, A. *Biostatistics for Epidemiologists*. Lewis Publishers, Inc., Boca Raton, FL, 1993, p. 214.

Alberts, B., D. Bray, J. Lewis, M. Raff, K. Roberts, and J.D. Watson. *Molecular Biology of the Cell*. Garland Publishing , Inc., New York, 1983, p. 1146.

Allen, H.E., R.H. Hall, and T.D. Brisbin. Metal speciation. Effects on aquatic toxicity. *Environ. Sci. Technol.* 14, pp. 441–442, 1980.

Aloj Totaro, E., F.A. Pisani, and P. Glees. The role of copper level in the formation of neuronal lipofuscin in the spinal ganglia of *Torpedo m. Mar. Environ. Res.* 15, pp. 153–163, 1985.

Aloj Totaro, E., F.A. Pisanti, P. Glees, and A. Continillo. The effect of copper pollution on mitochondrial degeneration. *Mar. Environ. Res.* 18, pp. 245–253, 1986.

Amiard–Triquet, C., D. Pain, G. Mauvais, and L. Pinault. Lead Poisoning in Waterfowl: Field and Experimental Data, in *Impact of Heavy Metals on the Environment*, Vernet, J.–P., Ed., Elsevier Science Publishers B.V., Amsterdam, 1992.

Amiro, B. D. Radiological dose conversion factors for generic non–human biota used for screening potential ecological impacts. *J. Environ. Radioact.* 35, pp. 37–51, 1997.

Anderson, D.P. Immunological indicators: Effects of environmental stress on immune protection and disease outbreaks. *Am. Fish. Soc. Symp.* 8, pp. 38–50, 1990

Anderson, E.V. Phasing lead out of gasoline: Hard knocks for lead alkyls producers. *Chem. Eng. News.* Feb. 6, pp. 12–16, 1978.

Anderson, P.D. and L.J. Weber. Toxic response as a quantitative function of body size. *Toxicol. Appl. Pharmacol.* 33, pp. 471–483, 1975.

Anderson, S. L. and F. L. Harrison. *Effects of Radiation on Aquatic and Radiological Methodologies for Effects Assessment*. EPA Report 5201-85-016, U.S. Environmental Protection Agency, Washington, DC, 1986, p. 128.

Andrew, R.W., K.E. Biesinger, and G.E. Glass. Effects of inorganic complexing on the toxicity of copper in *Daphnia magna*. *Water Res.* 11, pp. 309–315, 1977.

Ankley, G.T., G.L. Phipps, E.N. Leonard, D.A. Benoit, V.R. Mattson, P.A. Kosian, A.M. Cotter, J.R. Dierkes, D.J. Hansen, and J.D. Mahony. Acid–volatile sulfide as a factor mediating cadmium and nickel bioavailability in contaminated sediments. *Environ. Toxicol. Chem.* 10, pp. 1299–1307, 1991.

Anonymous. Heart disease, cancer linked to trace metals. *Chem. Eng. News*. May 3, pp. 24–27, 1976.

Anonymous. Hooker settle on Hyde Park dump cleanup. *Chem. Eng. News*. Jan. 26, pp. 10, 1981.

Anonymous. Most of Love Canal habitable, EPA says. *Chem. Eng. News.* July 19, pp. 6, 1982.

Anonymous. Lead use in gasoline: EPA proposes 91% cut by 1986. *Chem. Eng. News.* Aug. 6, pp. 4, 1984a.

Anonymous. India's chemical tragedy: Death toll at Bhopal still rising. *Chem. Eng. News.* Dec. 10, pp. 6–7, 1984b.

Anonymous. Global climate warming: Trace gases other than CO_2 play role. *Chem. Eng. News.* May 6, pp. 6–7, 1985.

Anonymous. Appendix II – Method 1311 toxicity characteristic leaching procedure. *Federal Register* 55, pp. 11863–11875, 1990.

Anspaugh, L.R., R.J. Catlin and M. Goldman. The global impact of the Chernobyl reactor accident. *Science (WASH DC)* 242, pp. 1513–1519, 1988.

APHA. *Standard Methods for the Examination of Water and Wastewater*, 15th ed., American Public Health Association, Washington, DC, 1981, p. 1134.

Apple, G.J., W.G. Hunter, and S. Bisgaard. Scientific Data and Environmental Regulation, in *Statistics and the Law*, DeGroot, M.H., S.E. Fienberg, and J.B. Kadane, Eds., John Wiley & Sons, Inc., New York, 1986.

Armitage, P. and I. Allen. Methods of estimating the LD 50 in quantal response data. *J. Hyg.* 48, pp. 298–322, 1950.

ASTM. Standard guide for conducting the frog embryo teratogenesis assay – *Xenopus* (FETAX), in *Annual Book of ASTM Standards*, American Society for Testing and Materials, Philadelphia, PA, 1993.

Atchison, G.J., M.G. Henry, and M.B. Sandheinrich. Effects of metals on fish: A review. *Environ. Biol. Fishes* 18, pp. 11–25, 1987.

Atchison, G.J., M.B. Sandheinrich, and M.D. Bryan. Effects of Environmental Stressors on Interspecific Interactions of Aquatic Animals, in *Ecotoxicology. A Hierarchical Treatment*, Newman, M.C. and C.H. Jagoe, Eds., CRC Press, Inc., Boca Raton, FL, 1996.

Aust, A.E. Mutations and Cancer, in *Genetic Toxicology*, Li, A.P. and R.H. Heflich, Eds., CRC Press, Inc., Boca Raton, FL, 1991.

Austin, A. and J. Deniseger. Periphyton community changes along a heavy metals gradient in a long narrow lake. *Environ. Exp. Bot.* 25, pp. 41–52, 1985.

Austin, A., J. Deniseger, and M.J.R. Clark. Lake algal populations and physico–chemical changes after 14 years input of metallic wastes. *Water Res.* 19, pp. 299–308, 1985.

Avery, T.E. and G.L. Berlin. *Interpretation of Aerial Photographs.* Macmillian Publishing Co., New York, 1985, p. 554.

Azenha, M., M.T. Vasconcelos, and J.P.S. Cabral. Organic ligands reduce copper toxicity in *Pseudomonas syringae*. *Environ. Toxicol. Chem.* 14, pp. 369–373, 1995.

Babukutty, Y. and J. Chacko. Chemical partitioning and bioavailability of lead and nickel in an estuarine system. *Environ. Toxicol. Chem.* 14, pp. 427–434, 1995.

Baes, C.F., Jr., and S.B. McLaughlin. Trace elements in tree rings: Evidence of recent and historical air pollution. *Science (WASH DC)* 224, pp. 494–497, 1984.

Baird, C. *Environmental Chemistry.* W.H. Freeman and Co., New York, 1995, p. 484.

Baker, A.J.M. and P.L. Walker. Physiological responses of plants to heavy metals and the quantification of tolerance and toxicity. *Chem. Speciation Bioavail.* 1, pp. 7–18, 1989.

Baker, C.E. and P.B. Dunaway. Retention of [134]Cs as an index to metabolism in the cotton rat (*Sigmodon hispidus*). *Health Phys.* 16, pp. 227–230, 1969.

Baker, L.A., A.T. Herlihy, P.R. Kaufmann and J.E. Eilers. Acidic lakes and streams in the United States: The role of acidic deposition. *Science (WASH DC).* 252, pp. 1151–1154, 1991.

Baker, R., B. Lavie, and E. Nevo. Natural selection for resistance to mercury pollution. *Experientia* 41, pp. 697–699, 1985.

Baker, R.J., R.A. Van Den Bussche, A.J. Wright, L.E. Wiggins, M.J. Hamilton, E.P. Reat, M.H. Smith, M.D. Lomakin, and R.K. Chesser. High levels of genetic change in rodents of Chernobyl. *Nature (LOND.)* 380, pp. 707–708, 1996.

Bantle, J.A. FETAX – A Developmental Toxicity Assay Using Frog Embryos, in *Fundamentals of*

Aquatic Toxicology. Effects, Environmental Fate, and Risk Assessment, 2nd Ed., Rand, G.M., Ed., Taylor & Francis, Washington, DC, 1995.

Bantle, J.A. and T.D. Sabourin. Standard guide for conducting the frog embryo teratogenesis assay – *Xenopus* (FETAX). American Society for Testing and Materials, *Am. Soc. Test. Mat. Spec. Pub.* E1439–91, pp. 1–11, 1991.

Barber, M.C., L.A. Suarez, and R.R. Lassiter. Modeling bioconcentration of nonpolar organic pollutants by fish. *Environ. Toxicol. Chem.* 7, pp. 545–558, 1988.

Barnaby, F. Chernobyl: The consequences to Europe. *Ambio* 15, pp. 332–334, 1986.

Barnthouse, L. W. Issues in ecological risk assessment: the CRAM perspective. *Risk Analysis* 14, pp. 251–256, 1994.

Barnthouse, L.W., G.W. Suter II, A.E. Rosen, and J.J. Beauchamp. Estimating responses of fish populations to toxic contaminants. *Environ. Toxicol. Chem.* 6, pp. 811–824, 1987.

Barrie, L.A. and J.M. Hales. The spatial distributions of precipitation acidity and major ion wet deposition in North America during 1980. *Tellus* 36B, pp. 333–335, 1984.

Barron, M.G. Bioaccumulation and Bioconcentration in Aquatic Organisms, in *Handbook of Ecotoxicology*, D.J. Hoffman, B.A. Rattner, G.A., Burton, Jr., and J. Cairns, Jr., Eds., CRC Press, Inc., Boca Raton, FL, 1995.

Barron, M.G., G.R. Stehly, and W.L. Hayton. Pharmacokinetic modeling in aquatic animals. I. Models and concepts. *Aquat. Toxicol. (AMST)* 18, pp. 61–86, 1990.

Bartholomew, G.A. The Roles of Physiology and Behaviour in the Maintenance of Homeotasis in the Desert Environment, in *Symposia of the Society for Experimental Biology, No. 18*, Academic Press, New York, 1964.

Baskin, Y. Ecolgists dare to ask: How much does diversity matter? *Science (WASH DC)* 264, pp. 202–203, 1994.

Battaglia, B., P.M. Bisol, V.U. Fossato, and E. Rodino. Studies on the genetic effects of pollution in the sea. *Rapp. P. V. Reun. Cons. Int. Explor. Mer.* 179, pp. 267–274, 1980.

Bechmann, R.K. Use of life tables and LC50 tests to evaluate chronic and acute toxicity effects of copper on the marine copepod *Tisbe furcata* (Baird). *Environ. Toxicol. Chem.* 13, pp. 1509–1517, 1994.

Becker, P.H., D. Henning, and R.W. Furness. Differences in mercury contamination and elimination during feather development in gull and tern broods. *Arch. Environ. Contam. Toxicol.* 27, pp. 162–167, 1994.

Beeby, A. Toxic Metal Uptake and Essential Metal Regulation in Terrestrial Invertebrates: A Review, in *Metal Ecotoxicology. Concepts and Applications*, Newman, M.C. and A.W. McIntosh, Eds., Lewis Publishers, Chelsea, MI, 1991.

Benson, A.A. and R.E. Summons. Arsenic accumulation in Great Barrier Reef invertebrates. *Science (WASH DC)* 211, pp. 482–483, 1981.

Bercovitz, K. and D. Laufer. Lead Release from Human Trabecular Bone, in *Impact of Heavy Metals on the Environment*, Vernet, J.–P., Ed., Elsevier Science Publishers B.V., Amsterdam, 1992.

Bergeron, J.M., D. Crews, and J.A. McLachlan. PCBs as environmental estrogens: Turtle sex determination as a biomarker of environmental contamination. *Environ. Health Perspect.* 102, pp. 780–781, 1994.

Berglind, R. Combined and separate effects of cadmium, lead and zinc in ALAD activity, growth and hemoglobin content in *Daphnia magna*. *Environ. Toxicol. Chem.* 5, pp. 989–995, 1986.

Berkson, J. Why I prefer logits to probits. *Biometrics* 7, pp. 327–339, 1951.

Berkson, J. Maximum likelihood and minimum χ^2 estimates of the logistic function. *J. Am. Stat. Assoc.* 50, pp. 130–162, 1955.

Beyer, W.N. A reexamination of biomagnification of metals in terrestrial food chains. *Environ. Toxicol. Chem.* 5, pp. 863–864, 1986.

Bezel, V.S. and V.N. Bolshakov. Population Ecotoxicology of Mammals, in *Bioindications of Chemical and Radioactive Pollution*, Krivolutsky, D.A., Ed., CRC Press, Inc., Boca Raton, FL, 1990.

Biesinger, K.E. and G.M. Christensen. Effects of various metals on survival, growth, reproduction, and metabolism of *Daphnia magna*. *J. Fish. Res. Board Can.* 29, pp. 1691–1700, 1972.

Biggins, P.D.E. and R.M. Harrison. Chemical speciation of lead compounds in street dusts. *Environ. Sci. Technol.* 14, pp. 336–339, 1980.

Bishop, W.E. and A.W. McIntosh. Acute lethality and effects of sublethal cadmium exposure on ventilation frequency and cough rate of bluegill (*Lepomis macrochirus*). *Arch. Environ. Contam. Toxicol.* 10, pp. 519–530, 1981.

Blanck, H. A simple, community level, ecotoxicological test system using samples of periphyton. *Hydrobiologia* 124, pp. 251–261, 1985.

Blaylock, B.G. Radionuclide data bases available for bioaccumulation factors for freshwater biota. *Nucl. Saf.* 23, pp. 427–438, 1982.

Blaylock, B. G. and J. R. Trabalka. Evaluating the effects of ionizing radiation on aquatic organisms. *Adv. Radiat. Biol.* 7, p. 103, 1978.

Bliss, C.I. The calculation of the dosage–mortality curve. *Ann. Appl. Biol.* 22, pp. 134–307, 1935.

Bliss, C.I. The size factor in the action of arsenic upon silkworm larvae. *J. Exp. Biol.* 13, pp. 95–110, 1936.

Blok, J. and F. Balk. Environmental Regulation in the European Community, in *Fundamentals of Aquatic Toxicology. Effects, Environmental Fate, and Risk Assessment*, 2nd ed., Rand, G.M., Ed., Taylor & Francis, Washington, DC, 1995.

Blum, D.J.W. and R.E. Speece. Determining chemical toxicity to aquatic species. *Environ. Sci. Technol.* 24, pp. 284–293. 1990.

Bongers, T. The maturity index: An ecological measure of environmental disturbance based on nematode species composition. *Oecologia (BERL)* 83, pp. 14–19, 1990.

Booth, W. Postmortem on Three Mile Island. *Science (WASH DC)*. 238, pp. 1342–1345, 1987.

Borgmann, U. Metal Speciation and Toxicity of Free Ions to Aquatic Biota, in *Aquatic Toxicology*, Nriagu, J.O., Ed., John Wiley & Sons, New York, 1983.

Bornschein, R.L. and S.-R. Kuang. Behavioral Effects of Heavy Metal Exposure, in *Biological Effects of Heavy Metals*, Foulkes, E.C., Ed., CRC Press, Inc., Boca Raton, FL, 1990.

Borovec, J. Changes in incidence of carcinoma *in situ* after the Chernobyl disaster in Central Europe. *Arch. Environ. Contam. Toxicol.* 29, pp. 266–269, 1995.

Bortone, S.A., W.P. Davis and C.M. Bundrick. Morphological and behavioral characters in mosquitofish as potential bioindication of exposure to Kraft mill effluent. *Bull. Environ. Contam. Toxicol.* 43, pp. 370–377, 1989.

Bouquegneau, J.M. Evidence for the protective effect of metallothioneins against inorganic mercury injuries to fish. *Bull. Environm. Contam. Toxicol.* 23, pp. 218–219, 1979.

Boutron, C.F. and C.C. Patterson. The occurrence of lead in Antarctic recent snow, firn deposited over the last two centuries and prehistoric ice. *Geochim. Cosmochim. Acta* 47, pp. 1355–1368, 1983.

Bowling, J.W., G.J. Leversee, P.F. Landrum, and J.P. Giesy. Acute mortality od anthracene–contaminated fish exposed to sunlight. *Aquat. Toxicol. (AMST)* 3, pp. 79–90, 1983.

Boyden, C.R. Trace element content and body size in molluscs. *Nature (LOND)* 251, pp. 311–314, 1974.

Boyden, C.R. Effect of size upon metal content of shellfish. *J. Mar. Biol. Assoc. UK* 57, pp. 675–714, 1977.

Bradley, R.W. and J.B. Sprague. The influence of pH, water hardness, and alkalinity on the acute lethality of zinc to rainbow trout (*Salmo gairdneri*). *Can. J. Fish. Aquat. Sci.* 42, pp. 731–736, 1985.

Branches, F.J.P., T.B. Erickson, S.E. Aks and D.O. Hryhorczuk. The price of gold: Mercury exposure in the Amazonian rain forest. *J. Toxicol. Clin. Toxicol.* 31, pp. 295–306, 1993.

Brezonik, P.L., S.O. King, and C.E. Mach. The Influence of Water Chemistry on Trace Metal Bioavailability and Toxicity to Aquatic Organisms, in *Metal Ecotoxicology. Concepts and Applications*, Newman, M.C. and A.W. McIntosh, Eds., Lewis Publishers, Chelsea, MI, 1991.

Bricelj, V.M., A.E. Bass, and G.R. Lopez. Absorption and gut passage time of microalgae in a suspension feeder: An evaluation of the ^{51}Cr:^{14}C twin tracer technique. *Mar. Ecol. Prog. Ser.* 17, pp. 57–63, 1984.

Bridgman, H. *Global Air Pollution. Problems for the 1990s.* John Wiley & Sons, New York, 1994, p. 261.

Broad, W.J. Sir Isaac Newton: Mad as a hatter. *Science (WASH DC).* 213, pp. 1341–1344, 1981.

Broderius, S.J., L.L. Smith, Jr., and D.T. Lind. Relative toxicity of free cyanide and dissolved sulfide forms to the fathead minnow (*Pimephales promelas*). *J. Fish. Res. Board Can.* 34, pp. 2323–2332, 1977.

Broman, D., C. Näf, C. Rolff, Y. Zebühr, B. Fry, and J. Hobbie. Using ratios of stable nitrogen to estimate bioaccumulation and flux of polychlorinated dibenzo–p–dioxins (PCDDs) and dibenzofurans (PCDFs) in two food chains from the Northern Baltic. *Environ. Toxicol. Chem.* 11, pp. 331–345, 1992.

Brouwer, A., A.J. Murk, and J.H. Koeman. Biochemical and physiological approaches in ecotoxicology. *Functional Ecol.* 4, pp. 75–281, 1990.

Brown, B.E. Lead detoxification by a copper–tolerant isopod. *Nature (LOND)* 276, pp. 388–390, 1978.

Brown, T.A. and A. Shrift. Selenium: Toxicity and tolerance in higher plants. *Biol. Rev.* 57, pp. 59–84, 1982.

Bruggeman, W.A., L.B.J.M. Martron, D. Kooiman, and O. Hutzinger. Accumulation and elimination kinetics of di–, tri– and tetra chlorobiphenyls by goldfish after dietary and aqueous exposure. *Chemosphere* 10, pp. 811–832, 1981.

Brumley, C.M., V.S. Haritos, J.T. Ahokas, and D.A. Holdway. Validation of biomarkers of marine pollution exposure in sand flathead using Aroclor 1254. *Aquat. Toxicol. (AMST)* 31, pp. 249–262, 1995.

Brunst, V. V. Effects of ionizing radiation on the development of amphibians. *Q. Rev. Biol.* 40, p. 1, 1965.

Bryan, G.W. and P.E. Gibbs. Impact of Low Concentrations of Tributyltin (TBT) on Marine Organisms: A Review, in *Metal Ecotoxicology. Concepts and Applications,* Newman, M.C. and A.W. McIntosh, Eds., Lewis Publishers, Chelsea, MI, 1991.

Buikema, A.L., B.R. Niederlehner, and J. Cairns, Jr. Biological monitoring. Part IV – Toxicity testing. *Water Res.* 16, pp. 239–262, 1982.

Bunce, N. *Environmental Chemistry.* Wuerz Publishing Ltd., Winnipeg, Canada, 1991, p. 339.

Bunzl, K., W. Schimmack, S. V. Krouglov and R. M. Alexakhin. Changes with time in the migration of radiocesium in the soil, as observed near Chernobyl and in Germany, 1986–1994. *Sci. Total Environ.* 175, pp. 49–56, 1995.

Burger, J., M.H. Lavery, and M. Gochfeld. Temporal changes in lead levels in common tern feathers in New York and relationship of field levels to adverse effects in the laboratory. *Environ. Toxicol. Chem.* 13, pp. 581–586, 1994.

Cabana, G. and J.B. Rasmussen. Modelling food chain structure and contaminant bioaccumulation using stable nitrogen isotopes. *Nature (LOND)* 372, pp. 255–257, 1994.

Cabana, G., A. Tremblay, J. Kalff, and J.B. Ramussen. Pelagic food chain structure in Ontario lakes: A determinant of mercury in lake trout (*Salvelinus namaycush*). *Can. J. Fish. Aquat. Sci.* 51, pp. 381–389, 1994.

Cade, T.J., J.L. Lincer, C.M. White, D.G. Roseneau, and L.G. Swartz. 1971. DDE residues and eggshell changes in Alaskan falcons and hawks. *Science (WASH DC).* 172, pp. 955–957, 1971.

Cairns, J., Jr. Heated Waste–Water Effects on Aquatic Ecosystems, in *Thermal Ecology II*, Esch, G.W. and R.W. McFarlane, Eds., National Technical Information Center, Springfield, VA, 1976.

Cairns, J., Jr. The myth of the most sensitive species. *Bioscience* 36, pp. 670–672, 1986.

Cairns, J., Jr. Will there ever be a field of landscape toxicology? *Environ. Toxicol. Chem.* 12, pp. 609–610, 1993.

Cairns, J., Jr., and D.I. Mount. Aquatic toxicology. *Environ. Sci. & Technol.* 24, pp. 154–161, 1990.

Cairns, J., Jr., and J.R. Pratt. Biotic Impoverishment: Effects of Anthropogenic Stress, in *The Earth in Transition: Patterns and Processes of Biotic Impoverishment*, G.M. Woodwell, Ed., Cambridge University Press, Cambridge, 1990.

Cairns, J., Jr., J.R. Pratt, B.R. Niederlehner, and P.V. McCormick. A simple cost–effective multi-species toxicity test using organisms with a cosmopolitan distribution. *Environ. Monit. Assess.* 6, pp. 207–220, 1986.

Calabrese, E.J., M.E. McCarthy, and E. Kenyon. The occurrence of chemically induced hormesis. *Health Phys.* 52, pp. 531–541, 1987.

Camner, P., T.W. Clarkson, and G.F. Nordberg. Route of Exposure, Dose and Metabolism of Metals, in *Handbook on the Toxicology of Metals*, Friberg, L., G.F. Nordberg, and V.B. Vouk, Eds., Elsevier/North–Holland Biomedical Press, Amsterdam, 1979.

Campbell, F.L. Relative susceptibility to arsenic in successive instars of the silkworm. *J. General Physiol.* 9, pp. 727–733, 1926.

Campbell, P.G.C., A.G. Lewis, P.M. Chapman, A.A. Crowder, W.K. Fletcher, B. Imber, S.N. Luoma, P.M. Stokes, and M. Winfrey. *Biologically Available Metals in Sediments*. NRCC No. 27694, NRCC/CNRC Publications, Ottawa, Canada, p. 298, 1988.

Campbell, P.G.C. and A. Tessier. Ecotoxicology of Metals in the Aquatic Environment: Geochemical Aspects, in *Ecotoxicology: A Hierarchical Treatment*, Newman, M.C. and Jagoe, C.H., Eds., CRC Press, Inc., Boca Raton, FL, 1996.

Carlson, A.R., G.L. Phipps, V.R. Mattson, P.A. Kosian, and A.M. Cotter. The role of acid–volatile sulfide in determining cadmium bioavailability and toxicity in freshwater sediments. *Environ. Toxicol. Chem.* 10, pp. 1309–1319, 1991.

Carlson, W. D. and F. X. Gassner. Effects of Ionizing Radiation on the Reproductive System, in *Proc. Int. Symp., Fort Collins, CO*. Pergamon Press, New York, 1964, p. 478.

Carrol, J.J., S.J. Ellis, and W.S. Oliver. Influences of hardness constituents on the acute toxicity of cadmium to brook trout (*Salvelinus fontinalis*). *Bull. Environ. Contam. Toxicol.* 22, pp. 575–581, 1979.

Carroll, L. *Through the Looking–Glass*, in *The Best of Lewis Carroll*. Castle, a division of Book Sales, Inc., Secaucus, NJ, p. 439

Carson, R., *Silent Spring*. Houghton–Mifflin Co., Boston, 1962, p. 368.

Casarett, A. P. *Radiation Biology*. Prentice–Hall, Inc., Englewood Cliffs, NJ, 1968, p. 368.

Casarett, L.J. and J. Doull. *Toxicology. The Basic Science of Poisons*. Macmillan Publishing Co., Inc., New York, 1975, p. 974.

Casti, J.L. *Paradigms Lost. Tackling the Unanswered Mysteries of Modern Science*. Avon Books, New York, 1989, p. 565.

Caswell, H., Demography Meets Ecotoxicology: Untangling the Population Level Effects of Toxic Substances, in *Ecotoxicology: A Hierarchical Treatment*, Newman, M.C. and C.H. Jagoe, Eds., CRC/Lewis Publishers, Boca Raton, FL, 1996.

Catallo, W.J. Ecotoxicology and wetland ecosystems: Current understanding and future needs. *Environ. Toxicol. Chem.* 12, pp. 2209–2224, 1993.

Chadwick, J.W. and S.P. Canton. Coal mine drainage on a lotic ecosystem in Northwest Colorado, U.S.A. *Hydrobiologia* 107, pp. 25–33, 1983.

Chamberlin, T.C. The method of multiple working hypotheses. *J. Geol.* 5, pp. 837–848, 1897.

Chapman, G.A. Sea Urchin Sperm Cell Test, in *Fundamentals of Aquatic Toxicology. Effects, Environmental Fate, and Risk Assessment*, 2nd Ed., Rand, G.M., Ed., Taylor & Francis, Washington, DC, 1995.

Chapman, P.M., R.S. Caldwell, and P.F. Chapman. A warning: NOECs are inappropriate for regulatory use. *Environ. Toxicol. Chem.* 15, pp. 77–79, 1996, Ch. 10.

Chattopadhyay, J. Effect of toxic substances on a two–species competitive system. *Ecol. Model.* 84, pp. 287–289, 1996.

Chernyak, S.M., L.L. McConnell, and C.P. Rice. Fate of some chlorinated hydrocarbons in arctic and far eastern ecosystems in the Russian Federation. *Sci. Total Environ.* 160/161, pp. 75–85, 1995.

Cherry, D.S. and J. Cairns, Jr., Biological monitoring. Part V – Preference and avoidance studies. *Water Res*. 16, pp. 263–301, 1982.

Chew, R.D. and M. A. Hamilton. Toxicity curve estimation: Fitting a compartment model to median survival times. *Trans. Am. Fish. Soc*. 114, pp. 403–412, 1985.

Chiou, C.T. Partition coefficients of organic compounds in lipid–water systems and correlations with fish bioconcentration factors. *Environ. Sci. Technol*. 19, pp. 57–62, 1985.

Choppin, G.R. and J. Rydberg, *Nuclear Chemistry. Theory and Applications*. Pergamon Press, Oxford, 1980, p. 667.

Christensen, E.R. Dose–response functions in aquatic toxicity testing and the Weibull model. *Water Res*. 18, pp. 213–221, 1984.

Christensen, E.R, and N. Nyholm. Ecotoxicological assays with algae: Weibull dose–response curves. *Environ. Sci. Technol*. 18, pp. 713–718, 1984.

Clark, K.E. and D. Mackay. Dietary uptake and biomagnification of four chlorinated hydrocarbons by guppies. *Environ. Toxicol. Chem*. 10, pp. 1205–1217, 1991.

Clark, R. H. Managing radiation risks. Presented at a workshop on: *Pathway Analysis and Risk Assessment for Environmental Compliance and Dose Reconstruction,* Kiawah Island, SC, Nov. 6–10, 1995.

Clayton, J.R., Jr., S.P. Pavlou, and N. F. Breitner. Polychlorinated biphenyls in coastal marine zooplankton: Bioaccumulation by equilibrium partitioning. *Environ. Sci. Technol*. 11, pp. 676–682, 1977.

Clements, W.H. and P.M. Kiffney. Assessing contaminant effects at higher levels of biological organization. *Environ. Toxicol. Chem*. 13, pp. 357–359, 1994.

Cockerham , L.G. and B.S. Shane. *Basic Environmental Toxicology*. CRC Press, Inc., Boca Raton, FL, 1994, p. 627.

Cohen, B.L. Perspective on occupational mortality risks. *Health Phys*. 40, pp. 703–724, 1981.

Cohen, B.L. A test of the linear–no threshold theory of radiation carcinogenesis. *Environ. Res*. 53, pp. 193–220, 1990.

Cohen, B.L. and I.–S. Lee. A catalog of risks. *Health Phys*. 36, pp. 707–722, 1979.

Comins, H.N. The development of insecticide resistance in the presence of migration. *J. Theor. Biol*. 64, pp. 177–197, 1977.

Connell, D.W. *Bioaccumulation of Xenobiotic Compounds*. CRC Press, Inc., Boca Raton, FL, 1990, p. 219.

Connell, D.W. and D.W. Hawker. Use use of polynomial expressions to describe the bioconcentration of hydrophobic chemicals by fish. *Ecotoxicol. Environ. Saf*. 16, pp. 242–257, 1988.

Connolly, J.P. and C.J. Pedersen. A thermodynamic–based evaluation of organic chemical accumulation in aquatic organisms. *Environ. Sci. Technol*. 22, pp. 99–103, 1988.

Cooke, A.S. Shell thinning in avian eggs by environmental pollutants. *Environ. Pollut*. 4, pp. 85–152, 1973.

Cooke, A.S. Egg shell characterisitcs of gannets *Sula bassana*, shags *Phalacrocorax aristotelis* and great black–backed gulls *Larus marinus* exposed to DDE and other environmental pollutants. *Environ. Pollut*. 19, pp. 47–65, 1979.

Cooney, R.V. and A.A. Benson. Arsenic metabolism in *Homarus americanus*. *Chemosphere*. 9, pp. 335–341, 1980.

Cooper, E.L. *Comparative Immunology*. Prentice–Hall, Inc., Englewood Cliffs, NJ, 1976, p. 338.

Cordasco, E.M., S.L. Demeter, and C. Zenz. *Environmental Respiratory Diseases*. Van Nostrand Reinhold, New York, 1995, p. 619.

Corn, M. Corporations viewed as environmental bad guys. *Chem. Eng. News*. May 3, pp. 47–48, 1982.

Correa, M. Physiological effects of metal toxicity on the tropical freshwater shrimp *Macrobrachium carcinus* (Linneo, 1758). *Environ. Pollut*. 45, pp. 149–155, 1987.

Correa, M. and H.I. Garcia. Physiological responses of juvenile white mullet, *Mugil curema*, exposed to benzene. *Bull. Environ. Contam. Toxicol*. 44, pp. 428–434, 1990.

Cosgrove, G. E. Reptilian radiobiology. *J. Am. Vet. Med. Assoc*. 159, p. 1678, 1971.

Cossa, D., E. Bourget, D. Pouliot, J. Piuze, and J.P. Chanut. Geographical and Seasonal Variations in the Relationship Between Trace Metal Content and Body Weight in *Mytilus edulis*. *J. Mar. Biol. Assoc. UK* 58, pp. 7–14, 1980.

Couillard, Y., P.G.C. Campbell, and A. Tessier. Response of metallothionein concentrations in a freshwater bivalve (*Anodonta grandis*) along an environmental cadmium gradient. *Limnol. Oceanogr.* 38, pp. 299–313, 1993.

Cowling, E.B. Acid precipitation in historical perspective. *Environ. Sci. & Technol.* 16, pp. 110A–123A, 1982.

Cowling, E.B. and R.A. Linthurst. The acid precipitation phenomenon and its ecological consequences. *Bioscience.* 31, pp. 649–654, 1981.

Craig, E.A. The heat shock response. *CRC Critical Reviews in Biochemistry* 18, pp. 239–280, 1985.

Craig, E.A. Chaperones: Helpers along the pathways to protein folding. *Science (WASH DC)* 260, pp. 1902–1903, 1993.

Crawford, M. Landmark ozone treaty negotiated. *Science (WASH DC).* 237, pp. 1557–1558, 1987.

Crecelus, E.A., J.T. Hardy, C.I. Bobson, R.L. Schmidt, C.W. Apts, J.M. Gurtisen, and S.P. Joyce. Copper bioavailability to marine bivalves and shrimp: Relationship to cupric ion activity. *Mar. Environ. Res.* 6, pp. 13–26, 1982.

Cremers, A., A. Elsen, P. DePreter and A. Maes. Quantitative analysis of radiocaesium retention in soils. *Nature (LOND)* 335, pp. 247–249, 1988, Ch. 14.

Cronan, C.S. and C.L. Schofield. Aluminum leaching response to acid precipitation: Effects on high–elevation watersheds in the Northeast. *Science (WASH DC)* 204, pp. 304–306, 1979.

Cunningham, P.A., S.L. Smith, J.P. Tippett, and A. Greene. A national fish consumption advisory data base: A step toward consistency. *Fisheries (BETHESDA)* 19, pp. 14–23, 1994.

Curtsinger, J.W., H.H. Fukui, D.R. Townsend and J.W. Vaupel. Demography of genotypes: Failure of the limited life–span paradigm in *Drosophila melanogaster. Science (WASH DC).* 258, pp. 461–463, 1992.

Cushing, C.E. and D.G. Watson. Cycling of Zinc–65 in a Simple Food Web, in *Proceedings of the Third National Symposium on Radioecology, Oak Ridge, Tennessee*, 1971. The Atomic Energy Commision, Oak Ridge National Laboratory, and the Ecological Society of America, Oak Ridge, TN, pp. 318–322, 1971.

Daniels, R.E. and J.D. Allan. Life table evaluation of chronic exposure to a pesticide. *Can. J. Fish. Aquat. Sci.* 38, pp. 485–494, 1981.

Danzmann, R.G., M.M. Feruson, F.W. Allendorf, and K.L. Knudsen. Heterozygosity and developmental rate in a strain of rainbow trout (*Salmo gairdneri*). *Evolution* 40, pp. 86–93, 1986.

Davies, P.H., J.P. Goettl, Jr., J.R. Sinley, and N.F. Smith. Acute and chronic toxicity of lead to rainbow trout *Salmo gairdneri*, in hard and soft water. *Water Res.* 10, pp. 199–206, 1976.

Davis, J.J. and R.F. Foster. Bioaccumulation of radioisotopes through aquatic food chains. *Ecology* 39, pp. 530–535, 1958.

Day, K. and N.K. Kaushik. An assessment of the chronic toxicity of the synthetic pyrethroid, Fenvalerate, to *Daphnia galeata mendotae*, using life tables. *Environ. Pollut.* 44, pp. 13–26, 1987.

Deevey, Jr., E.S. Life tables for natural populations of animals. *Q. Rev. Biol.* 22, pp. 283–314, 1947.

de Lacerda, L.D., W.C. Pfeiffer, A.T. Ott and E.G. da Silveira. Mercury contamination in the Madeira River, Amazon – Hg inputs to the environment. *Biotropica.* 21, pp. 91–93, 1989.

Diamond, J.M., M.J. Parson, D. Gruber. Rapid detection of sublethal toxicity using fish ventilatory behavior. *Environ. Toxicol. Chem.* 9, pp. 3–11, 1990.

Diamond, S.A., M.C. Newman, M. Mulvey, P.M. Dixon, and D. Martinson. Allozyme genotype and time to death of mosquitofish, *Gambusia affinis* (Baird and Girard), during acute exposure to inorganic mercury. *Environ. Toxicol. Chem.* 8, pp. 613–622, 1989.

Dickson, D. Details of 1957 British nuclear accident withheld to avoid endangering U.S. ties. *Science (WASH DC).* 239, pp. 137, 1988.

Di Giulio, R.T., W.H. Benson, B.M. Sanders, and P.A. Van Veld. Biochemical mechanisms: Metabolism, adaptation, and toxicity, in *Fundamentals of Aquatic Toxicology. Effects, Environ-*

mental Fate, and Risk Assessment, 2nd Ed., Rand, G.M., Ed., Taylor & Francis, Washington, DC, 1995.

Di Giulio, R.T., P.C. Washburn, R.J. Wenning, G.W. Winston, and C.S. Jewell. Biochemical responses in aquatic animals: A review of determinants of oxidative stress. *Environ. Toxicol. Chem.* 8, pp. 1103–1123, 1989.

Dillon, T.M. and M.P. Lynch. Physiological Responses as Determinants of Stress in Marine and Estuarine Organisms, in *Stress Effects on Natural Ecosystems*, Barrett, G.W. and R. Rosenberg, Eds., John Wiley & Sons Ltd., 1981.

Di Toro, D.M., J.D. Mahony, D.J. Hansen, K.J. Scott, M.B. Hicks, S.M. Mayr, and M.S. Redmond. Toxicity of cadmium in sediments: The role of acid volatile sulfide. *Environ. Toxicol. Chem.* 9, pp. 1487–1502, 1990.

Di Toro, D.M., C.S. Zarba, D.J. Hansen, W.J. Berry, R.C. Swartz, C.E. Cowan, S.O. Pavlou, H.E. Allen, N.A. Thomas, and P.R. Paquin. Technical basis for establishing sediment quality criteria for nonionic organic chemicals using equilibrium partitioning. *Environ. Toxicol. Chem.* 10, pp. 1541–1583, 1991.

Dixon, D.G. and J.B. Sprague. Acclimation to copper by rainbow trout (*Salmo gairdneri*) – A modifying factor in toxicity. *Can. J. Fish. Aquat. Sci.* 38, pp. 880–888, 1981.

Dixon, D.G. and J.B. Sprague. Acclimation–induced changes in toxicity of arsenic and cyanide in rainbow trout, *Salmo gairdneri* Richardson. *J. Fish Biol.* 18, pp. 579–589, 1981.

Dixon, D.R. and K.R. Clarke. Sister chromatid exchange: A sensitive method for detecting damage caused by exposure to environmental mutagens in the chromosomes of adult *Mytilus edulis*. *Mar. Biol. Lett.* 3, pp. 163–172, 1982.

Dixon, P.M. and M.C. Newman. Analyzing Toxicity Data Using Statistical Models of Time–to–Death: An Introduction, in *Metal Ecotoxicology. Concepts and Applications*, Newman, M.C. and A.W. McIntosh, Eds., Lewis Publishers, Chelsea, MI, 1991.

Dodge, E.A. and T.L. Theis. Effect of chemical speciation on the uptake of copper by *Chironomous tentans*. *Environ. Sci. Technol.* 13, pp. 1287–1288, 1979.

Dolphin, R. 1959 Lake County Mosquito Abatement District Gnat Research Program. Clear Lake Gnat (*Chaoborus astictopus*). *Proceedings of 27th Annual Conference of the California Mosquito Control Association*, pp. 47–48, 1959.

Donkin, S.G. and D.B. Dusenbery. Using the *Caenorhabditis elegans* soil toxicity test to identify factors affecting toxicity of four metal ions in intact soil. *Water Air Soil Pollut.* 86, pp. 359–373, 1994.

Donnelly, K.C., C.S. Anderson, G.C. Barbee, and D.J. Manek. Soil Toxicology, in *Basic Environmental Toxicology*, Cockerham, L.G. and B.S. Shane, CRC Press, Inc., Boca Raton, FL, 1994.

Dopp, E., C.M. Barker, D. Schiffmann, and C.L. Reinisch. Detection of micronuclei in hemocytes of *Mya arenaria*: Association with leukemia and induction with an alkylating agent. *Aquat. Toxicol. (AMST)* 34, pp. 31–45, 1996.

Doust, J.L., M. Schmidt, and L.L. Doust. Biological assessment of aquatic pollution: A review, with emphasis on plants as biomonitors. *Biol. Rev.* 69, pp. 147–186, 1994.

Downs, T.D. and R.F. Frankowski. Influence of repair processes on dose–response models. *Drug Metab. Rev.* 13, pp. 839–852, 1982.

Driscoll, C.T., V. Blette, C. Yan, C.L. Schofield, R. Munson, and J. Holsapple. The role of dissolved organic carbon in the chemistry and bioavailability of mercury in remote Adirondack lakes. *Water Air Soil Pollut.* 80, pp. 499–508, 1995.

Drummond, R.A., G.F. Olson, and A.R. Batterman. Cough response and uptake of mercury by brook trout, *Salvelinus fontinalis*, exposed to mercuric compounds at different hydrogen–ion concentrations. *Trans. Am. Fish. Soc.* 2, pp. 244–249, 1974.

Duce, R.A. and W.H. Zoller. Atmospheric trace metals at remote northern and southern hemisphere sites: Pollution or natural? *Science (WASH DC)* 187, pp. 59–61, 1975.

Duffus, J.H., *Environmental Toxicology*. John Wiley & Sons, New York, 1980, p. 164.

Dunson, W.A. and J. Travis. The role of abiotic factors in community organization. *Am. Nat.* 138, pp. 1067–1091, 1991.

Dwyer, F.J., C.J. Schnitt, S.E. Finger, and P.M. Mehrle. Biochemical changes in longear sunfish, *Lepomis megalotis*, associated with lead, cadmium and zinc from mine tailings. *J. Fish Biol.* 33, p. 307–317, 1988.

Eberhardt, L.L. Relationship of cesium–137 half–life in humans to body weight. *Health Phys.* 13, pp. 88–90, 1967.

Eckerman, K. F., A. B. Wolbrast and A. C. B. Richardson. *Limiting Values of Radionuclide Intake and Air Concentration and Dose Conversion Factors for Inhalation, Submersion, and Ingestion.* Federal Guidance Report, No. 11, EPA–520/1–88–020, U.S. Environmental Protection Agency, 1988, p. 225.

Eckerman, K. F. and J. C. Ryman. *External Exposure to Radionuclides in Air, Water and Soil.* Federal Guidance Report, No. 12, EPA 402–R–93–081, U.S. Environmental Protection Agency, 1993, p. 235.

Edmonds, J.S. and K.A. Francesconi. Isolation and identification of arsenobetaine from the American lobster, *Homarus americanus. Chemosphere.* 10, pp. 1041–1044, 1981.

Edwards, M. Pollution in the former U.S.S.R. Lethal legacy. *Nat. Geogr.* 186, pp. 70–99, 1994.

Ehrlich, P.R. and A.H. Ehrlich. *Extinction, the Causes and Consequences of the Disappearance of Species.* Random House, New York, 1981, p. 305.

Eichhorn, G.L., J.J. Butzow, P. Clark, Y.A. Shin. Studies on Metal Ions and Nucleic Acids, in *Effects of Metals on Cells, Subcellular Elements, and Macromolecules*, Maniloff, J., J.R. Coleman, and M.W. Miller, Eds., Charles C. Thomas Publisher, Springfield, IL, 1970.

Eichhorn, G.L. Active Sites of Biological Macromolecules and Their Interaction with Heavy Metals, in *Ecological Toxicology. Effects of Heavy Metal and Organohalogen Compounds*, McIntyre, A.D. and C.F. Mills, Eds., Plenum Press, New York, 1975.

Eisenbud, M. *Environmental Radioactivity.* Academic Press. New York, 1973.

Ellgehausen, H., J.A. Guth, and H.O. Essner. Factors determining the bioaccumulation potential of pesticides in the individual compartment of aquatic food chains. *Ecotoxicol. Environ. Saf.* 4, pp. 134–157, 1980.

Elliott, J.M. Tolerance and resistance to thermal stress in juvenile Atlantic salmon, *Salmo salar. Freshwater Biol.* 25, pp. 61–70, 1991.

Ellis, D., *Environments at Risk. Case Histories of Impact Assessment.* Springer–Verlag, Berlin, 1989, p. 325.

Elsom, D. 1987. *Atmospheric Pollution. Causes, Effects and Control Policies.* Basil Blackwell, Inc., New York, 1987, p. 322.

Ember, L.R. Environmental lead: Insidious health problem. *Chem. Eng. News.* June 23, pp. 28–35, 1980.

Ember, L.R. EPA study backs cut in lead use in gas. *Chem. Eng. News.* April 9, pp. 18, 1984.

Endler, J.A. *Natural Selection in the Wild.* Princeton University Press, Princeton, NJ, 1986, p. 336.

EPA. *Water Quality Standards Handbook, December 1983.* NTIS, Springfield, VA, 1983, p. 66.

EPA. *Methods for Measuring the Acute Toxicity of Effluents to Freshwater and Marine Organisms, PB85–205383 March 1985.* NTIS, Springfield, VA, 1985a, p. 216.

EPA. *Ambient Water Quality Criteria for Cadmium – 1984, PB85–227031 January 1985.* NTIS, Springfield, VA, 1985b, p. 127.

EPA. *Ambient Water Quality Criteria for Copper – 1984, PB85–227023, January 1985.* NTIS, Springfield, VA, 1985c, p. 142.

EPA. *Ambient Water Quality Criteria for Lead – 1984, PB85–227437, January 1985.* NTIS, Springfield, VA, 1985d, p. 81.

EPA. *Guidelines for Deriving Numerical National Water Quality Criteria for the Protection of Aquatic Organisms and Their Uses, PB85–227049, January 1985.* NTIS, Springfield, VA, 1985e, p. 98.

EPA. *The Enhanced Stream Water Quality Models QUAL2E and QUAL2E–UNCAS: Documentation and User Manual, EPA/600/3–87/007, May 1987.* NTIS, Springfield, VA, 1987a, p. 189.

EPA. *Ambient Water Quality Criteria for Zinc – 1987, PB87–153581, February 1987.* NTIS, Springfield, VA, 1985d, p. 214.

EPA. Radiation Protection Guidance to Federal Agencies for Occupational Exposure. *Federal Register* 52(17), p. 2822, 1987.

EPA. *Short–Term Methods for Estimating the Chronic Toxicity of Effluents and Receiving Waters to Marine and Estuarine Organisms, PB89–220503 May 1988.* NTIS, Springfield, VA, 1988a, p. 415.

EPA. *Ambient Water Quality Criteria for Aluminum – 1988, PB88–245998 August 1988.* NTIS, Springfield, VA, 1988b, p. 47.

EPA. *Short–Term Methods for Estimating the Chronic Toxicity of Effluents and Receiving Waters to Freshwater Organisms, EPA/600/4–89/001 March 1989.* NTIS, Springfield, VA, 1989a, p. 249.

EPA. *Short–Term Methods for Estimating the Chronic Toxicity of Effluents and Surface Waters to Freshwater Organisms. Supplement, PB90–145764 September 1989.* NTIS, Springfield, VA, 1989b, p. 262.

EPA. *Ecological Assessment of Hazardous Waste Sites, EPA 600/3–89/013 March 1989.* NTIS, Springfield, VA, 1989c, p. 260.

EPA. *Risk Assessment Guidance for Superfund. Volume I. Human Health Evaluation Manual (Part A). Interim Final, EPA 540/1–89/002 December 1989.* NTIS, Springfield, VA, 1989d, p. 290.

EPA. *Risk Assessment Guidance for Superfund, Volume II: Environmental Evaluation Manual, EPA 540 1–89 001.* NTIS, Springfield, VA, 1989e, p. 57.

EPA. *Summary Report on Issues in Ecological Risk Assessment, EPA/625/3–91/018 February 1991.* NTIS, Springfield, VA, 1991a, p. 46.

EPA. *MINTEQA2/PRODEFA2, A Geochemical Assessment Model for Environmental Systems: Version 3.0 User's Manual, EPA/600/3–91/021 March 1991.* NTIS, Springfield, VA, 1991b, p. 106.

EPA. *Risk Assessment Guidance for Superfund, Volume I: Human Health Evaluation Manual/Supplemental Guidance–Standard Default Exposure Factors.* OSWER Directive 9285.6–03, 1991c.

EPA. *Ecological Risk Assessment Guidance for Superfund: Process for Designing and Conducting Ecological Risk Assessment, Review Draft, September 26, 1994.*

EPA. *Proposed Guidelines for Ecological Risk Assessment, EPA/630/R–95/002B August 1996.* NTIS, Springfield, VA, 1996, p. 247.

Erickson, R.J. and J.M. McKim. A simple flow–limited model for exchange of organic chemicals at fish gills. *Environ. Toxicol. Chem.* 9, pp. 159–165, 1990.

Erickson, J.M., M. Rahire, and J.–D. Rochaix. Herbicide resistance and cross–resistance: Changes at three distinct sites in the herbicide–binding protein. *Science (WASH DC)* 228, pp. 204–207, 1985.

Evans, D.H. The fish gill: Site of action and model for toxic effects of environmental pollutants. *Environ. Health Perspect.* 71, pp. 47–58, 1987.

Evans, H.J. Leukaemia and radiation. *Nature (LOND)* 345, pp. 16–17, 1990.

Evans, M.S., G.E. Noguchi, and C.P. Rice. The biomagnification of polychlorinated biphenyls, toxaphene, and DDT compounds in a Lake Michigan offshore food web. *Arch. Environ. Contam. Toxicol.* 20, pp.87–93, 1991.

Exeley, C., J.S. Chappell, and J.D. Birchall. A mechanism of acute aluminum toxicity in fish. *J. Theor. Biol.* 151, pp. 417–428, 1991.

Fabacher, D.L. and H. Chambers. Rotenone tolerance in mosquitofish. *Environ. Pollut.* 3, 139–141, 1972.

Fagerström, T. Body weight, metabolic rate, and trace substance turnover in animals. *Oecologia (BERL)* 29, pp. 99–104, 1977.

Fairchild, J.F., T. Boyle, W.R. English, and C. Rabeni. Effects of sediment and contaminated sediment on structural and functional components of experimental stream ecosystems. *Water Air Soil Pollut.* 36, pp. 271–293, 1987.

Ferson, S. and H.R. Akçakaya. *Modeling Fluctuations in Age–structured Populations.* Exeter Software, Inc., Seatauket, NY, 1991, p. 146.

Finney, D.J. *Statistical Method in Biological Assay*. Charles Griffin & Company Ltd., London, 1971, p. 668.

Fisher, N.S., J.–L. Teyssié, S. Krishnaswami, and M. Baskaran. Accumulation of Th, Pb, U, and Ra in marine phytoplankton and its geochemical significance. *Limnol. Oceanogr.* 32, pp. 131–142, 1987.

Fisher, R.A., A.S. Corbet, and C.B. Williams. The relation between the number of species and the number of individuals in a random sample of an animal population. *J. An. Ecol.* 12, pp. 42–58, 1943.

Forbes, V.E. and T.L. Forbes, *Ecotoxicology in Theory and Practice*. Chapman & Hall, London, 1994, p. 247.

Ford, J. The Effects of Chemical Stress on Aquatic Species Composition and Community Structure, in *Ecotoxicology: Problems and Approaches*, Levin, S.A., M.A. Harwell, J.R. Kelly, K.D. Kimball, Eds., Springer–Verlag, New York, 1989.

Forni, A. Chromosomal Effects of Lead. A Critical Review, in *Reviews on Environmental Health, Vol. III*, James, G.V., Ed., Freund Publishing House, Ltd., Tel–Aviv, Israel, 1980.

Foster, R.B. Environmental Legislation, in *Fundamentals of Aquatic Toxicology*, Rand, G.M. and S.R. Petrocelli, Eds., Hemisphere Publishing Corp., Washington, DC, 1985.

Fox, G.A., S.W. Kennedy, R.J. Norstrom, and D.C. Wigfield. Porphyria in herring gulls: A biochemical response to chemical contamination of Great Lakes food chains. *Environ. Toxicol. Chem.* 7, pp. 831–839, 1988.

Frankel, E.G. *Ocean Environmental Management. A Primer on the Role of the Oceans and How to Maintain their Contributions to Life on Earth*. Prentice Hall PTR, Englewood Cliffs, NJ, 1995, p. 381.

Franson, J.C., L. Sileo, and N.J. Thomas. Causes of Eagle Deaths, in *Our Living Resources*, LaRoe, E.T., G.S. Farris, C.E. Puckett, P.D. Doran, and M.J. Mac, Eds., U.S. Dept. of the Interior – National Biological Service, Washington, DC, 1995.

Freedman, W. Regulation in the public interest, *Federal Statutes on Environmental Protection*. Quorum Books, New York, 1987, p. 174.

French, N.R. Comparison of Radioisotope Assimilation by Granivorous and Herbivorous Mammals, in *Radioecological Concentration Processes, Proceedings of an International Symposium Held in Stockholm, 1966*, B. Åberg and F.P. Hungate, Eds., Pergamon Press, Inc., New York, 1967, pp. 665–673.

Fromm, P.O. A review of some physiological and toxicological responses of freshwater fish to acid stress. *Env. Biol. Fishes* 5, pp. 79–93, 1980.

Fromm, P.O. and J.R. Gillette. Effect of ambient ammonia on blood ammonia and nitrogen excretion of rainbow trout (*Salmo gairdneri*). *Comp. Biochem. Physiol.* 26, pp. 887–896, 1968.

Fry, B. Stable isotope diagrams of freshwater food webs. *Ecology* 72, pp. 2293–2297, 1991.

Fry, B. Food web structure on Georges Bank from stable C, N, and S isotopic compositions. *Limnol. Oceanogr.* 33, pp. 1182–1190, 1988.

Fry, D.M. and C.K. Toone. DDT–induced feminization of gull embryos. *Science (WASH DC)* 213, pp. 922–924, 1981.

Gächter, R. and W. Geiger. MELIMEX, an experimental heavy metal pollution study: Behavior of heavy metals in an aquatic food chain. *Schweiz. Z. Hydrol.* 41, pp. 277–290, 1979.

Gad, S.C. Statistical analysis of behavioral toxicology data and studies. *Arch. Toxicol. Suppl.* 5, pp. 256–266, 1982.

Gaddum, J.H. Bioassays and mathematics. *Pharacol. Rev.* 5, pp. 87–134, 1953.

Gallegos, A.F. and F.W. Whicker. Radiocesium Retention by Rainbow Trout as Affected by Temperature and Weight, in *Proceedings of the Third National Symposium on Radioecology*, Oak Ridge National Laboratories, Oak Ridge, TN, pp. 361–371, 1971.

Gallopin, G. and S. Öberg. Quality of life, in *An Agenda of Science for Environment and Development into the 21st Century*, Dooge, J.C.I., G.T. Goodman, J.W.M. la Rivière, J. Marton–Lefèvre, T. O'Riordan, and F. Praderie, Eds., Cambridge University Press, Cambridge, 1992.

Galtsoff, P.S. *The American Oyster Crassostrea virginica Gmelin*, Fishery Bull. of the Fish and Wildlife Service Vol. 64, U.S. Printing Office, Washington, 1964, p. 480.

Gardner, M.J., M.P. Snee, A.J. Hall, C.A. Powell, S. Downes, and J.D. Terrell. Results of case–control study of leukaemia and lymphoma among young people near Sellafield nuclear plant in West Cumbria. *Br. Med. J.* 300, pp. 423–434, 1990.

Garvey, J.S. Metallothonein: A Potential Biomonitor of Exposure to Environmental Toxins, in *Biomarkers of Environmental Contamination*, McCarthy, J.F. and L.R. Shugart, Eds., Lewis Publishers, Boca Raton, FL, 1990.

Gaylor, D.W., F.F. Kadlubar, and F.A. Beland. Application of biomarkers to risk assessment. *Environ. Health Perspect.* 98, pp. 139–141, 1992.

Geisel, T.S. and A.S. Geisel., *The Lorax*. Random House, New York, 1971, p. 70.

George, S.G. Enzymology and Molecular Biology of Phase II Xenobiotic–Conjugating Enzymes in Fish, in *Aquatic Toxicology. Molecular, Biochemical and Cellular Perspectives*, Malins, D.C. and G.K. Ostrander, Eds., CRC Press, Inc., Boca Raton, FL, 1994.

Gerhart, D.Z., S.M. Anderson, and J. Richter. Toxicity bioassays with periphyton communities: Design of experimental streams. *Water Res.* 11, pp. 567–570, 1977.

Geyer, H., D. Sheehan, D. Kotzias, D. Freitag and F. Korte. Prediction of ecotoxicological behavior of chemicals: relationship between physiochemical properties and bioaccumulation of organic compounds in the mussel. *Chemosphere* 11, pp. 1121–1134, 1982.

Giattina, J.D. and R.R. Garton. A review of the preference–avoidance responses of fishes to aquatic contaminants. *Residue Rev.* 87, pp. 43–90, 1983.

Gibaldi, M. *Biopharmaceutics and Clinical Pharmacokinetics*. Lea and Febiger, Philadelphia, PA, 1991, p. 406.

Gibaldi, M. and D. Perrier. *Pharmacokinetics*, 2nd ed. Marcel Dekker, Inc., New York, 1982, p. 494.

Gibbs, M.H., L.F. Wicker, and A.J. Stewart. A method for assessing sublethal effects of contaminants in soils to the earthworm, *Eisenia foetida*. *Environ. Toxicol. Chem.* 15, pp. 360–368, 1996.

Giesy, J.P. Cadmium inhibition of leaf decomposition in an aquatic microcosm. *Chemosphere* 6, pp. 467–475, 1978.

Giesy, J.P., S.R. Denzer, C.S. Duke, and G.W. Dickson. Phosphoadenylate concentrations and energy charge in two freshwater crustaceans: Responses to physical and chemical stressors. *Verh. Internat. Verein. Limnol.* 21, pp. 205–220, 1981.

Giesy, J.P. and R.A. Hoke. Freshwater sediment quality criteria: Toxicity bioassessment, in *Sediments: Chemistry and Toxicity of In–Place Pollutants*, Baudo, R., J.P. Giesy, and H. Muntau, Eds., Lewis Publishers, Inc., Chelsea, MI, 1990.

Gillespie, R.B. Allozyme Frequency Variation as an Indicator of Contaminant–Induced Impacts in Aquatic Populations, in *Techniques in Aquatic Toxicology*, Ostrander, G.K., Ed., CRC Press, Inc., Boca Raton, FL, 1996.

Gillespie, R.B. and S.I. Guttman. Effects of contaminants on the frequencies of allozymes in populations of the central stoneroller. *Environ. Toxicol. Chem.* 8, pp. 309–317, 1989.

Gillett, J.W. The Role of Terrestrial Microcosms and Mesocosms in Ecotoxicological Research, in *Ecotoxicology: Problems and Approaches*, Levin, S.A., M.A. Harwell, J.R. Kelly, and K.D. Kimball, Eds., Springer–Verlag, New York, 1989.

Glass, N.R., D.E. Arnold, J.N. Galloway, G.R. Hendrey, J.J. Lee, W.W. McFee, S.A. Norton, C.F. Powers, D.L. Rambo and C.L. Schofield. Effects of acidic precipitation. *Environ. Sci. & Technol.* 16, pp. 163A–169A, 1982.

Gobas, F.A.P.C. and D. Mackay. Dynamics of hydrophobic organic chemical bioconcentration in fish. *Environ. Toxicol. Chem.* 6, pp. 495–504, 1987.

Gobas, F.A.P.C., J.R. McCorquodale, and G.D. Haffner. Intestinal absorption and biomagnification of organochlorines. *Environ. Toxicol. Chem.* 12, pp. 567–576, 1993.

Goksøyr, A. and L. Förlin. The cytochrome P–450 system in fish, aquatic toxicology and environmental monitoring. *Aquat. Toxicol. (AMST)* 22, pp. 287–312, 1992.

Goldberg, E.D. The mussel watch concept. *Environ. Monit. Assess.* 7, pp. 91–103, 1986.

Goldsborough, L.G. and G.G.C. Robinson. Changes in periphytic algal community structure as a consequence of short herbicide exposures. *Hydrobiologia* 139, pp. 177–192, 1986.

Golley, F.B. *A History of the Ecosystem Concept in Ecology. More Than the Sum of the Parts.* Yale University, New Haven, CT, 1993, p. 254.

Goodyear, C.P. A simple technique for detecting effects of toxicants or other stresses on a predator–prey interaction. *Trans. Am. Fish. Soc.* 101, pp. 367–370, 1972.

Gore, A. *Earth in the Balance. Ecology and the Human Spirit.* Penguin Books USA, Inc., New York, 1992, p. 407.

Gorman, M., *Environmental Hazards. Marine Pollution.* ABC–CLIO, Inc., Santa Barbara, CA, 1993, p. 252.

Gowen, J. W. and J. Stadler. Acute Irradiation Effects on Reproductivity of Different Strains of Mice, in *Effects of Ionizing Radiation on the Reproductive System,* Carlson, W. D. and F. X. Gassner, Eds., Pergamon Press, New York, 1964, p. 478.

Graham, J.H., J.M. Emlen, and D.C. Freeman. Developmental stability and its applications in ecotoxicology. *Ecotoxicology* 2, pp. 175–184, 1993b.

Graham, J.H., D.C. Freeman, and J.M. Emlen. Developmental Stability: A Sensitive Indicator of Populations Under Stress, in *Environmental Toxicology and Risk Assessment, ASTM STP 1179,* Landis, W.G., J.S. Hughes, and M.A. Lewis, Eds., American Society for Testing and Materials, Philadelphia, PA, 1993a.

Graham, J.H., K.E. Roe, and T.B. West. Effects of lead and benzene on the developmental stability of *Drosophilia melanogaster. Ecotoxicology* 2, pp. 185–195, 1993b.

Grandy, N.J. Role of the OECD in Chemicals Control and International Harmonization of Testing Methods, in *Fundamentals of Aquatic Toxicology. Effects, Environmental Fate, and Risk Assessment,* 2nd ed., Rand, G.M., Ed., Taylor & Francis, Washington, DC, 1995.

Graney, R.L., Jr., D.S. Cherry, and J. Cairns, Jr. The influence of substrate, pH, diet and temperature upon cadmium accumulation in the asiatic clam (*Corbicula fluminea*) in laboratory artificial streams. *Water Res.* 18, pp. 833–842, 1984.

Gray, J.S. Pollution–induced changes in populations. *Phil. Trans. Roy. Soc. Lond.* 286, pp. 545–561, 1979.

Gray, R.H. Fish behavior and environmental assessment. *Environ. Toxicol. Chem.* 9, pp. 53–67, 1990.

Green, A.S. and G.T. Chandler. Life–table evaluation of sediment–associated chlorpyrifos chronic toxicity to the benthic copepod, *Amphiascus tenuiremis. Arch. Environ. Contam. Toxicol.* 31, pp. 77–83, 1996.

Grill, E., E.–L. Winnacker, and M.H. Zenk. Phytochelatins: The principal heavy–metal complexing peptides in higher plants. *Science (WASH DC)* 230, pp. 674–676, 1985.

Grosch, D.S. *Biological Effects of Radiations.* Blaisdell Publishing Co., Waltham, MA, 1965, p. 293.

Haasch, M.L., R. Prince, P.J. Wejksnora, K.R. Cooper, and J.J. Lech. Caged and wild fish: Induction of hepatic cytochrome P–450 (CYP1A1) as an environmental monitor. *Environ. Toxicol. Chem.* 12, pp. 885–895, 1993.

Hall, E. J. *Radiobiology for the Radiologist.* Harper & Row Publishers, San Francisco, CA, 1978, p. 460.

Hall, R.J., Impact of Pesticides on Bird Populations, in *Silent Spring Revisited,* Marco, G.J., R.M. Hollingworth and W. Durham, Eds., American Chemical Society, Washington, DC, 1987.

Ham, L., R. Quinn, and D. Pascoe. Effects of cadmium on the predator–prey interaction between the turbellarian *Dendrocoelum lacteum* (Müller, 1774) and the isopod crustacean *Asellus aquaticus* (L.). *Arch. Environ. Contam. Toxicol.* 29, pp. 358–365, 1995.

Hamilton, M.A., R.C. Russo, and R.V. Thurston. Trimmed Spearman–Karber method for estimating median lethal concentrations in toxicity bioassays. *Environ. Sci. Technol.* 11, pp. 714–719, 1977.

Hamilton, S.J. and P.M. Mehrle. Metallothionein in fish: Review of its importance in assessing stress from metal contaminants. *Trans. Amer. Fish. Soc.* 115, pp. 596–609, 1986.

Hansen, L.G. and B.S. Shane. Xenobiotic Metabolism, in *Basic Environmental Toxicology*, Cockerham, L.G. and B.S. Shane, Eds., CRC Press, Inc., Boca Raton, FL, 1994.

Harrison, F.L. and I.M. Jones. An *in vivo* sister–chromatid exchange assay in the larvae of the mussel *Mytilus edulis*: Response to 3 mutagens. *Mutat. Res.* 105, pp. 235–242, 1982.

Hatakeyama, S. and M. Yasuno. Effects of cadmium on the periodicity of parturition and brood size of *Moina macrocopa* (Cladocera). *Environ. Pollut.* 26, pp. 111–120, 1981.

Haux, C. and L. Förlin. Biochemical methods for detecting effects of contaminants on fish. *Ambio* 17, pp. 376–380, 1988.

Hayton, W.L. Pharmacokinetic parameters for interspecies scaling using allometric techniques. *Health Phys.* 57 (Sup. 1), pp. 159–164, 1989.

Heading, R.C., J. Nimmo, L.F. Prescott, and P. Tothill. The dependence of paracetamol absorption on the rate of gastric emptying. *Br. J. Pharmacol.* 47, pp. 415–421, 1973.

Heagler, M.G., M.C. Newman, M. Mulvey, and P.M. Dixon. Allozyme genotype in mosquitofish, *Gambusia holbrooki*: Temporal stability, concentration effects and field verification. *Environ. Toxicol. Chem.* 12, pp. 385–395, 1993.

Hedtke, S.F. Structure and function of copper–stressed aquatic microcosms. *Aquat. Toxicol. (AMST)* 5, pp. 227–244, 1984.

Hennig, H.F.–K.O. Metal–binding proteins as metal pollution indicators. *Environ. Health Perspect.* 65, pp. 175–187, 1986.

Henny, C.J. and G.B. Herron. DDE, selenium, mercury, and white–faced ibis reproduction at Carson Lake, Nevada. *J. Wildl. Manage.* 53, pp. 1032–1045, 1989.

Henry, M.G. and G.J. Atchison. Influence of social rank on the behavior of bluegill, *Lepomis macrochirus* Rafinesque, exposed to sublethal concentrations od cadmium and zinc. *J. Fish. Biol.* 15, pp. 309–315, 1979.

Henry, M.G. and G.J. Atchison. Metal Effects on Fish Behavior – Advances in Determining the Ecological Significance of Responses, in *Metal Ecotoxicology. Concepts & Applications*, Lewis Publishers, Inc., Chelsea, MI, 1991.

Herricks, E.E. and J. Cairns, Jr., Biological monitoring. Part III – Receiving system methodology based on community structure. *Water Res.* 16, pp. 141–153, 1982.

Hesslein, R.H., M.J. Capel, D.E. Fox, and K.A. Hallard. Stable isotopes of sulfur, carbon, and nitrogen as indicators of trophic level and fish migration in the Lower Mackenzie River Basin, Canada. *Can. J. Fish. Aquat. Sci.* 48, pp. 2258–2265, 1991.

Heusner, A.A. What does the power function reveal about structure and function in animals of different size. *Annual Rev. Physiol.* 49, pp. 121–133, 1987.

Heylin, M. Bhopal. *Chem. Eng. News.* Feb. 11, pp. 14–15, 1985.

Hickey, C.W., D.S. Roper, P.T. Holland, and T.M. Trower. Accumulation of organic contaminants in two sediment–dwelling shellfish with contrasting feeding modes: Deposit– (*Macomona liliana*) and filter–feeding (*Austrovenus stutchburyi*). *Arch. Environ. Contam. Toxicol.* 29, pp. 221–231, 1995.

Hickey, D.A. and T. McNeilly. Competition between metal tolerant and normal plant populations: A field experiment on normal soil. *Evolution* 29, pp. 458–464, 1975.

Hickey, J.J. and D.W. Anderson. Chlorinated hydrocarbons and eggshell changes in raptorial and fish–eating birds. *Science (WASH DC).* 162, pp. 271–273, 1968.

Hightower, L.E. Heat shock, stress proteins, chaperons, and proteotoxicity. *Cell* 66, pp. 191–197, 1991.

Hileman, B. Acid fog. *Environ. Sci. Technol.* 17, pp. 117A–120A, 1983.

Hileman, B. Recent reports on the greenhouse effect. *Environ. Sci. Technol.* 18, pp. 454–455, 1984.

Hileman, B. Global warming. *Chem. Eng. News.*, March 13, pp. 25–44, 1989.

Hill, A.B. The environment and disease: Association or causation? *Proc. R. Soc. Med.* 58, pp. 295–300, 1965.

Hinton, D.E. Cells, Cellular Responses, and Their Markers in Chronic Toxicity of Fishes, in *Aquatic Toxicology. Molecular, Biochemical and Cellular Perspectives*, Malins, D.C. and G.K. Ostrander, Eds., CRC Press, Inc., Boca Raton, FL, 1994.

Hinton, D.E. and D.J. Laurén. Integrative histopathological approaches to detecting effects of environmental stressors on fishes. *Amer. Fish. Soc. Symp.* 8, pp. 51–66, 1990a.

Hinton, D.E. and D.J. Laurén. Liver Structural Alterations Accompanying Chronic Toxicity in Fishes: Potential Biomarkers of Exposure, in *Biomarkers of Environmental Contamination*, McCarthy, J.F. and L.R. Shugart, Eds., Lewis Publishers, Boca Raton, FL, 1990b.

Hirao, Y. and C.C. Patterson. Lead aerosol pollution in the High Sierra overrides natural mechanisms which exclude lead from a food chain. *Science (WASH DC)* 184, pp. 989–992, 1974.

Hobson, J.F. and W.J. Birge. Acclimation–induced changes in toxicity and induction of metal-lothionein–like proteins in the fathead minnow following sublethal exposure to zinc. *Environ. Toxicol. Chem.* 8, pp. 157–169, 1989.

Hoekstra, J.A. and P.H. Van Ewijk. Alternatives for the No–Observed–Effect–Level. *Environ. Toxicol. Chem.* 12, pp. 187–194, 1993.

Hoffman, D.J., B.A. Rattner, G.A. Burton, Jr., and J. Cairns, Jr., *Handbook of Ecotoxicology*. CRC Press, Boca Raton, FL, 1995, p. 755.

Hoffman, F. O. Presented at a workshop on: *Pathway Analysis and Risk Assessment for Environmental Compliance and Reconstruction,* Kiawah Island, SC, 1991.

Holl, K.D. and J. Cairns, Jr. Landscape indicators in ecotoxicology, in *Handbook of Ecotoxicology*, Hoffman, D.J., B.A. Rattner, G.A. Burton, Jr., J. Cairns, Jr., Eds., Lewis Publishers, Boca Raton, FL, 1995.

Holloway, G.J., R.M. Sibly, and S.R. Povey. Evolution in toxin–stressed environments. *Functional Ecol.* 4, pp. 289–294, 1990.

Hopkin, S.P. Ecophysiological Strategies of Terrestrial Arthropods for Surviving Heavy Metal Pollution, in *Proceedings of the 3rd European Congress of Entomology, Amsterdam, 1986*. Nederlandse Entomologische Vereniging, Amsterdam, The Netherlands, pp. 263–266, 1986.

Hopkin, S.P., *Ecophysiology of Metals in Terrestrial Invertebrates*. Elsevier Applied Science, London, 1989, p. 366.

Hopkin, S.P. Ecological implications of "95% protection levels" for metals in soil. *Oikos* 66, pp. 137–141, 1993.

Hopkin, S.P. *In Situ* Biological Monitoring of Pollution in Terrestrial and Aquatic Ecosystems, in *Handbook of Ecotoxicology*, Calow, P., Ed., Blackwell Scientific Publications, London, 1993.

Hopkin, S.P., C.A.C. Hames, and A. Dray. X–ray microanalytical mapping of the intracellular distribution of pollutant metals. *Microscopy and Analysis* November 1989, pp. 1–4, 1989.

Hopkin, S.P. and J.A. Nott. Some observations on concentrically structured, intracellular granules in the hepatopancreas of the shore crab, *Carcinis maenas* (L.). *J. Mar. Biol. Assoc. UK* 59, pp. 867–877, 1979.

Howard, B., P.C.H. Mitchell, A. Ritchie, K. Simkiss, and M. Taylor. The composition of invertebrate granules from metal–accumulating cells of the common garden snail (*Helix aspersa*). *Biochem. J.* 194, pp. 507–511, 1981.

Howell, W.M., D.A. Black, and S.A. Bortone. Abnormal expression of secondary sex characteristics in a population of mosquitofish, *Gambusia affinis holbrooki*: Evidence for environmentally–induced masculinization. *Copeia* 1980, pp. 676–681, 1980.

Howson, C. and P. Urbach. *Scientific Reasoning. The Bayesian Approach.* Open Court Publishing Co., La Salle, IL, 1989, p. 312.

Huckabee, J.W., J.W. Elwood, and S.G. Hildebrand. Accumulation of Mercury in Freshwater Biota, in *The Biogeochemistry of Mercury in the Environment*, Nriagu, J.O., Ed., Elsevier/North–Holland Biomedical Press, Amsterdam, 1979.

Hughes, R.M. and D.P. Larsen, Ecoregions: An approach to surface water protection. *J. Water Pollut. Control Fed.* 60, pp. 486–493, 1988.

Hulett, Jr., L.D., A.J. Weinberger, K.J. Northcutt and M. Ferguson. Chemical species in fly ash from coal–burning power plants. *Science (WASH DC)* 210, pp. 1356–1358, 1980.

Hunt, E.G. and A.I. Bischoff. Inimical effects on wildlife of periodic DDD applications to Clear Lake. *Calif. Fish Game.* 46, pp. 91–106, 1960.

Hunt, G.L. and M.W. Hunt, Jr. Female–female pairing in western gulls (*Larus occidentalis*) in southern California. *Science (WASH DC)* 196, pp. 1466–1467, 1977.

Hursh, J.B., M.R. Greenwood, T.W. Clarkson, J. Allen, and S. Demuth. The effect of ethanol on the fate of mercury vapor inhaled by man. *J. Pharmacol. Exp. Ther.* 214, pp. 520–527, 1980.

Huxley, J.S. Relative growth and form transformation. *Proc. R. Soc. London B. Biol.* 137, pp. 465–470, 1950.

IAEA (International Atomic Energy Agency). *Effects of Ionizing Radiation on Plants and Animals at Levels Implied by Current Radiation Protection Standards.* IAEA, Vienna, Austria, 1992, p. 73.

Ibsen, I. *Peer Gynt.* New American Library of World Literature, Inc., New York, 1964, p. 253.

ICRP (International Commission on Radiological Protection). *Recommendations of the International Commission on Radiological Protection.* ICRP Publication 26, Annals of the ICRP 1 ICRP, Pergamon Press, New York, 1977, p. 53.

ICRP (International Commission on Radiological Protection). *Radionuclide Transformations: Energy and Intensity of Emissions.* ICRP Publication 38, Annals of the ICRP, Vols. 11–13, Pergamon Press, New York, 1983, p. 1250.

ICRP (International Commission on Radiological Protection). *1990 Recommendations of the International Commission on Radiological Protection.* ICRP Publication 60, Annals of the ICRP 21, Pergamon Press, New York, 1991, p. 211.

Jagoe, C.H. Responses at the Tissue Level: Quantitative Methods in Histopathology Applied to Ecotoxicology, in Newman, M.C. and C.H. Jagoe, *Ecotoxicology: A Hierarchical Treatment,* CRC Press, Inc., Boca Raton, FL, 1996.

Jagoe, C.H., A. Faivre, and M.C. Newman. Morphological and morphometric changes in the gills of mosquitofish (*Gambusia holbrooki*) after exposure to mercury (II). *Aquat. Toxicol. (AMST)* 34, pp. 163–183, 1996.

Jagoe, R. and M.C. Newman. Bootstrap estimation of community NOEC values. *Ecotoxicology* 6, pp. 293–306, 1997.

Janes, N. and R.C. Playle. Modeling silver binding to gills of rainbow trout (*Oncorhynchus mykiss*). *Environ. Toxicol. Chem.* 14, pp. 1847–1858, 1995.

Janicki, R.H. and W.B. Kinter. DDT: Disrupted osmoregulatory events in the intestine of the eel *Anguilla rostrata* adapted to seawater. *Science (WASH DC)* 173, pp. 1146–1148, 1971.

Jobling, S., D. Sheahan, J.A. Osborne, P. Matthiessen, and J.P. Sumpter. Inhibition of testicular growth in rainbow trout (*Oncorhychus mykiss*) exposed to estrogenic alkylphenolic chemicals. *Environ. Toxicol. Chem.* 15, pp. 194–202, 1996.

Johansson–Sjöbeck, M.–L. and Å. Larsson. The effect of cadmium on the hematology and on the activity of δ–aminolevulinic acid dehydratase (ALA–D) in blood and hematopoietic tissues of the flounder, *Pleuronectes flesus* L. *Environ. Res.* 17, pp. 191–204, 1978.

Johansson–Sjöbeck, M.–L. and Å. Larsson. Effects of inorganic lead on delta–aminolevulinic acid dehydratase activity and hemotological variables in the rainbow trout, *Salmo gairdnerii.* *Arch. Environ. Contam. Toxicol.* 8, pp. 419–431, 1979.

Johns, J. E. and J. R. Cunningham. *The Physics of Radiology.* Charles C. Thomas, Springfield, IL, 1980, p. 800.

Jones, K.A. and T.J. Hara. Behavioral alterations in Arctic char (*Salvelinus alpinus*) briefly exposed to sublethal chlorine levels. *Can. J. Fish. Aquat. Sci.* 45, pp. 749–752, 1988.

Jones, N.J. and J.M. Parry. The detection of DNA adducts, DNA base changes and chromosome damage for the assessment of exposure to genotoxic pollutants. *Aquat. Toxicol. (AMST)* 22, pp. 323–344, 1992.

Jørgensen, S.E., *Modelling in Ecotoxicology.* Elsevier, New York, 1990, p. 360.

Josephson, J.R. and S.G. Josephson. *Abductive Inference. Computation, Philosophy, Technology.* Cambridge University Press, Cambridge, UK, 1996, p. 306.

Jung, R.E. and C.H. Jagoe. Effects of low pH and aluminum on body size, swimming performance, and susceptibility to predation of green tree frog (*Hyla cinerea*) tadpoles. *Can. J. Zool.* 73, pp. 2171–2183, 1995.

Kac, M. Some mathematical models in science. *Science (WASH DC)* 166, pp. 695–699, 1969.

Kahn, B. and K.S. Turgeon. The bioaccumulation factor for phosphorus–32 in edible fish tissue. *Health Phys.* 46, pp. 321–333, 1984.

Kaiser, K.L., M.B. McKinnon, D.H. Stendahl, and W.B. Pett. Response threshold levels of selected organic compounds for rainbow trout (*Oncorhynchus mykiss*). *Environ. Toxicol. Chem.* 14, pp. 2107–2113, 1995.

Kammenga, J.E., M. Busschers, N.M. Van Straalen, P.C. Jepson, and J. Bakker. Stress induced fitness is not determined by the most sensitive life–cycle trait. *Functional Ecol.* 10, pp. 106–111, 1996.

Kania, H.K. and J. O'Hara. Behavioral alterations in a simple predator–prey system due to sublethal exposure to mercury. *Trans. Am. Fish. Soc.* 103, pp. 134–136, 1974.

Karin, M. and H.R. Herschman. Induction of metallothionein in HeLa cells by dexamethasone and zinc. *European J. Biochemistry* 113, pp. 267–272, 1981.

Kaufman, L.H. Stream Aufwuchs accumulation: Disturbance frequency and stress resistance and resilience. *Oecologia (BERL)* 52, pp. 57–63, 1982.

Keklak, M.M., M.C. Newman, and M. Mulvey. Enhanced uranium tolerance of an exposed population of the Eastern moquitofish (*Gambusia holbrooki* Girard 1859). *Arch. Environ. Contam. Toxicol.* 27, pp. 20–24, 1994.

Kelce, W.R., C.R. Stone, S.C. Laws, L.E. Gray, J.A. Kemppainen, and E.M. Wilson. Persistent DDT metabolite p,p'–DDE is a potent androgen receptor antagonist. *Nature (LOND)* 375, pp. 581–585, 1995.

Kerr, R.A. Ozone hole bodes ill for the globe. *Science (WASH DC)* 241, pp. 785–786, 1988a.

Kerr, R.A. Is there life after climate change? *Science (WASH DC)* 242, pp. 1010–1013, 1988b.

Kerr, R.A. New assaults seen on Earth's ozone shield. *Science (WASH DC)* 255, pp. 797–798, 1992.

Kerr, R.A. Ozone–destroying chlorine tops out. *Science (WASH DC)* 271, p. 32, 1996.

Kersting, K. Normalizing ecosystem strain: A system parameter for analysis of toxic stress in (micro–) ecosystems. *Ecol. Bull.* 36, pp. 150–153, 1984.

Kettlewell, H.B.D. Selection experiments on industrial melanism in the *Lepidoptera*. *Heredity* 9, pp. 323–342, 1955.

Khalil, A.M. Chromosome aberrations in blood lymphocytes from petroleum refinery workers. *Arch. Environ. Contam. Toxicol.* 28, pp. 236–239, 1995.

Khalil, M.A.K. and R.A. Rasmussen. Carbon monoxide in the earth's atmosphere: Increasing trend. *Science (WASH DC)* 224, pp. 54–56, 1984.

Khayat, A.I. and Z.A. Shaikh. Dose–effect relationship between ethyl alcohol pretreatment and retention and tissue distribution of mercury vapor in rats. *J. Pharmacol. Exp. Ther.* 223, pp. 649–653, 1982.

Khera, K.S. Teratogenic and Genetic Effects of Mercury, in *The Biogeochemistry of Mercury in the Environment*, J.O. Nriagu, Ed., Elsevier/North–Holland Biomedical Press, Amsterdam, 1979.

Kidd, K.A., R.H. Hesslein, R.J.P. Fudge, and K.A. Hallard. The influence of trophic level as measured by δ ^{15}N on mercury concentrations in freshwater organisms. *Water Air Soil Pollut.* 80, pp. 1011–1015, 1995.

Kidd, K.A., D.W. Schindler, R.H. Hesslein, and D.C.G. Muir. Correlation between stable nitrogen isotope ratios and concentrations of organochlorines in biota from a freshwater food web. *Sci. Total Environ.* 160/161, pp. 381–390, 1995.

King, J.J., *The Environmental Dictionary*, 2nd ed. Executive Enterprises Publications Co., Inc., New York, 1993, p. 977.

Klaassen, C.D., M.O. Amdur, and J. Doull (Eds). *Casarett and Doull's Toxicology. The Basic Science of Poisons*. Macmillan Publishing Co., New York, NY, 1987, p. 974.

Klaverkamp, J.F., M.D. Dutton, H.S. Majewski, R.V. Hunt, and L.J. Wesson. Evaluating the Effectiveness of Metal Pollution Controls in a Smelter by Using Metallothionein and Other Biochemical Responses in Fish, in *Metal Ecotoxicology. Concepts and Applications*, Newman, M.C. and A.W. McIntosh, Eds., Lewis Publishers, Chelsea, MI, 1991.

Kleinow, K.M., M.J. Melancon, and J.J. Lech. Biotransformation and induction: Implications for toxicity, bioaccumulation and monitoring of environmental xenobiotics in fish. *Environ. Health Perspect.* 71, pp. 105–119, 1987.

Klerks, P.L. and J.S. Weis. Genetic adaptation to heavy metals in aquatic organisms: A review. *Environ. Pollut.* 45, pp. 173–205, 1987.

Kling, G.W., B. Fry, and W.J. O'Brien. Stable isotopes and planktonic trophic structure in Arctic lakes. *Ecology* 73, pp. 561–566, 1992.

Knoll, G. F. *Radiation Detection and Measurement.* John Wiley & Sons, New York, NY, 1989.

Knoph, M.B. and Y.A. Olsen. Subacute toxicity of ammonia to Atlantic salmon (*Salmo salar* L.) in seawater: Effects on water and salt balance, plasma cortisol and plasma ammonia levels. *Aquat. Toxicol. (AMST)* 30, pp. 295–310, 1994.

Koehn, R.K. and B.L. Bayne. Towards a physiological and genetical understanding of the energetics of the stress response. *Biol. J. Linn. Soc.* 37, pp. 157–171, 1989.

Koehn, R.K. and P.M. Gaffney. Genetic heterozygosity and growth rate in *Mytilus edulis. Mar. Biol.* 82, pp. 1–7, 1984.

Koivisto, S. and M. Ketola. Effects of copper on life–histroy traits of *Daphnia pulex* and *Bosmina longirostris. Aquat. Toxicol. (AMST)* 32, pp. 255–269, 1995.

Kolaja, G.J. and D.E. Hinton. DDT–Induced Reduction in Eggshell Thickness, Weight, and Calcium is Accompanied by Calcium ATPase Inhibition, in *Animals as Monitors of Environmental Pollutants*, National Academy of Sciences, Washington, DC, 1979.

Kolkwitz, R. and M. Marsson. Ökologie der pflanzlichen Saprobien. *Ber. Dtsch. Bot. Ges.* 26, pp. 505–519, 1908.

Kondo, S. *Health Effects of Low–level Radiation.* Kinki University Press, Osaka, Japan, 1993, p. 213.

Kooijman, S.A.L.M. Parametric analyses of mortality rates in bioassays. *Water Res.* 15, pp. 107–119, 1981.

Kooijman, S.A.L.M. *Dynamic Energy Budgets in Biological Systems.* Cambridge University Press, Cambridge, 1993, p. 350.

Kooijman, S.A.L.M. and J.J.M. Bedaux. *The Analysis of Aquatic Toxicity Data.* VU University Press, Amsterdam, 1996, p. 149.

Kopp, R.L., S.I. Guttman, and T.E. Wissing. Genetic indicators of environmental stress in central mudminow (*Umbra limi*) populations exposed to acid deposition in the Adirondack Mountains. *Environ. Toxicol. Chem.* 11, pp. 665–676, 1992.

Koss, G., E. Schuler, B. Arndt, J. Seidel, S. Seubert, and A. Seubert. A comparative toxicological study of pike (*Esox lucius* L.) from the River Rhine and River Lahn. *Aquat. Toxicol. (AMST)* 8, pp. 1–9, 1986.

Kotz, S. and N.L. Johnson. *Encyclopedia of Statistical Sciences*, Volume 8. John Wiley & Sons, Inc., New York, p. 798, 1988.

Kramer, V.J. and M.C. Newman. Inhibition of glucosephosphate isomerase allozymes of the mosquitofish, *Gambusia holbrooki*, by mercury. *Environ. Toxicol. Chem.* 13, pp. 9–14, 1994.

Kramer, V.J., M.C. Newman, M. Mulvey, and G.R. Ultsch. Glycolysis and Krebs cycle metabolites in mosquitofish, *Gambusia holbrooki*, Girard 1859, exposed to mercuric chloride: Allozyme genotype effects. *Environ. Toxicol. Chem.* 11, pp. 357–364, 1992.

Krantzberg, G. Spatial and temporal variability in metal bioavailability and toxicity of sediment from Hamilton Harbour, Lake Ontario. *Environ. Toxicol. Chem.* 13, pp. 1685–1698, 1994.

Krause, P.R. Effects of an oil production effluent on gametogenesis and gamete performance in the purple sea urchin (*Strongylocentrotus purpuratus* Stimpson). *Environ. Toxicol. Chem.* 13, pp. 1153–1161, 1994.

Kuhn, T.S., *The Structure of Scientific Revolutions.* The University of Chicago, Chicago, IL, 1970, p. 210.

Lamb, T., J.W. Bickham, J.W. Gibbons, M.J. Smolen, and S. McDowell. Genetic damage in a population of slider turtles (*Trachemys scripta*) inhabiting a radioactive reservoir. *Arch. Environ. Contam. Toxicol.* 20, pp. 138–142, 1991.

Landis, W.G. and M.–H. Yu, *Introduction to Environmental Toxicology.* Lewis Publishers, Boca Raton, FL, 1995, p. 328.

Landrum, P.F., G.A. Harkey, and J. Kukkonen. Evaluation of Organic Contaminant Exposure in Aquatic Organisms: The Significance of Bioconcentration and Bioaccumulation, in *Ecotoxicology: A Hierarchical Treatment*, Newman, M.C. and Jagoe, C.H., Eds., CRC Press, Inc., Boca Raton, FL, 1996.

Landrum, P.F., H. Lee II, and M.J. Lydy. Toxicokinetics in aquatic systems: Model comparisons and use in hazard assessment. *Environ. Toxicol. Chem.* 11, pp. 1709–1725, 1992.

Landrum, P.F. and M.J. Lydy, personal communication during toxicokinetics short course, Nov. 3, 1991, Society of Environmental Toxicology and Chemistry 12th Annual Meeting, 1991.

Landrum, P.F. and J.A. Robbins. Bioavailability of Sediment–Associated Contaminants to Benthic Invertebrates, in *Sediments: Chemistry and Toxicity of In–Place Pollutants*, Baudo, R., J. Giesy, and H. Muntau, Eds., Lewis Publishers, Inc., Chelsea, MI, 1990.

Landrum, P.F. and C.R. Stubblefield. Role of respiration in the accumulation of organic xenobiotics by the amphipod, *Pontoporeia hoyi. Environ. Toxicol. Chem.* 10, pp. 1019–1028, 1991.

Lange, B.W. *Drosophila melanogaster* metallothionein genes: Selection for duplications? Ph.D. Dissertation, Duke University, Durham, NC, 1989.

Langston, W.J. and G.R. Burt. Bioavailability and effects of sediment–bound TBT in deposit–feeding clams, *Scrobicularia plana. Mar. Environ. Res.* 32, pp. 61–77, 1991.

Lapp, R. E. and H. L. Andrews. *Nuclear Radiation Physics*. Prentice–Hall, Inc., Englewood Cliffs, NJ, 1972.

Larson, R.A. and E.J. Weber. *Reaction Mechanisms in Environmental Organic Chemistry*. CRC Press, Inc., Boca Raton, FL, 1994, p. 433.

Larsson, Å., B.–E. Bengtsson, and C. Haux. Disturbed ion balance in flounder, *Platichthys flesus* L. exposed to sublethal levels of cadmium. *Aquat. Toxicol. (AMST)* 1, pp. 19–35, 1981.

Laskowski, R. Are the top carnivores endangered by heavy metal biomagnification? *Oikos* 60, pp. 387–390, 1991.

Laurén, D.J. and D.G. McDonald. Effects of copper on branchial ionoregulation in the rainbow trout, *Salmo gairdneri* Richardson. Modulation by water hardness and pH. *J. Comp. Physiol. B.* 155, pp. 635–644, 1985.

La Via, M.F. and R.B. Hill, Jr., *Principles of Pathobiology*, Oxford University Press, New York, 1971, p. 281.

Lavie, B. and E. Nevo. Heavy metal selection of phosphoglucose isomerase allozymes in marine gastropods. *Mar. Biol.* 71, pp. 17–22, 1982.

Lavie, B. and E. Nevo. Genetic selection of homozygote allozyme genotypes in marine gastropods exposed to cadmium pollution. *Sci. Total Environ.* 57, pp. 91–98, 1986.

Laxen, D.P.H. and R.M. Harrison. The highway as a source of water pollution: An appraisal with the heavy metal lead. *Water Res.* 11, pp. 1–11, 1977.

Lech, J.J. and M.J. Vodicnik. Biotransformation, in *Fundamentals of Aquatic Toxicology*, Rand, G.M. and S.R. Petrocelli, Eds., Hemisphere Publishing Corp., Washington, DC, 1985.

Lee, G. Lawmakers move to check CFC phaseout. *Washington Post.* Sept. 21, p. A13, 1995.

Lee, C., M.C. Newman, and M. Mulvey. Time to death of mosquitofish (*Gambusia holbrooki*) during acute inorganic mercury exposure: Population structure effects. *Arch. Environ. Contam. Toxicol.* 22, pp. 284–287, 1992.

Leggett, R.W. Predicting the retention of Cs in individuals. *Health Phys.* 50, pp. 747–759, 1986.

Leland, H.V. and J.L. Carter. Effects of copper on species composition of periphyton in a Sierra Nevada, California, stream. *Freshwater Biol.* 14, p. 281–296, 1984.

Leland, H.V. and J.S. Kuwabara. Trace Elements, in *Fundamentals of Aquatic Toxicology*, Rand, G.M. and S.R. Petrocelli, Eds., Hemisphere Publishing Corp., Washington, DC, 1985.

Lepkowski, W. Bhopal. Indian city begins to heal but conflicts remain. *Chem. Eng. News.* Dec. 2, pp. 18–32, 1985.

Levin, S.A., M.A. Harwell, J.R. Kelly, and K.D. Kimball, *Ecotoxicology: Problems and Approaches*. Springer–Verlag, New York, 1989, p. 547.

Liber, K., N.K. Kaushik, K.R. Solomon, and J.H. Carey. Experimental designs for aquatic mesocosm studies: A comparison of the "ANOVA" and "Regression" design for assessing the impact of tetrachlorophenol on zooplankton populations in limnocorrals. *Environ. Toxicol. Chem.* 11, pp. 61–77, 1992.

Lide, D.R., *CRC Handbook of Chemistry and Physics, 73rd Ed.* CRC Press, Boca Raton, FL, 1992, p. I–33.

Likens, G.E. Acid precipitation. *Chem. Eng. News.* Nov. 22, pp. 29–44, 1976.

Likens, G.E. and F.H. Borman. Acid rain: A serious environmental problem. *Science (WASH DC).* 184, pp. 1176–1179, 1974.

Lindberg, S.E., R.C. Harris and R.R. Turner. Atmospheric deposition of metals to forest vegetation. *Science (WASH DC)* 215, pp. 1609–1612, 1982.

Lindqvist, L. and M. Block. Excretion of cadmium and zinc during moulting in the grasshopper *Omocestus viridulus* (Orthoptera). *Environ. Toxicol. Chem.* 13, pp. 1669–1672, 1994.

Lindzen, R.S. Global warming. The origin and nature of the alleged scientific consensus. *CATO Review of Business & Government* Spring, pp. 87–98, 1992.

Lindzen, R.S. On the scientific basis for global warming scenarios. *Environ. Pollut.* 83, pp. 125–134, 1994.

Lipnick, R.L. A perspective on quantitative structure–activity relationships in ecotoxicology. *Environ. Toxicol. Chem.* 4, pp. 255–257, 1985.

Lipnick, R.L. Structure–Activity Relationships, in *Fundamentals of Aquatic Toxicology. Effects, Environmental Fate, and Risk Assessment*, 2nd Ed., Rand, G.M., Ed., Taylor & Francis, Washington, DC, 1995.

Litchfield, J.T. A method for rapid graphic solution of time–per cent effect curves. *J. Pharm. Exp. Ther.* 97, pp. 399–408, 1949.

Litchfield, J.T. and F. Wilcoxon. A simplified method of evaluating dose–effect experiments. *J. Pharm. Exp. Ther.* 96, pp. 99–113, 1949.

Little, E.E. and S.E. Finger. Swimming behavior as an indicator of sublethal toxicity in fish. *Environ. Toxicol. Chem.* 9, pp. 13–19, 1990.

Lloyd, R. and D.W.M. Herbert. The influence of carbon dioxide on the toxicity of un–ionized ammonia to rainbow trout *(Salmo gairdneri* Richardson). *Ann. Appl. Biol.* 48, pp. 399–404, 1960.

Lloyd, R. and L.D. Orr. The diuretic response by rainbow trout to sub–lethal concentrations of ammonia. *Water Res.* 3, pp. 335–344, 1969.

Lovelock, J.E. Gaia as seen through the atmosphere. *Atmos. Environ.* 6, pp. 579–580, 1972.

Lovelock, J.E., *The Ages of Gaia. A Biography of Our Living Earth*. Oxford University Press, Oxford, 1988, p. 252.

Lovelock, J.E., *Healing Gaia. Practical Medicine for the Planet*. Gaia Books Limited, New York, 1991, p. 192.

Luoma, J.R. New effect of pollutants: Hormone mayhem. *New York Times*, May 24, 1992.

Luoma, S.N. Can we determine the biological availability of sediment–bound trace elements? Hydrobiolia. 176/177, pp. 379–396, 1989.

Luoma, S.N. and G.W. Bryan. Factors controlling the availability of sediment–bound lead to the estuarie bivalve, *Scrobicularia plana. J. Mar. Biol. Assoc.* UK 58, pp. 793–802, 1978.

Lyon, R., M. Taylor, and K. Simkiss. Ligand activity in the clearance of metals from the blood of the crayfish (*Austropotamobius pallipes*). *J. Exp. Biol.* 113, pp. 19–27, 1984.

Mackay, D. Correlation of bioconcentration factors. *Environ. Sci. Technol.* 16, pp. 274–278, 1982.

Mackay, D. *Multimedia Environmental Models. The Fugacity Approach*. Lewis Publishers, Inc., Chelsea, MI, p. 257, 1991.

Mackay, D. and S. Paterson. Fugacity revisited. The fugacity approach to environmental transport. *Environ. Sci. Technol.* 116, pp. 654A–660A, 1982.

Mackay, D. and F. Wania, Transport of contaminants to the Arctic: Partitioning processes and models. *Sci. Total Environ.* 160/161, pp. 25–38, 1995.

Mackenthun, K.M. and J.I. Bregman. *Environmental Regulations Handbook*. Lewis Publishers, Boca Raton, FL, 1992, p. 297.

Macklis, R.M. The great radium scandal. *Sci. Am.* August 1993, pp. 94–99, 1993, Ch. 3.

Magurran, A.E. *Ecological Diversity and Its Measurement*. Princeton University Press, Princeton, NJ, 1988, p. 179.

Malins, D.C. Identification of hydroxyl radical–induced lesions in DNA base structure: Biomarkers with a putative link to cancer development. *J. Toxicol. Environ. Health* 40, pp. 247–261, 1993.

Mallatt, J. Fish gill structural changes induced by toxicants and other irritants: A statistical review. *Can. J. Fish. Aquat. Sci.* 42, pp. 630–648, 1985.

Mallet, J. The evolution of insecticide resistance: Have the insects won? *Trends Ecol. Evol.* 4, pp. 336–340, 1989.

Manahan, S.E. *Fundamentals of Environmental Chemistry*. Lewis Publishers, Chelsea, MI, 1993, p. 844.

Mance, G. *Pollution Threat of Heavy Metals in Aquatic Environments*. Elsevier Applied Science, London, 1987, p. 372.

Marco, G.J., R.M. Hollingworth and W. Durham, *Silent Spring Revisited*. American Chemical Society, Washington, DC, 1987, p. 214.

Margoshes, M. and B.L. Vallee. A cadmium protein from equine kidney cortex. *J. Am. Chem. Soc.* 79, pp. 4813–4814, 1957.

Margulis, L. and J.E. Lovelock, Gaia and Geognosy, in *Global Ecology. Towards a Science of the Biosphere*, Rambler, M.B., L. Margulis, and R. Fester, Eds., Academic Press, Inc., Boston, 1989.

Marking, L.L. Toxicity of Chemical Mixtures, in Rand, G.M., and Petrocelli, S.R., Eds., *Fundamentals of Aquatic Toxicology*. Hemisphere Publishing Corp., Washington, DC, 1985.

Marking, L.L. and V.K. Dawson. Method for assessment of toxicity or efficacy of mixtures of chemicals. *U.S. Fish Wildl. Serv. Invest. Fish Control* 67, pp. 1–8, 1975.

Maroni, G., J. Wise, J.E. Young, and E. Otto. Metallothionein gene duplications and metal tolerance in natural populations of *Drosophila melanogaster*. *Genetics* 117, pp. 739–744, 1987.

Marshall, E. EPA may allow more lead in gasoline. *Science (WASH DC)*. 215, pp. 1375–1378, 1982.

Marshall, J.S. The effects of continuous gamma radiation on the intrinisc rate of natural increase of *Daphnia pulex*. *Ecology* 43, pp. 598–607, 1962.

Martínez–Jerónimo, F., R. Villaseñor, F. Espinosa, and G. Rios. Use of life–tables and application factors for evaluating chronic toxicity of Kraft mill wastes on *Daphnia magna*. *Bull. Environ. Contam. Toxicol.* 50, pp. 377–384, 1993.

Marubini, E. and M.G. Valsecchi. *Analysing Survival Data from Clinical Trials and Observational Studies*. John Wiley & Sons, Inc., New York, 1995, p. 414.

Mason, A.Z. and J.A. Nott. The role of intracellular biomineralized granules in the regulation and detoxification of metals in gastropods with special reference to the marine prosobranch *Littorina littorea*. *Aquat. Toxicol. (AMST)* 1, pp. 239–256, 1981.

Mason, A.Z., K. Simkiss, and K.P. Ryan. The ultrastructural localization of metals in specimens of *Littorina littorea* collected from clean and polluted sites. *J. Mar. Biol. Assoc. UK* 64, pp. 699–720, 1984.

Masood, E. Climate panel confirms human role in warming, fights off oil states. *Nature (LOND)*, p. 524, 1995.

Matthews, R.A., W.G. Landis, and G.B. Matthews. The community conditioning hypothesis and its application to environmental toxicology. *Environ. Toxicol. Chem.* 15, pp. 597–603, 1996.

Maugh, T.H. It isn't easy being king. *Science (WASH DC)*. 203, pp. 637, 1974.

Maurer, B.A. and R.D. Holt. Effects of chronic pesticide stress on wildlife populations in complex landscapes: Processes at multiple scales. *Environ. Toxicol. Chem.* 15, pp. 420–426, 1996.

May, R.M. Biological populations with nonoverlapping generations: Stable points, stable cycles, and chaos. *Science (WASH DC)* 186, pp. 645–647, 1974.

May, R.M. *Theoretical Ecology. Principles and Applications*. W.B. Saunders, Philadelphia, PA, 1976a, p. 317.

May, R.M. Simple mathematical models with very complicated dynamics. *Nature (LOND.)* 261, pp. 459–467, 1976b.

McBee, K., J.W. Bickham, K.W. Brown, and K.C. Donnelly. Chromosomal aberrations in native small mammals (*Peromyscus leucopus* and *Sigmodon hispidus*) at a petrochemical waste disposal site: I. Standard karyology. *Arch. Environ. Contam. Toxicol.* 16, pp. 681–688, 1987.

McCarthy, J.F. and L.R. Shugart, *Biomarkers of Environmental Contamination*. Lewis Publishers, Chelsea, MI, 1990, p. 457.

McCarthy, J.R., R.S. Halbrook, and L.R. Shugart. *Conceptual Strategy for Design, Implementation, and Validation of a Biomarker–based Biomonitoring Capability*. Oak Ridge National Laboratory, Oak Ridge, TN, 1991, p. 85.

McCormick, P.V., J. Cairns, Jr., S.E. Belanger, and E.P. Smith. Response of protistan assemblages to a model toxicant, the surfactant C12–TMAC (dodecyl trimethyl ammonium chloride), in laboratory streams. *Aquat. Toxicol. (AMST)* 21, pp. 41–70, 1991.

McFarlane, G.A. and W.G. Franzin. Elevated heavy metals: A stress on a population of white suckers, *Catostomus commersoni*, in Hamell Lake, Saskatchewan. *J. Fish. Res. Board Can.* 35, pp. 963–970, 1978.

McGregor, G.I. *Environmental Law and Enforcement*. Lewis Publishers, Boca Raton, FL, 1994, p. 239.

McIntosh. A. Trace Metals in Freshwater Sediments: A Review of the Literature and an Assessment of Research Needs, in *Metal Ecotoxicology. Concepts and Applications*, Newman, M.C. and A.W. McIntosh, Eds., Lewis Publishers, Chelsea, MI, 1991.

McKim, J.M. Early Life Stage Tests, in Rand, G.M., and Petrocelli, S.R., Eds., *Fundamentals of Aquatic Toxicology*. Hemisphere Publishing Corp., Washington, DC, 1985.

McLachlan, J.A. Functional toxicology: A new approach to detect biologically active xenobiotics. *Environ. Health Perspect.* 101, pp. 386–387, 1993.

McNeill, K.G. and G.A.D. Trojan. The cesium–potassium discrimination ratio. *Health Phys.* 4, pp. 109–112, 1960.

Medawar, P.B., *The Art of the Soluble*. Methuen Co., London, 1967, p. 160.

Medawar, P.B., *Pluto's Republic*. Oxford University Press, Oxford, 1982, p. 351.

Medvedev, Z.A., *The Ural and Chernobyl Nuclear Accidents*, seminar presented September 5, 1995 at the University of Georgia's Savannah River Ecology Laboratory.

Mehrle, P.M. and F.L. Mayer. Biochemistry/Physiology, in *Fundamentals of Aquatic Toxicology. Methods and Applications*, Rand, G.M. and S.R. Petrocelli, Eds., Hemisphere Publishing Corp., Washington, DC, 1985.

Melancon, M.J. Bioindicators Used in Aquatic and Terrestrial Monitoring, in *Handbook of Ecotoxicology*, Hoffman, D.J., B.A. Rattner, G.A. Burton, Jr., and J. Cairns, Jr., Eds., CRC Press, Inc., Boca Raton, FL, 1995.

Mellinger, P. J. and V. Schultz. Ionizing radiation and wild birds: A review. *Crit. Rev. Environ. Control* 5, p. 397, 1975.

Mertz, W. The essential trace elements. *Science (WASH DC)* 213, pp. 1332–1338, 1981.

Metalli, P. and E. Ballardin. Radiobiology of *Artemia*: Radiation effects and ploidy. *Curr. Top. Radiat. Res. Q.* 7, p. 181, 1972.

Meyers, T.R. and J.D. Hendricks. Histopathology, in *Fundamentals of Aquatic Toxicology. Methods and Applications*, Rand, G.M. and S.R. Petrocelli, Eds., Hemisphere Publishing Corp., Washington, DC, 1985.

Meyers–Schöne, L.R. Shugart, J.J. Beauchamp, and B.T. Walton. Comparison of two freshwater turtle species as monitors of radionuclide and chemical contamination: DNA damage and residue analysis. *Environ. Toxicol. Chem.* 12, pp. 1487–1496, 1993.

Michener, W.K., J.W. Brunt, and S.G. Stafford. *Environmental Information Management and Analysis: Ecosystem to Global Scales*. Taylor & Francis, Ltd., London, 1994, p. 555.

Millar, I.B. and P.A. Cooney. Urban lead—A study of environmental lead and its significance to school children in the vicinity of a major trunk road. *Atmos. Environ.* 16, pp. 615–620, 1982.

Miller, R.G., Jr. *Beyond ANOVA, Basics of Applied Statistics*. John Wiley & Sons, New York, NY, 1986.

Milodowski, A.E., J.M. West, J.M. Pearce, E.K. Hyslop, I.R. Basham, and P.J. Hooker.

Uranium–mineralized microorganisms associated with uraniferous hydrocarbons in southeast Scotland. *Nature (LOND)*. 347, pp. 465–467, 1990.

Minagawa, M. and E. Wada. Stepwise enrichment of ^{15}N along food chains: Further evidence and the relation between δ ^{15}N and animal age. *Geochim. Cosmochim. Acta* 48, pp. 1135–1140, 1984.

Mishima, J. and E.P. Odum. Excretion rate of Zn^{65} by *Littorina irrorata* in relation to temperature and body size. *Limnol. Oceanogr.* 8, pp. 39–44, 1963.

Molander, S., H. Blanck, and M. Söderström. Toxicity assessment by pollution–induced community tolerance (PICT), and identification of metabolites in periphyton communities after exposure to 4,5,6–trichloroguaiacol. *Aquat. Toxicol. (AMST)* 18, pp. 115–136, 1990.

Møller, V., V.E. Forbes, and M.H. DePledge. Influence of acclimation and exposure temperature on the acute toxicity of cadmium to the freshwater snail *Potamophygus antipodarum* (Hydrobiidae). *Environ. Toxicol. Chem.* 13, pp. 1519–1524, 1994.

Monk, C.D. Effects of short–term gamma irradiation on an old field. *Radiat. Bot.* 6, pp. 329–335, 1966.

Monkhouse, F.J. *A Dictionary of Geography*. Aldine Publishing Co., Chicago, IL, 1965, p. 344.

Moolgavkar, S.H. Carcinogenesis modeling: From molecular biology to epidemiology. *Ann. Rev. Public Health* 1986, pp. 151–169, 1986.

Moore, M.J. and M.S. Myers. Pathobiology of Chemical–Associated Neoplasia in Fish, in *Aquatic Toxicology. Molecular, Biochemical and Cellular Perspectives*, Malins, D.C. and G.K. Ostrander, Eds., CRC Press, Inc., Boca Raton, FL, 1994.

Moriarty, F. *Ecotoxicology. The Study of Pollutants in Ecosystems*. Academic Press, Inc., London, 1983, p. 233.

Morton, B. The tidal rhythm and rhythm of feeding and digestion in *Cardium edule. J. Mar. Biol. Assoc. U.K.* 50, pp. 499–512, 1970.

Mossman, K. L., M. Goldman, F. Masse, W. Mills, K. Schiager, and R. Vetter. Radiation risk in perspective. *Health Physics Newsletter*, March 1996. Health Physics Society, McLean, VA.

Mount, D.I. and C.E. Stephan. A method for establishing acceptable toxicant limits for fish – Malathion and 2,4–D. *Trans. Am. Fish. Soc.* 96, pp. 185–193, 1967.

Muir, D.C.G., N.P. Grift, W.L. Lockhart, P. Wilkinson, B.N. Billeck, and G.J. Brunskill. Spatial trends and historical profiles of organochlorine pesticides in Arctic lake sediments. *Sci. Total Environ.*160/161, pp. 447–457, 1995.

Muir, D.C.G., R.J. Norstrom, and M. Simon. Organochlorine contaminants in Arctic marine food chains: Accumulation of specific polychlorinated biphenyls and chlordane–related compounds. *Environ. Sci. Technol.* 22, pp. 1071–1079, 1988.

Mulholland, P.J., A.V. Palumbo, J.W. Elwood, and A.D. Rosemond. *J. N. Am. Benthol. Soc.* 6, pp. 147–158, 1987.

Müller, H. and G. Pröhl. ECOSYS–87: A dynamic model for the assessment of the radiological consequences of nuclear accidents. *Health Phys.* 64, pp. 232–252, 1993.

Mulvey, M., M.C. Newman, A. Chazal, M.M. Keklak, M.G. Heagler, and L.S. Hales, Jr. Genetic and demographic responses of mosquitofish (*Gambusia holbrooki* Girard 1859) populations stressed by mercury. *Environ. Toxicol. Chem.* 14, pp. 1411–1418, 1995.

Mulvey, M. and S.A. Diamond. Genetic Factors and Tolerance Acquisition in Populations Exposed to Metals and Metalloids, in *Metal Ecotoxicology. Concepts and Applications*, Newman, M.C. and A.W. McIntosh, Eds., Lewis Publishers, Chelsea, MI, 1991.

Munkittrick, K.R. and D.G. Dixon. Growth, fecundity, and energy stores of white suckers (*Catostomus commersoni*) from lakes containing elevated levels of copper and zinc. *Can. J. Fish. Aquat. Sci.* 45, pp. 1355–1365, 1988.

Murdoch, M.H. and P.D.N. Hebert. Mitochondrial DNA diversity of brown bullhead from contaminated and relatively pristine sites in the Great Lakes. *Environ. Toxicol. Chem.* 8, pp. 1281–1289, 1994.

Murphy, D.L. and J.W. Gooch. Accumulation of *cis* and *trans* chlordane by channel catfish during dietary exposure. *Arch. Environ. Contam. Toxicol.* 29, pp. 297–301, 1995.

Nagel, E., *The Structure of Science. Problems in the Logic of Scientific Explanation.* Harcourt, Brace and World, Inc., New York, 1961, p. 618.

NAS/NRC (National Academy of Sciences/National Research Council, Committee on the Biological Effects of Ionizing Radiations). *Health Risks of Exposure to Low Levels of Ionizing Radiation, BEIR V.* National Academy Press, Washington, DC, 1990, p. 754.

National Research Council. *Risk Assessment in the Federal Government: Managing the Process.* National Academy Press, Washington, DC, 1983.

NCRP (National Council on Radiation Protection and Measurements). *Exposure of the Population in the United States and Canada from Natural Background Radiation.* NCRP Report No. 94, Bethesda, MD, 1987, p. 212.

NCRP (National Council on Radiation Protection and Measurements). *Radiological Assessment: Predicting the Transport, Bioaccumulation, and Uptake by Man of Radionuclides Released to the Environment.* NCRP Report No. 76, Bethesda, MD, 1984, 304.

NCRP (National Council on Radiation Protection and Measurements). *Screening Techniques for Determining Compliance with Environmental Standards.* NCRP Commentary No. 3, Bethesda, MD, 1989, 134.

NCRP (National Council on Radiation Protection and Measurements). *Effects of Ionizing Radiation on Aquatic Organisms.* NCRP Report No. 109, p. 115, Bethesda, MD, 1991.

NCRP (National Council on Radiation Protection and Measurements). *Risk Estimates for Radiation Protection.* NCRP Report No. 115, Bethesda, MD, 1993, p. 148.

NCRP (National Council on Radiation Protection and Measurements). *A Guide for Uncertainty Analysis in Dose and Risk Assessments Related to Environmental Contamination.* NCRP Commentary No. 14, Bethesda, MD, 1996, p.54.

Neathery, M.W. and W.J. Miller. Metabolism and toxicity of cadmium, mercury, and lead in animals: A review. *J. Dairy Sci.* 58, 1767–1781, 1975, p. 211.

Neely, W.B., D.R. Bronson and G.E. Blau. Partition coefficient to measure bioconcentration potential of organic chemicals in fish. *Environ. Sci. Technol.* 8, pp. 1113–1115, 1974.

Neter, J., W. Waserman, and M.H. Kutner. *Applied Linear Statistical Models. Regression, Analysis of Variance, and Experimental Design*, 3rd ed., Richard D. Irwin, Inc., Homewood, IL, 1990, p. 1181.

Neuhold, J.M. The relationship of life history attributes to toxicant tolerance in fishes. *Environ. Toxicol. Chem.* 6, pp. 709–716, 1987.

Neville, C.M. and P.G.C. Campbell. Possible mechanisms of aluminum toxicity in dilute, acidic environment to fingerlings and older life stages of salmonids. *Water Air Soil Pollut.* 42, pp. 311–327, 1988.

Nevo, E., T. Perl, A. Beiles and D. Wool. Mercury selection of allozyme genotypes in shrimps. *Experientia* 37, pp. 1152–1154, 1981.

Nevo, E., T. Shimony and M. Libni. Thermal selection of allozyme polymorphisms in barnacles. *Nature (LOND)* 267, pp. 699–701, 1977.

Nevo, E., T. Shimony and M. Libni. Pollution selection of allozyme polymorphisms in barnacles. *Experientia* 34, pp. 1562–1564, 1978.

Newman, M.C. A statistical bias in the derivation of hardness–dependent metals criteria. *Environ. Toxicol. Chem.* 10, pp. 1295–1297, 1991.

Newman, M.C. Correction of prediction bias in least–squares linear regression methods used to fit power and exponential relationships in environmental chemistry and toxicology. *Environ. Toxicol. Chem.* 12, pp. 1129–1133, 1993.

Newman, M.C., *Quantitative Methods in Aquatic Ecotoxicology.* Lewis Publishers, Boca Raton, FL, 1995, p. 426.

Newman, M.C., Ecotoxicology as a Science, in *Ecotoxicology: A Hierarchical Treatment*, Newman, M.C. and C.H. Jagoe, Eds., CRC/Lewis Publishers, Boca Raton, FL, 1996.

Newman, M.C., Alberts, J.J., and V.A. Greenhut. Geochemical factors complicating the use of *aufwuchs* to monitor bioaccumulation of arsenic, cadmium, chromium, copper and zinc. *Water. Res.* 19, pp. 111–128, 1985.

Newman, M.C. and M.S. Aplin. Enhancing toxicity data interpretation and prediction of ecological risk with survival time modeling: An illustration using sodium chloride toxicity to mosquitofish (*Gambusia holbrooki*). *Aquat. Toxicol. (AMST)* 23, pp. 85–96, 1992.

Newman, M.C., S.A. Diamond, M. Mulvey, and P. Dixon. Allozyme genotpye and time to death of mosquitofish, *Gambusia affinis* (Baird and Girard) during acute toxicant exposure: A comparison of arsenate and inorganic mercury. *Aquat. Toxicol. (AMST)* 15, pp. 141–156, 1989.

Newman, M.C. and P.M. Dixon. Ecologically Meaningful Estimates of Lethal Effect in Individuals, in *Ecotoxicology: A Hierarchical Treatment*, Newman, M.C. and C.H. Jagoe, Eds., CRC/Lewis Publishers, Boca Raton, FL, 1996.

Newman, M.C. and M.G. Heagler. Allometry of Metal Bioaccumulation and Toxicity, in *Metal Ecotoxicology. Concepts and Applications*, Newman, M.C. and A.W. McIntosh, Eds., Lewis Publishers, Chelsea, MI, 1991.

Newman, M.C. and C.H. Jagoe. Ligands and the Bioavailability of Metals in Aquatic Environments, in *Bioavailability. Physical, Chemical and Biological Interactions*, Hamelink, J.L., P.F. Landrum, H.L. Bergman and W.H. Benson, Eds., CRC Press, Inc, Boca Raton, FL, 1994.

Newman, M.C. and R.H. Jagoe. Bioaccumulation models with time lags: Dynamics and stability criteria. *Ecol. Model.* 1424, pp. 281–286, 1996.

Newman, M.C. and C.H. Jagoe, Eds. *Ecotoxicology: A Hierarchical Treatment*. CRC/Lewis Publishers, Boca Raton, FL, 1996.

Newman, M.C., M.M. Keklak, and M.S. Doggett. Quantifying animal size effects on toxicity: A general approach. *Aquat. Toxicol. (AMST)* 28, pp. 1–12, 1994.

Newman, M.C. and J.T. McCloskey. Predicting relative toxicity and interactions of divalent metal ions: Microtox® bioluminescence assay. *Environ. Toxicol. Chem.* 15, pp. 275–281, 1966a.

Newman, M.C. and J.T. McCloskey. Time-to–event analyses of ecotoxicology data. *Ecotoxicology* 5, pp. 187–196, 1996b.

Newman, M.C. and A.W. McIntosh. Slow accumulation of lead from contaminated food sources by the freshwater gastropods, *Physa integra* and *Campeloma decisum*. *Arch. Environ. Contam. Toxicol.* 12, pp. 685–692, 1983.

Newman, M.C. and A.W. McIntosh. Appropriateness of *aufwuchs* as a monitor of bioaccumulation. *Environ. Pollut.* 60, pp. 83–100, 1989.

Newman, M.C. and A.W. McIntosh, Eds. *Metal Ecotoxicology. Concepts and Applications.* Lewis Publishers, Chelsea, MI, 1991.

Newman, M.C., McIntosh, A.W., and V.A. Greenhut. Geochemical factors complicating the use of *aufwuchs* as a biomonitor for lead levels in two New Jersey reservoirs. *Water Res.* 17, pp. 625–630, 1983.

Newman, M.C. and S.V. Mitz. Size dependence of zinc elimination and uptake from water by mosquitofish *Gambusia affinis* (Baird and Girard). *Aquat. Toxicol. (AMST)* 12, pp. 17–32, 1988.

Newman, M.C., M. Mulvey, A. Beeby, R.W. Hurst, and L. Richmond. Snail (*Helix aspersa*) exposure history and possible adaptation to lead as reflected in shell composition. *Arch. Environ. Contam. Toxicol.* 27, pp. 346–351, 1994.

Nichols, J.W., J.M. McKim, M.E. Andersen, M.L. Gargas, H.J. Clewell, III, and R.J. Erickson. A physiologically based toxicokinetic model for the uptake and disposition of waterborne organic chemicals in fish. *Toxicol. Appl. Pharmacol.* 106, pp. 433–447, 1990.

Nieboer, E. and D.H.S. Richardson. The replacement of the nondescript term "heavy metals" by a biologically and chemically significant classification of metal ions. *Environ. Pollut.* 1B, pp. 3–26, 1980.

Niederlehner, B.R., J.R. Pratt, A.L. Buikema, Jr., and J. Cairns, Jr. Laboratory tests evaluating the effects of cadmium on freshwater protozoan communities. *Environ. Toxicol. Chem.* 4, pp. 155–165, 1985.

Nihlgård, B. The ammonium hypothesis – An additional explanation to the forest dieback in Europe. *Ambio* 14, pp. 2–8, 1985.

Nikinmaa, K. How does environmental pollution affect red cell function in fish? *Aquat. Toxicol. (AMST)* 22, pp. 227–238, 1992.

Nissen, P. and A.A. Benson. Arsenic metabolism in freshwater and terrestrial plants. *Physiol. Plant.* 54, pp. 446–450, 1982.

Norton, S.B., D.J. Rodier, J.H. Gentile, W.H. Van Der Schalie, W.P. Wood, and M.W. Slimak. A framework for ecological risk assessment at the EPA. *Environ. Toxicol. Chem.* 11, pp. 1663–1672, 1992.

Nott, J.A. and A. Nicolaidou. Bioreduction of zinc and manganese along a molluscan food chain. *Comp. Biochem. Physiol.* 104A, pp. 235–238, 1993.

O'Brien, R. D. and L. S. Wolfe. Nongenetic Effects of Radiation, in *Radiation, Radioactivity, and Insects.* Academic Press, New York, 1964.

Odin, M., A. Feurtet–Mazel, F. Ribeyre, and A. Boudou. Actions and interactions of temperature, pH and photoperiod on mercury bioaccumulation by nymphs of the burrowing mayfly *Hexagenia rigida,* from the sediment contamination source. *Environ. Toxicol. Chem.* 13, pp. 1291–1302, 1994.

Odum, E.P., *Fundamentals of Ecology.* W.B. Saunders Co., Philadelphia, PA, 1971, p. 574.

Odum, E.P. Trends expected in stressed ecosystems. *Bioscience* 35, pp. 419–422, 1985.

Odum, E.P., Preface, in *Ecotoxicology: A Hierarchical Treatment,* Newman, M.C. and C.H. Jagoe, Eds., CRC/Lewis Publishers, Boca Raton, FL, 1996.

Ogilvie, D.M. and D.L. Miller. Duration of a DDT–induced shift in the selected temperature of Atlantic salmon (*Salmo salar*). *Bull. Environ. Contam. Toxicol.* 16, pp. 86–89, 1976.

Ohlendorf, H.M., D.J. Hoffman, M.K. Saiki, and T.A. Aldrich. Embryonic mortality and abnormalities of aquatic birds: Apparent impacts of selenium from irrigation drainwater. *Sci. Total Environ.* 52, pp. 49–63, 1986.

Öhman, L.–O. and S. Sjöberg. Thermodynamic Calculations with Special Reference to the Aqueous Aluminum System, in *Metal Speciation: Theory, Analysis and Application,* Kramer, J.R. and H.E. Allen, Eds., Lewis Publishers, Inc., Chelsea, MI, 1988.

Olivieri, G., J. Bodycote, and S. Wolff. Adaptive response of human lymphocytes to low concentrations of radioactive thymidine. *Science (WASH DC)* 223, pp. 594–597, 1984.

Omernik, J.M., Ecoregions in the conterminous United States. *Annals Assoc. Am. Geographers.* 77, pp. 118–125, 1987.

Pacyna, J.M. The origin of Arctic air pollutants: Lessons learned and future research. *Sci. Total Environ.* 160/161, pp. 39–53, 1995.

Pagenkopf, G.K. Gill surface interaction model for trace–metal toxicity to fishes: Role of complexation, pH, and water hardness. *Environ. Sci. Technol.* 17, pp. 342–347, 1983.

Pain, D.J. Lead in the Environment, in *Handbook of Ecotoxicology,* Hoffman, D.J., B.A. Rattner, G.A. Burton, Jr., and J. Cairns, Jr., Eds., CRC Press, Inc., Boca Raton, FL, 1995.

Pao, E. M., K. H. Fleming, P. M. Gueuther, and S. J. Mickle. *Food Commonly Eaten by Individuals: Amount per Day and per Eating Occasion.* U.S. Department of Agriculture, Springville, VA, 1982, p. 431.

Parke, D. Cytochrome P–450 and the detoxication of environmental chemicals. *Aquat. Toxicol. (AMST)* 1, pp. 367–376, 1981.

Parsons, P.A. Inherited stress resistance and longevity: A stress theory of ageing. *Heredity* 75, pp. 216–221, 1995.

Partridge, G.G. Relative fitness of genotypes in a population of *Rattus norvegicus* polymorphic for warfarin resistance. *Heredity* 43, pp. 239–246, 1979.

Paterson, S. and D. Mackay. A steady–state fugacity–based pharmacokinetic model with simultaneous multiple exposure routes. *Environ. Toxicol. Chem.* 6, pp. 395–408, 1987.

Patrick, R. Use of algae, especially diatoms, in the assessment of water quality. *American Society for Testing and Materials, Special Technical Publication* 528, pp. 76–95, 1973.

Peakall, D. *Animal Biomarkers as Pollution Indicators.*Chaman & Hall, London, 1992, p. 291.

Peltier, W.H. and C.I. Weber. *Methods for Measuring the Acute Toxicity of Effluents to Freshwater and Marine Organisms,* EPA/600/4–85/013. Environmental Monitoring and Support Laboratory, EPA, Cincinnati, OH, 1985.

Pendleton, R.C., R.D. Lloyd, C.W. Mays, and B.W. Church. Trophic level effect on the accumulation of caesium–137 in cougars feeding on mule deer. *Nature (LOND)* 204, pp. 708–709, 1964.

Peoples, S.A. The Metabolism of Arsenic in Man and Animals, in *Arsenic. Industrial, Biomedical, Environmental Perspectives*, Lederer, W.H. and R.J. Fensterheim, Eds., Van Nostrand Reinhold Co., Inc., New York, 1983.

Peters, T. The effect of x–rays on fertilized egg cells of *Triton clopostris* with special emphasis on the effects of small doses and less damage. *Strahlentherapie.* 112, pp. 525–542, 1960.

Phillips, D.J.H. The use of biological indicator organisms to monitor trace metal pollution in marine and estuarine environments – A review. *Environ. Pollut.* 13, pp. 281–317, 1977.

Piatt, J.F., C.J. Lensink, W. Butler, M. Kendziorek, and D.R. Nysewander. Immediate impact of the 'Exxon Valdez' oil spill on marine birds. *Auk.* 10, pp. 387–397, 1990.

Pielou, E.C. *Population and Community Ecology. Principles and Methods.* Gordon and Breach Science Publishers, New York, 1974, p. 286.

Pimentel, D. and C.A. Edwards. Pesticides and ecosystems. *Bioscience* 32, pp. 595–600, 1982.

Piszkiewicz, D. *Kinetics of Chemical and Enzyme–Catalyzed Reactions.* Oxford University Press, New York, 1977, p. 235.

Platt, J.R. Strong inference. *Science (WASH DC).* 146, pp. 347–353, 1964.

Playle, R.C. and C.M. Wood. Water chemistry changes in the gill microenviroment of rainbow trout: Experimental observations and theory. *J. Comp. Physiol.* 159B, pp. 527–537, 1989.

Playle, R.C. and C.M. Wood. Mechanisms of aluminum extraction and accumulation at the gills of rainbow trout, *Oncorhynchus mykiss* (Walbaum) fingerlings. *Aquat. Toxicol. (AMST)* 21, pp. 267–278, 1991.

Polikarpov, G.G. *Radioecology of Aquatic Organisms.* North–Holland Publishing Co., Amsterdam, 1966, p. 314.

Pontasch, K.W., B.R. Niederlehner, and J. Cairns, Jr. Comparisons of single–species, microcosm and field responses to a complex effluent. *Environ. Toxicol. Chem.* 8, pp. 521–532, 1989.

Posthuma, L., R.F. Hogervorst, E.N.G. Joose, and N.M. Van Straalen. Genetic variation and co-variation for characteristics associated with cadmium tolerance in natural populations of the springtail *Orchesella cincta* (L.). *Evolution* 47, pp. 619–631, 1993.

Postma, J.F., S. Mol, H. Larsen, and W. Admiraal. Life–cycle changes and zinc shortage in cadmium–tolerant midges, *Chironomus riparius* (Diptera), reared in the absence of cadmium. *Environ. Toxicol. Chem.* 14, pp. 117–122, 1995.

Postma, J.F., A. van Kleunen, and W. Admiraal. Alterations in life–history traits of *Chironomus riparius* (Diptera) obtained from metal contaminated rivers. *Arch. Environ. Contam. Toxicol.* 29, pp. 469–475, 1995.

Pratt, J.R., and J. Cairns. Ecotoxicology and the Redundancy Problem: Understanding Effects on Community Structure and Function, in *Ecotoxicology: A Hierarchical Treatment*, Newman, M.C. and C.H. Jagoe, Eds., CRC/Lewis Publishers, Boca Raton, FL, 1996.

Preston, F.W. The commonness, and rarity, of species. *Ecology* 29, pp. 254–283, 1948.

Price, W.J. *Analytical Atomic Absorption Spectrometry.* Heyden & Son, Ltd., London, 1972, p. 239.

Pulliam, H.R., G.W. Barrett, and E.P. Odum. Bioelimination of Tracer [65]Zn in Relation to Metabolic Rates in Mice, in *Symposium on Radioecology,* Nelson, D.J. and F.C. Evans, Eds., University of Michigan, Ann Arbor, MI, 1967.

Pulliam, H.R. and B.J. Danielson. Source, sinks, and habitat selection: A landscape perspective on population dynamics. *Am. Nat.* 137, Supplement, pp. S50–S66, 1991.

Putka, G. Research on lead poisoning is questioned. *Wall Street Journal.* March 6, pp. B1, 1992.

Pynnönen, K., D.A. Holwerda, and D.I. Zandee. Occurrence of calcium concretions in various tissues of freshwater mussels, and their capacity for cadmium sequestration. *Aquat. Toxicol. (AMST)* 10, pp. 101–114, 1987.

Quine, W.V. *Methods of Logic.* Harvard University Press, Cambridge, MA, 1982, p. 333.

Rago, P.J. and R.M. Dorazio. Statistical inference in life–table experiments: The finite rate of increase. *Can. J. Fish. Aquat. Sci.* 41, pp. 1361–1374, 1984.

Ramade, F. *Ecotoxicology.* John Wiley & Sons, New York, 1987, p. 272.

Rand, G.M. Behavior, in Rand, G.M., and Petrocelli, S.R., Eds., *Fundamentals of Aquatic Toxicology*. Hemisphere Publishing Corp., Washington, DC, 1985.

Rand, G.M. and S.R. Petrocelli, Eds. *Fundamentals of Aquatic Toxicology*. Hemisphere Publishing Corp., Washington, DC, 1985, p. 666.

Rasmussen, J.B., D.J. Rowan, D.R.S. Lean, and J.H. Carey. Food chain structure in Ontario lakes determines PCB levels in lake trout (*Salvelinus namaycush*) and other pelagic fish. *Can. J. Fish. Aquat. Sci.* 47, pp. 2030–2038, 1990.

Ratcliffe, D.A. Decrease in eggshell weight in certain birds of prey. *Nature (LOND).* 215, pp. 208–210, 1967.

Ratcliffe, D.A. Changes attributable to pesticides in egg breakage frequency and eggshell thickness in some British birds. *J. Appl. Ecol.* 7, pp. 67–107, 1970.

Rau, G.H. Low $^{15}N/^{14}N$ in hydrothermal vent animals: Ecological implications. *Nature (LOND)* 289, pp. 484–485, 1981.

Rauser, W.E. and E.B. Dumbroff. Effects of excess cobalt, nickel and zinc on the water relations of *Phaseolus vulgaris*. *Environ. Exp. Bot.* 21, pp. 249–255, 1981.

Ravera, O. The ecological effects of acid deposition. Part 1. An introduction. *Experientia* 42, pp. 329–330, 1986.

Ray, S., D.W. McLeese, and M.R. Peterson. Accumulation of copper, zinc, cadmium and lead from two contaminated sediments by three marine invertebrates – A laboratory study. *Bull. Environ. Contam. Toxicol.* 26, pp. 315–322, 1981.

Regoli, F. and G. Principato. Glutathione, glutathione–dependent and antioxidant enzymes in mussel, *Mytilus galloprovincialis*, exposed to metals under field and laboratory conditions: Implications for the use of biochemical biomarkers. *Aquat. Toxicol. (AMST)* 31, pp. 143–164, 1995.

Reichhardt, T. Environmental GIS: The world in a computer. *Environ. Sci. Technol.* 30, pp. 340A–343A, 1996.

Reichle, D.E. Relation of body size to food intake, oxygen consumption, and trace element metabolism in forest floor arthropods. *Ecology* 49, pp. 538–541, 1968.

Reichle, D.E., P.B. Dunaway, and D.J. Nelson. Turnover and concentration of radionuclides in food chains. *Nucl. Saf.* 11, pp. 43–55, 1970.

Reichle, D.E. and R.I. Van Hook, Jr. Radionuclide dynamics in insect food chains. *Manit. Entomol.* 4, pp. 22–32, 1970.

Reinert, R.E., L.J. Stone, and W.A. Willford. Effect of temperature on accumulation of methylmercuric chloride and p,p'–DDT by rainbow trout (*Salmo gairdneri*). *J. Fish. Res. Board Can.* 31, pp. 1649–1652, 1974.

Reinfelder, J.R. and N.S. Fisher. The assimilation of elements ingested by marine copepods. *Science (WASH DC).* 251, pp. 794–796, 1991.

Rench, J.D. Environmental Epidemiology, in *Basic Environmental Toxicology*, Cockerham, L.G. and Shane, B.S., CRC Press, Inc., Boca Raton, FL, 1994.

Reuther, R. Mercury accumulation in sediment and fish from rivers affected by alluvial gold mining in the Brazilian Madeira River basin, Amazon. *Environ. Monit. Assess.* 32, pp. 239–258, 1994.

Rhodes, R. *The Making of the Atomic Bomb.* Simon and Schuster, New York, 1986, p. 886.

Richards, C. and G. Host. Examining land use influences on stream habitats and macroinvertebrates: A GIS approach. *Water Resour. Bull.* 30, pp. 729–738, 1994.

Richards, C., G.E. Host and J.W. Arthur. Identification of predominant environmental factors structuring stream macroinvertebrate communities within a large agricultural catchment. *Freshwater Biol.* 29, pp. 285–294, 1993.

Roberts, D.K., T.C. Hutchinson, J. Paciga, A. Chattopadhyay, R.E. Jervis, and J. Van Loon. Lead contamination around secondary smelters: Estimation of dispersal and accumulation by humans. *Science (WASH DC).* 186, pp. 1120–1124, 1974.

Roberts, L. Hard choices ahead on biodiversity. *Science (WASH DC).* 241, pp. 1759–1761, 1988.

Robison, S.H., O. Cantoni, and M. Costa. Analysis of metal–induced DNA lesionsand DNA–repair replication in mammalian cells. *Mutat. Res.* 131, pp. 173–181, 1984.

Robohm, R.A. Paradoxical effects of cadmium exposure on antibacterial antibody responses in two fish species: Inhibition in cunners (*Tautogolabrus adspersus*) and enhancement in striped bass (*Morone saxatilis*). *Vet. Immunol. Immunopathol.* 12, pp. 251–262, 1986.

Roch, M., J.A. McCarter, A.T. Matheson, M.J.R. Clark, and R.W. Olafson. Hepatic metallothionein in rainbow trout (*Salmo gairdneri*) as an indicator of metal pollution in the Campbell River system. *Can. J. Fish. Aquat. Sci.* 39, pp. 1596–1601, 1982.

Roesijadi, G. Metallothioneins in metal regulation and toxicity in aquatic animals. *Aquat. Toxicol. (AMST)* 22, pp. 81–114, 1992.

Roesijadi, G. and W.E. Robinson. Metal Regulation in Aquatic Animals: Mechanisms of Uptake, Accumulation, and Release, in *Aquatic Toxicology. Molecular, Biochemical and Cellular Perspectives*, Malins, D.C. and G.K. Ostrander, Eds., CRC Press, Inc., Boca Raton, FL, 1994.

Rolff, C., D. Broman, C. Näf, and Y. Zebühr. Potential biomagnification of PCDD/Fs – New possibilities for quantitative assessment using stable isotope trophic position. *Chemosphere* 27, pp. 461–468, 1993.

Rose, K. S. B. Lower limits of radiosensitivity in organisms, excluding man. *J. Environ. Radioact.* 15, pp. 113–133, 1992.

Rousch, W. Putting a price tag on Nature's bounty. *Science (WASH DC)* 276, p. 1029, 1997.

Rowan, D.J. and J.B. Rasmussen. Bioaccumulation of radiocesium by fish: The influence of physicochemical factors and trophic structure. *Can. J. Fish. Aquat. Sci.* 51, pp. 2388–2410, 1994.

Roy, R. and P.G.C. Campbell. Survival time modeling of exposure of juvenile Atlantic salmon (*Salmo salar*) to mixtures of aluminum and zinc in soft water at low pH. *Aquat. Toxicol. (AMST)* 33, pp. 155–176, 1995.

Rueter, J.G., Jr., and F.M.M. Morel. The interaction between zinc deficiency and copper toxicity as it affects the silicic acid uptake mechanisms in *Thalassiosira pseudonana*. *Limnol. Oceanogr.* 26, pp. 67–73, 1981.

Rugh, R. and J. Wolff. Threshold x–irradiation sterilization of the ovary. *Fertil. Steril.* 8, pp. 428–430, 1957.

Rule, J.H. and R.W. Alden, III. Cadmium bioavailablity to three estuarine animals in relation to geochemical fractions to sediments. *Arch. Environ. Contam. Toxicol.* 19, pp. 878–885, 1990.

Ruohtula, M. and J.K. Miettinen. Retention and excretion of [203]Hg–labelled methylmercury in rainbow trout. *Oikos* 26, pp. 385–390, 1975.

Rupp, E. M. Age dependent values of dietary intake for assessing human exposures to environmental pollutants. *Health Phys.* 39, pp. 141–146, 1980.

Rupp, E. M., F. L. Miller, and C. F. Baes, III. Some results of recent surveys of fish and shellfish consumption by age and region of U.S. residents. *Health Phys.* 39, pp. 1965–1970, 1980.

Russell, R.W., R. Lazar, and G.D. Haffner. Biomagnification of organochlorines in Lake Erie white bass. *Environ. Toxicol. Chem.* 14, pp. 719–724, 1995.

Saavedra Alvarez, M.M. and D.V. Ellis. Widespread neogastropod inposex in the Northeast Pacific: Implications for TBT contamination surveys. *Mar. Pollut. Bull.* 21, pp. 244–247, 1990.

Sabins, Jr., F.F., *Remote Sensing. Principles and Interpretation.* W.H. Freeman and Co., New York, 1987, p. 449.

Sagan, L.A. What is hormesis and why haven't we heard about it before? *Health Phys.* 52, pp. 521–525, 1987.

Saiki, M.K. and R.S. Ogle. Evidence of impaired reproduction by Western mosquitofish inhabiting seleniferous agricultural drainwater. *Trans. Am. Fish. Soc.* 124, pp. 578–587, 1995.

Salbu, B., E. H. Oughton, A. V. Ratnikov, T. L. Zhigareva, S. V. Kruglov, K. V. Petrov, N. V. Grebenshakikova, S. K. Firsakova, N. P. Astasheva, N. A. Loshchilov, K. Hove, and P. Strand. The mobility of ^{137}Cs and ^{90}Sr in agricultural soils in the Ukraine, Belarus, and Russia, 1991. *Health Phys.* 67, pp. 518–528, 1994.

Salsburg, D.S. *Statistics for Toxicologists.* Marcel Dekker, Inc., New York, NY, 1986.

Samallow, P.B. and M.E. Soule. A case of stress related heterozygote superiority in nature. *Evolution* 37, pp. 646–649, 1983.

Sanborn, J.R., R.L. Metcalf, W.N. Bruce, and P.–Y. Lu. The fate of chlordane and toxaphene in a terrestrial–aquatic model ecosystem. *Environ. Entomol.* 5, pp. 533–538, 1976.

Sanders, B.M. Stress proteins: Potential as Multitiered Biomarkers, in *Biomarkers of Environmental Contamination*, McCarthy, J.F. and L.R. Shugart, Eds., Lewis Publishers, Boca Raton, FL, 1990.

Sanders, B.M. and S.D. Dyer. Cellular stress response. *Environ. Toxicol. Chem.* 13, pp. 1209–1210, 1994.

Sanders, B.M., K.D. Jenkins, W.G. Sunda, and J.D. Costlow. Free cupric ion activity in seawater: Effects on metallothionein and growth in crab larvae. *Science (WASH DC)* 222, pp. 53–55, 1983.

Sanders, B.M. and L.S. Martin. Stress proteins as biomarkers of contaminant exposure in archived environmental samples. *Sci. Total Environ.* 139/140, pp. 459–470, 1993.

Sandheinrich, M.B. and G.J. Atchison. Sublethal toxicant effects on fish foraging behavior: Empirical vs. mechanistic approaches. *Environ. Toxicol. Chem.* 9, pp. 107–119, 1990.

Sandstead, H.H. Interactions That Influence Bioavailability of Essential Metals to Humans, in *Metal Speciation: Theory, Analysis and Application*, Kramer, J.R. and H.E. Allen, Eds., Lewis Publishers, Inc., Chelsea, MI, 1988.

Sanzharova, N. I., S. V. Fesenko, R. M. Alexakhin, V. S. Anisimov, V. K. Kuznetsov and L. G. Chernyayeva. Changes in the forms of ^{137}Cs and its availability for plants as dependent on properties of fallout after the Chernobyl nuclear power plant accident. *Sci. Total Environ.* 154, pp. 9–22, 1994.

Sarokin, D. and J. Schulkin. The role of pollution in large–scale population disturbances. Part 1: Aquatic populations. *Environ. Sci. Technol.* 26, pp1476–1484, 1992.

SAS Institute Inc., *SAS® Technical Report P–200, SAS/STAT® Software: CALIS and LOGISTIC Procedures, Release 6.04*. SAS Institute Inc., Cary, NC, 1990, p. 236.

Scheninger, M.J., M.J. CeNiro, and H. Tauber. Stable nitrogen isotope ratios of bone collagen reflect marine and terrestrial components of prehistoric human diet. *Science (WASH DC)* 220, pp. 1381–1383, 1983.

Schimmack, W., K. Bunzl and L. Zelles. Initial rates of migration of radionuclides from the Chernobyl fallout in undisturbed soils. *Geoderma.* 44, pp. 211–218, 1989.

Schindler, D.W. Effects of acid rain on freshwater ecosystems. *Science (WASH DC)* 239, pp. 149–157, 1988.

Schindler, D.W. Ecosystems and Ecotoxicology: A Personal Perspective, in *Ecotoxicology: A Hierarchical Treatment*, Newman, M.C. and C.H. Jagoe, Eds., CRC/Lewis Publishers, Boca Raton, FL, 1996.

Schlueter, M.A., S.I. Guttman, J.T. Oris, and A.J. Bailer. Survival of copper–exposed juvenile fathead minnows (*Pimephales promelas*) differs among allozyme genotypes. *Environ. Toxicol. Chem.* 10, pp. 1727–1734, 1995.

Schmitt, C.J., M.L. Wildhaber, J.B. Hunn, T. Nash, M.N. Tieger, and B.L. Steadman. Biomonitoring of lead–contaminated Missouri streams with an assay for erythrocyte δ–aminolevulinic acid dehydratase activity in fish blood. *Arch. Environ. Contam. Toxicol.* 25, pp. 464–475, 1993.

Schnell, J.H. Some effects of neutron–gamma radiation on late summer bird populations. *Auk* 81, pp. 528–533, 1964.

Schnute, J.T. and L.J. Richards. A unified approach to the analysis of fish growth, maturity, and survivorship data. *Can. J. Fish. Aquat. Sci.* 47, pp. 24–40, 1990.

Schober, U. and W. Lampert. Effects of sublethal concentrations of the herbicide Atrazin® on growth and reproduction of *Daphnia pulex*. *Bull. Environ. Contam. Toxicol.* 17, pp. 269–277, 1977.

Schreck, C.B. Physiological, behavioral, and performance indicators of stress. *Am. Fish. Soc. Symp.* 8, pp. 29–37, 1990.

Schultz, I.R. and W.L. Hayton. Body size and the toxicokinetics of trifluralin in rainbow trout. *Toxicol. Appl. Pharmacol.* 129, pp. 138–145, 1994.

Schütt, P. and E.B. Cowling. Waldsterben, a general decline of forests in central Europe: Symptoms, development, and possible causes. *Plant Disease* 69, pp. 548–558, 1985.

Selye, H. *The Stress of Life*. McGraw–Hill Book Company, New York, 1956, p. 54.

Selye, H. The evolution of the stress concept. *Amer. Sci.* 61, pp. 692–699, 1973.

Semagin, V. N. Influence of prolonged low dose irradiation on brain of rat embryos. *Radiobiologiya* 15, pp. 583–588, 1986 (from Rose, 1992).

Settle, D.M. and C.C. Patterson. Lead in albacore: Guide to lead pollution in Americans. *Science (WASH DC).* 207, pp. 1167–1176, 1980.

Shane, B.S., Introduction to Ecotoxicology, in *Basic Environmental Toxicology,* Cockerham, L.G. and B.S. Shane, Eds., CRC Press, Boca Raton, FL, 1994.

Shea, D. Developing national sediment quality criteria. *Environ. Sci. Technol.* 22, pp. 1256–1261, 1988.

Shore, P.A., B.B. Brodie, and C.A.M. Hogben. The gastric secretion of drugs: A pH partition hypothesis. *J. Pharmacol. Exp. Ther.* 119, pp. 361–369, 1957.

Shugart, L.R. Quantitation of chemically induced damage to DNA of aquatic organisms by alkaline unwinding assay. *Aquat. Toxicol. (AMST)* 13, pp. 43–52, 1988.

Shugart, L.R. Environmental genotoxicology, in *Fundamentals of Aquatic Toxicology. Effects, Environmental Fate, and Risk Assessment*, 2nd ed., Rand, G.M., Ed., Taylor & Francis, Washington, DC, 1995.

Shugart, L.R. Molecular Markers to Toxic Agents, in *Ecotoxicology: A Hierarchical Treatment*, Newman, M.C. and Jagoe, C.H., Eds., CRC Press, Inc., Boca Raton, FL, 1996.

Shukla, K.K., C.S. Dombroski, and S.H. Cohn. Fallout ^{137}Cs levels in man over a 12 yr period. *Health Phys.* 24, pp. 555–557, 1973.

Sibly, R.M. Effects of Pollutants on Individual Life Histories and Population Growth Rates, in Newman, M.C. and C.H. Jagoe, *Ecotoxicology: A Hierarchical Treatment*, CRC Press, Inc., Boca Raton, FL, 1996.

Sibly, R.M. and P. Calow. A life–cycle theory of responses to stress. *Biol. J. Linn. Soc.* 37, pp. 101–116, 1989.

Simkiss, K. Calcium, pyrophosphate and cellular pollution. *Trends Biochem. Sci.* April 1981, pp. 1–3, 1981.

Simkiss, K., Lipid solubility of heavy metals in saline solutions. *J. Mar. Biol. Assoc. U.K.* 63, pp. 1–7, 1983.

Simkiss, K., Ecotoxicants at the Cell–Membrane Barrier, in *Ecotoxicology: A Hierarchical Treatment*, Newman, M.C. and C.H. Jagoe, Eds., CRC/Lewis Publishers, Boca Raton, FL, 1996.

Simkiss, K., S. Daniels, and R.H. Smith. Effects of population density and cadmium on growth and survival of blowflies. *Environ. Pollut.* 81, p. 41–45, 1993.

Simkiss, K. and M. Taylor. Cellular mechanisms of metal ion detoxification and some new indices of pollution. *Aquat. Toxicol. (AMST)* 1, pp. 279–290, 1981.

Simonich, S.I. and R.A. Hites. Global distribution of persistent organochlorine compounds. *Science (WASH DC).* 269, pp. 1851–1854, 1995.

Sinclair, W. K. *Science, Radiation Protection and the NCRP.* (Lauriston S. Taylor lectures in Radiation Protection and Measurements, Lecture 17). NCRP, Bethesda, MD, 1993, p. 56.

Skidmore, J.F. and P.W.A. Tovell. Toxic effects of zinc sulphate on the gills of rainbow trout. *Water Res.* 6, pp. 217–230, 1972.

Slater, T.F. Free–radical mechanisms in tissue injury. *Biochem. J.* 222, pp. 1–15, 1984. (CH. 6)

Slobodkin, L.B. and D.E. Dykhuizen, Applied Ecology, Its Practice and Philosophy, in *Integrated Environmental Management*, Cairns, J., Jr., and T.V. Crawford, Eds., Lewis Publishers, Chelsea, MI, 1991.

Smith, A.L. and R.H. Green. Uptake of mercury by freshwater clams (Family Unionidae). *J. Fish. Res. Board Can.* 32, pp. 1297–1303, 1975.

Smith, R.J. The risks of living near Love Canal. *Science (WASH DC).* 217, pp. 808–811, 1982.

Smith, R.L. *Elements of Ecology*, 2nd ed. Harper & Row, Publishers, New York, 1986, p. 677.

Smith, W.H. *Air Pollution and Forests*. Springer–Verlag, Inc., New York, 1981, p. 379.

Smith, W.E. and A.M. Smith, *Minamata*. Holt, Rinehart and Winston, New York, 1975, p. 192.

Soimasuo, R., I. Jokinen, J. Kukkoen, T. Petänen, T. Ristola, and A. Oikari. Biomarker responses along a pollution gradient: Effects of pulp and paper mill effluents on caged whitefish. *Aquat. Toxicol. (AMST)* 31, pp. 329–345, 1995.

Sorkhoh, N., R. Al–Hasan, S. Radwan and T. Höpner. Self–cleaning of the Gulf. *Nature (LOND)*. 359, p. 109, 1992.

Spacie, A. and J.L. Hamelink, Bioaccumulation, in *Fundamentals of Aquatic Toxicology*, Rand, G.M. and S.R. Petrocelli, Eds., Hemisphere Publishing Corp., Washington, DC, 1985.

Spacie, A., L.S. McCarty, and G.M. Rand. Bioaccumulation and Bioavailability in Multiphase Systems, in *Fundamentals of Aquatic Toxicology*, 2nd ed., Rand, G.M., Ed., Taylor & Francis, Washington, 1995.

Sparks, A.K. *Invertebrate Pathology. Noncommunicable Diseases*. Academic Press, Inc., New York, 1972, p. 387.

Sparrow, A. H., J. P. Binnington, and V. Pond. *Bibliography on the Effects of Ionizing Radiations on Plants, 1896–1955*. U.S. AEC Rep. BNL–504(L–103), Brookhaven National Laboratory, Upton, Long Island, NY, 1958, p. 222.

Spies, R.B., D.W. Rice, Jr., P.J. Thomas, J.J. Stegeman, J.N. Cross, and J.E. Hose. A field test of correlates of poor reproductive success and genetic damage in contaminated populations of starry flounder, *Platichthys stellatus*. *Mar. Env. Res.* 28, pp. 542–543, 1989.

Spitzer, P.R., R.W. Risebrough, W. Walker II, R. Hernandez, A. Poole, D. Puleston, and I.C.T. Nisbet. Productivity of ospreys in Connecticut–Long Island increases as DDE residues decline. *Science (WASH DC)*. 202, pp. 333–335, 1978.

Sprague, J.B. Measurement of pollutant toxicity to fish. I. Bioassay methods for acute toxicity. *Water Res.* 3, pp. 793–821, 1969.

Sprague, J.B. Measurement of pollutant toxicity to fish. II. Utilizing and applying bioassay results. *Water Res.* 4, pp. 3–32, 1970.

Sprague, J.B. Measurement of pollutant toxicity to fish. III. Sublethal effects and "safe" concentrations. *Water Res.* 5, pp. 245–266, 1971.

Sprague, J.B. Current status of sublethal tests of pollutants on aquatic organisms. *J. Fish. Res. Board Can*. 33, pp. 1988–1992, 1976.

Squibb, .S. and B.A. Fowler. Relationship between metal toxicity to subcellular systems and the carcinogenic response. *Environ. Health Perspect.* 40, pp. 181–188, 1981.

Stacell, M. and D.G. Huffman. Oxytetracycline–induced photosensitivity of channel catfish. *Prog. Fish–Cult.* 56, pp. 211–213, 1994.

Stebbing, A.R.D. Hormesis – The stimulation of growth by low levels of inhibitors. *Sci. Total Environ.* 22, pp. 213–234, 1982.

Steedman, R.J. Modification and assessment of an index of biotic integrity to quantify stream quality in southern Ontario. *Can. J. Fish. Aquat. Sci.* 45, pp. 492–501, 1988.

Stegeman, J.J. and M.E. Hahn. Biochemistry and Molecular Biology of Monoxygenases: Current Perspectives on Forms, Functions, and Regulation of Cytochrome P450 in Aquatic Species, in *Aquatic Toxicology. Molecular, Biochemical and Cellular Perspectives*, Malins, D.C. and G.K. Ostrander, Eds., CRC Press, Inc., Boca Raton, FL, 1994.

Stein, J.E., W.L. Reichert, M. Nishimoto, and U. Varanasi. Overview of studies on liver carcinogenesis in English sole from Puget Sound; Evidence for a xenobiotic chemical etiology II: Biochemical studies. *Sci. Total Environ.* 94, pp. 51–69, 1990.

Stenehjem, M. Indecent exposure. *Nat. Hist.* 9/90, pp. 6–21, 1990.

Stephan, C.E. Methods for calculating an LC50, in *Aquatic Toxicology and Hazard Evaluation*. *ASTM STP 634*, Mayer, F.L. and J.L. Hamelink, Eds., American Society for Testing and Materials, Philadelphia, PA, 1977.

Stephan, C.E. and J.W. Rogers. Advantages of Using Regression Analysis to Calculate Results of Chronic Toxicity Tests, in *Aquatic Toxicology and Hazard Assessment: Eighth Symposium*. *ASTM STP 891*, Bahner, R.C. and Hansen, D.J., Eds., American Society for Testing and Materials, Philadelphia, PA, 1985.

Stone, R. Can a father's exposure lead to illness in his children? *Science (WASH DC)* 258, pp. 31, 1992.

Strong, C.R. and S.N. Luoma. Variations in the correlation of body size with concentrations of Cu and Ag in the bivalve *Macoma balthica*. *Can. J. Fish. Aquat. Sci.* 38, pp. 1059–1064, 1981.

Stumm, W. and J.J. Morgan. *Aquatic Chemistry. An Introduction Emphasizing Chemical Equilibrium in Natural Waters*. John Wiley and Sons, Inc., New York, 1970, p. 583.

Stumm, W., L. Sigg, and J.L. Schnoor. Aquatic chemistry of acid deposition. *Environ. Sci. Technol.* 21, pp. 8–13, 1987.

Sullivan, J.F., G.J. Atchison, D.J. Kolar, and A.W. McIntosh. Changes in predator-prey behavior of fathead minnows (*Pimephales promelas*) and largemouth bass (*Micropterus salmoides*) caused by cadmium. *J. Fish. Res. Board Can.* 35, pp. 446–451, 1978.

Suter, G.W., II, *Ecological Risk Assessment*. Lewis Pubishers, Boca Raton, FL, 1993, p. 538.

Tachikawa, M. and R. Sawamura. The effects of salinity on pentachlorophenol accumulation and elimination by killifish (*Oryzias latipes*). *Arch. Environ. Contam. Toxicol.* 26, pp. 304–308, 1994.

Tackett, S.L. Lead in the environment: Effects of human exposure. *Am. Lab. (FAIRFIELD CONN)* July, pp. 32–41, 1987.

Tagatz, M.E. Effect of mirex on predator–prey interaction in an experimental estuarine ecosystem. *Trans. Am. Fish. Soc.* 4, pp. 546–549, 1976.

Taub, F.B. Standardized aquatic microcosms. *Environ. Sci. Technol.* 23, pp. 1064–1066, 1989.

Taylor, E.J., J.E. Morrison, S.J. Blockwell, A. Tarr, and D. Pasoe. Effects of lindane on the predator–prey interaction between *Hydra oligactis* Pallas and *Daphnia magna* Strauss. *Arch. Environ. Contam. Toxicol.* 29, pp. 291–26, 1995.

Teasdale, S. There Will Come Soft Rains (War Time), in *The Collected Poems of Sara Teasdale*, Macmillian Co., New York, 1937.

Terhaar, C.J., W.S. Ewell, S.P. Dziuba, W.W. White, and P.J. Murphy. A laboratory model for evaluating the behavior of heavy metals in an aquatic environment. *Water Res.* 11, pp. 101–110, 1977.

Tessier, A., P.G.C. Campbell, and M. Bisson. Sequential extraction procedure for the speciation of particulate trace metals. *Anal. Chem.* 51, pp. 844–851, 1979.

Tessier, A., P.G.C. Campbell, J.C. Auclair, and M. Bisson. Relationships between partitioning of trace metals in sediments and their accumulation in the tissues of the freshwater mollusc *Elliptio complanata* in a mining area. *Can. J. Fish. Aquat. Sci.* 41, pp. 1463–1472, 1984.

Tessier, L., G. Vaillancourt, and L. Pazdernik. Temperature effects on cadmium and mercury kinetics in freshwater molluscs under laboratory conditions. *Arch. Environ. Contam. Toxicol.* 26, pp. 179–184, 1994.

Thomann, R.V. Bioaccumulation model of organic chemical distribution in aquatic food chains. *Environ. Sci. Technol.* 23, pp. 699–707, 1989.

Thomas, P. Molecular and biochemical responses of fish to stressors and their potential use in environmental monitoring. *Am. Fish. Soc. Symp.* 8, pp. 9–28, 1990.

Thompson, H.M. Interactions between pesticides; a review of reported effects and their implications for wildlife risk assessment. *Ecotoxicology* 5, pp. 59–81, 1996.

Thompson, K.W., A.C. Hendricks, G.L. Nunn, and J. Cairns, Jr. Ventilatory responses of bluegill sunfish to sublethal fluctuating exposures to heavy metals (Zn^{++} and Cu^{++}). *Water Resour. Bull.* 19, pp. 719–727, 1983.

Thompson, R.D. The Changing Atmosphere and Its Impact on Planet Earth, in *Environmental Issues in the 1990s*, Mannion, A.M. and S.R. Bowlby, Eds.,John Wiley & Sons, Ltd., West Sussex, 1992.

Thoreau, H.D., *Essay on the Duty of Disobedience and Walden*. Reprinted 1968 by Lancer Books, Inc., New York, 1854, p. 445.

Thurston, R.V., R.C. Russo and G.A. Vonogradov. Ammonia toxicity to fishes. Effect of pH on the toxicity of the un-ionized ammonia species. *Environ. Sci. Technol.* 15, pp. 837–840, 1981.

Till, J. E. and H. R. Meyer. *Radiological Assessment: A Textbook on Environmental Dose Analysis*. U. S. Nuclear Regulatory Commission, Washington, DC and NTIS, Springfield, VA, 1983.

Tolmazin, D. Soviet environmental practices. *Science (WASH DC).* 221, pp. 1136, 1983.

Touart, L.W. The Federal Insecticide, Fungicide, and Rodenticide Act, in *Fundamentals of Aquatic Toxicology. Effects, Environmental Fate, and Risk Assessment*, 2nd ed., Rand, G.M., Ed., Taylor & Francis, Washington, DC, 1995.

Trabalka, J.R., L.D. Eyman and S.I. Aurbach. Analysis of the 1957–1958 Soviet nuclear accident. *Science (WASH DC).* 209, pp. 345–353, 1980.

Truhaut, R. Ecotoxicology: Objectives, principles and perspectives. *Ecotoxicol. Env. Saf.* 1, pp. 151–173, 1977.

Tucker, J.D., A. Auletta, M.C. Cimino, K.L. Dearfield, D. Jacobson–Kram, R.R. Tice, and A.V. Carrano. Sister–chromatid exchange: Second report of the Gene–Tox program. *Mutat. Res.* 297, pp. 101–180, 1993. ˘

Turner, F. B. Effects of continuous irradiation on animal populations. *Adv. Radiat. Biol.* 5, p. 83, 1975.

Tuurala, H. and A. Soivio. Structural and circulatory changes in the secondary lamallae of *Salmo gairdneri* gills after sublethal exposures to dehydroabietic acid and zinc. *Aquat. Toxicol. (AMST)* 2, pp. 21–29, 1982.

Ugedal, O., B. Jonsson, O. Njåstad, and R. Næumanň. Effects of temperature and body size on radiocaesium retention in brown trout, *Salmo trutta. Freshwater Biol.* 28, pp. 165–171, 1992.

Ugolini, F.C. and H. Spaltenstein. Pedosphere, in *Global Biogeochemical Cycles*, Butcher, S.S., R.J. Charlson, G.H. Orians, and G.V. Wolfe, Ed., Academic Press, London, 1992.

Ullrich, R. L., M. C. Jernigan, L. C. Satterfield and N. D. Bowles. Radiation carcinogenesis: Time–dose relationships. *Radiat. Res.* 111, pp. 179–184, 1987.

Ulmer, D.D. Effects of Metals on Protein Structure, in *Effects of Metals on Cells, Subcellular Elements, and Macromolecules*, Maniloff, J., J.R. Coleman, and M.W. Miller, Eds., Charles C. Thomas Publisher, Springfield, IL, 1970.

Underwood, A. J. and C. H. Peterson. Towards an ecological framework for investigating pollution. *Mar. Ecol. Prog. Ser.* 46, pp. 227–234, 1988.

UNSCEAR (United Nations Scientific Committee on the Effects of Atomic Radiation). *Genetic and Somatic Effects of Ionizing Radiation.* Publication E.86.IX.9, United Nations, NY, 1986, p. 366.

Van Beneden, R.J. and G.K. Ostrander. Expression of Oncogenes and Tumor Supressor Genes in Teleost Fishes, in *Aquatic Toxicology. Molecular, Biochemical and Cellular Perspectives*, Malins, D.C. and G.K. Ostrander, Eds., CRC Press, Inc., Boca Raton, FL, 1994.

van den Heuvel, M.R., L.S. McCarty, R.P. Lanno, B.E. Hickie, and D.G. Dixon. Effect of total body lipid on the toxicity and toxicokinetics of pentachlorophenol in rainbow trout (*Oncorhynchus mykiss*). *Aquat. Toxicol. (AMST)* 20, pp. 235–252, 1991.

Van Leeuwen, C.J., F. Moberts, G. Niebeek. Aquatic toxicological aspects of dithiocarbamates and related compounds. II. Effects on survival, reproduction and growth of *Daphnia magna. Aquat. Toxicol. (AMST.)* 7, pp. 165–175, 1985.

Van Straalen, N.M. and C.A. Denneman. Ecotoxicological evaluation of soil quality criteria. *Ecotoxicol. Env. Saf.* 18, pp. 241–251, 1989.

Van Veld, P.A., D.J. Westbrook, B.R. Woodin, R.C. Hale, C.L. Smith, R.J. Huggett, and J.J. Stegeman. Induced cytochrome P–450 in intestine and liver of spot (*Leiostomus xanthurus*) from a polycyclic aromatic hydrocarbon contaminated environment. *Aquat. Toxicol. (AMST)* 17, pp. 119–132, 1990.

Velasko, R. H., M. Belli, U. Sanasone, and S. Menegon. Vertical transport of radiocesium in surface soils: model implementation and dose–rate computation. *Health Phys.* 64, pp. 37–44, 1993.

Víg, É. and J. Nemcsók. The effects of hypoxia and paraquat on the superoxide dismutase activity in different organs of carp, *Cyprinus carpio* L. *J. Fish Biol.* 35, pp. 23–25, 1989.

Wagner, C. and H. Løkke. Estimation of ecotoxicological protection levels from NOEC toxicity data. *Water Res.* 25, pp. 1237–1242, 1991.

Wagner, J.G. *Fundamentals of Clinical Pharmacokinetics.* Drug Intelligence Publications, Inc., Hamilton, IL, 1975, p. 461.

Walker, B. Biodiversity and ecological redundancy. *Conserv. Biol.* 6, pp. 12–23, 1991.

Walker, C.H. Species differences in microsomal monooxygenase activity and their relationship to biological half–lives. *Drug Metab. Rev.* 7, pp. 295–323, 1978.

Walker, C.H., I. Newton, S.D. Hallam, and M.J.J. Ronis. Activities and toxicological significance of hepatic microsomal enzymes of the kestrel (*Falco tinnunculus*) and sparrowhawk (*Accipiter nisus*). *Comp. Biochem. Physiol.* 86C, pp. 379–382, 1987.

Wang, C.H., D.L. Willis, and W.D. Loveland. *Radiotracer Methodology in the Biological, Environmental, and Physical Sciences.* Prentice–Hall, Inc., Englewood Cliffs, NJ, 1975, p. 480.

Wangen, L.E. Elemental composition of size–fractionated aerosols associated with a coal–fired power plant plume and background. *Environ. Sci. Technol.* 15, pp. 1080–1088, 1981.

Wania, F. and D. Mackay. A global distribution model for persistent organic chemicals. *Sci. Total Environ.* 160/161, pp. 211–232, 1995.

Wania, F. and D. Mackay. Tracking the distribution of persistent organic pollutants. *Environ. Sci. Technol.* 30, pp. 390A–396A, 1996.

Warnau, M., G. Ledent, A. Temara, V. Alva, M. Jangoux, P. Dubois. Allometry of heavy metal bioconcentration in the Echinoid *Paracentrotus lividus. Arch. Environ. Contam. Toxicol.* 29, pp. 393–399, 1995.

Watkins, B. and K. Simkiss. The effect of oscillating temperatures on the metal ion metabolism of *Mytilus edulis. J. Mar. Biol. Assoc.* UK 68, pp. 93–100, 1988.

Webb, R.E. and F. Horsfall, Jr. Endrin resistance in the pine mouse. *Science (WASH DC)* 156, pp. 1762, 1967.

Weber, C.I., W.H. Peltier, T.J. Norberg–King, W.B. Horning, II., F.A. Kessler, J.R. Menkedick, T.W. Neiheisel, P.A. Lewis, D.J. Klemm, Q.H. Pickering, E.L. Robinson, J.M. Lazorchak, L.J. Wymer, and R.W. Freyberg. *Short–Term Methods for Estimating the Chronic Toxicity of Effluents and Receiving Waters to Freshwater Organisms*, EPA/600/4–89/001. Environmental Monitoring Systems Laboratory, EPA, Cincinnati, OH, 1989.

Weeks, J.M. and P.S. Rainbow. A dual–labelling technique to measure the relative assimilation efficiencies of invertebrates taking up trace metals from food. *Functional Ecol.* 4, pp. 711–717, 1990.

Weis, J.S. and J. Perlmutter. Effects of tributyltin on activity and burrowing behavior of the fiddler crab, *Uca pugilator. Estuaries* 10, pp. 342–346, 1987.

Weis, J.S. and P. Weis. Pollutants as developmental toxicants in aquatic organisms. *Environ. Health Perspect.* 71, pp. 77–85, 1987.

Weis, J.S. and P. Weis. Effects of environmental pollutants on early fish development. *CRC Critical Reviews in Aquatic Sciences* 1, pp. 45–73, 1989a.

Weis, J.S. and P. Weis. Tolerance and stress in a polluted environment. The case of the mummichog. *Bioscience* 39, pp. 89–95, 1989b.

Weis, J.S. and P. Weis. Effects of embryonic exposure to methylmercury on larval prey–capture ability in the mummichog, *Fundulus heteroclitus. Environ. Toxicol. Chem.* 14, pp. 153–156, 1995.

Welch, W.J. Mammalian Stress Response: Cell Physiology and Biochemistry of Stress Proteins, in *Stress Proteins in Biology and Medicine.* Morimoto, R.I., A. Tissieres, and C.C. Georgopolis, Eds., Cold Spring Harbor Press, Cold Spring Harbor, NY, 1990.

Wenning, R.J., R.T. Di Giulio, and E.P. Gallagher. Oxidant–mediated biochemical effects of paraquat in the ribbed mussel, *Geukensia demissa. Aquat. Toxicol. (AMST)* 12, pp. 157–170, 1988.

Wetzel, R.G. *Limnology.* Saunders College Publishing, Philadelphia, PA, 1982, p. 767.

Wheeler, D.L. An eclectic biologist argues that humans are not evolution's most important result; bacteria are. *Chron. Higher Education* Sept. 6, A23–A24, 1996.

Whicker, F. W. and L. Fraley, Jr. Effects of ionizing radiation on terrestrial plant communities. *Adv. Radiat. Biol.* 4, p. 317, 1974.

Whicker, F. W., T. G. Hinton, and D. J. Niquette. Health Risk to Hypothetical Residents of a Radioactively Contaminated Lake Bed, in *Proceedings of the Environmental Remediation Conference,* Augusta, GA, US DOE, October 24–28, 1993.

Whicker, F. W. and V. Schultz. *Radioecology: Nuclear Energy and the Environment.* CRC Press, Boca Raton, FL, 1982, Vol. I, p. 212; Vol. II., p. 228.

Whicker, F. W. and T. B. Kirchner. PATHWAY: A dynamic food–chain model to predict radionuclide ingestion after fallout deposition. *Health Phys.* 52, pp. 717–737, 1987.

White, R.M. The great climate debate. *Sci. Am.* 263, pp. 36–43, 1990.

Wichterman, R. Biological effects of ionizing radiations on protozoa: Some discoveries and unsolved problems. *BioScience* 22, p. 281, 1972.

Wilkinson, K.J. and P.G.C. Campbell. Aluminum bioconcentration at the gill surface of juvenile Atlantic salmon in acidic media. *Environ. Toxicol. Chem.* 12, pp. 2083–2095, 1993.

Williams, L.G. and D.I. Mount. Influence of zinc on periphytic communities. *Am. J. Bot.* 52, pp. 26–34, 1965.

Williamson, P. Use of ^{65}Zn to determine the field metabolism of the snail *Cepaea nemoralis* L. *Ecology* 56, pp. 1185–1192, 1975.

Wilson, J.B. The cost of heavy–metal tolerance: An example. *Evolution* 42, pp. 408–413, 1988.

Winge, D.R. and M. Brouwer. Discussion summary. Techniques and problems in metal–binding protein chemistry and implications for proteins in nonmammalian organisms. *Environ. Health Perspect.* 65, pp. 211–214, 1986.

Winner, R.W., M.W. Boesel, and M.P. Farrell. Insect community structure as an index of heavy–metal pollution in lotic ecosystems. *Can. J. Fish. Aquat. Sci.* 37, pp. 647–655, 1980.

Winner, R.W., T. Keeling, R. Yeager, and M.P. Farrell. Effect of food type on the acute and chronic toxicity of copper to *Daphnia magna*. *Freshwater. Biol.* 7, pp. 343–349, 1977.

Winston, G.W. and R.T. Di Giulio. Prooxidant and antioxidant mechanisms in aquatic organisms. *Aquat. Toxicol. (AMST)* 19, pp. 137–161, 1991.

Wirgin, I.I., C. Grunald, S. Courtenay, G–.L. Kreamer, W.L. Reichert, and J.E. Stein. A biomarker approach to assessing xenobiotic exposure in Atlantic tomcod from the North American Atlantic Coast. *Environ. Health Perspect.* 102, pp. 764–770, 1994.

Wise, D., J.D. Yarbrough, and R.T. Roush. Chromosomal analysis of insecticide resistant and susceptible mosquitofish. *J. Hered.* 77, pp. 345–348, 1986.

Witters, H.E. Acute acid exposure of rainbow trout, *Salmo gairdneri* Richardson: Effects of aluminum and calcium on ion balance and haematology. *Aquat. Toxicol. (AMST)* 8, pp. 197–210, 1986.

Wofford, H.W. and P. Thomas. Effect of xenobiotics on peroxidation of hepatic microsomal lipids from striped mullet (*Mugil cephalus*) and Atlantic croaker (*Micropogonus undulatus*). *Mar. Environ. Res.* 2, pp. 285–289, 1988.

Wolff, S. Are radiation–induced effects hormetic? *Science (WASH DC)* 245, pp. 575–621, 1989.

Woltering, D.M., J.L. Hedtke, and L.J. Weber. Predator–prey interactions of fishes under the influence of ammonia. *Trans. Am. Fish. Soc.* 107, pp. 500–504, 1978.

Wong, C.K. and P.K. Wong. Life table evaluation of the effects of cadmium exposure on the freshwater cladoceran *Moina macrocopa*. *Bull. Environ. Contam. Toxicol.* 44, pp. 135–141, 1990.

Wood, J.M. and H.–K. Wang. Microbial resistance to heavy metals. *Environ. Sci. Technol.* 17, pp. 582A–590A, 1983.

Woods, J.S., M.D. Martin, C.A. Naleway, and D. Echeverria. Urinary porphyrin profiles as a biomarker of mercury exposure: Studies on dentists with occupational exposure to mercury vapor. *J. Toxicol. Environ. Health.* 40, pp. 235–246, 1993.

Woodward, L.A., M. Mulvey, and M.C. Newman. Mercury contamination and population–level responses in chironomids: Can allozyme polymorphism indicate exposure? *Environ. Toxicol. Chem.* 15, pp. 1309–1316, 1996.

Woodwell, G.M. Effects of ionizing radiation on terrestrial ecosystems. *Science (WASH DC).* 138, pp. 572–577, 1962.

Woodwell, G.M. The ecological effects of radiation. *Sci. Am.* 208, pp. 2–11, 1963.

Woodwell, G.M. Toxic substances and ecological cycles. *Sci. Am.* 216, pp. 24–31, 1967.

Woodwell, G.M. The carbon dioxide question. *Sci. Am.* 238, pp. 34–43, 1978.

Wren, C.D. and H.R. MacCrimmon. Comparative bioaccumulation of mercury in two adjacent freshwater ecosystems. *Water Res.* 20, pp. 763–769, 1986.

Wren, C.D., H.R. MacCrimmon, and B.R. Loescher. Examination of bioaccumulation and biomagnification of metals in a Precambrian Shield lake. *Water Air Soil Pollut.* 19, pp. 277–291, 1983.

Wright, P.A. and C.D. Zamuda. Copper accumulation by two bivalue molluscs. Salinity effect is independent of cupric ion activity. *Mar. Environ. Res.* 23, pp. 1–14, 1987.

Wu, L., J. Chen, K.K. Tanji, and G.S. Banuelos. Distribution and biomagnification of selenium in a restored upland grassland contaminated by selenium from agricultural drain water. *Environ. Toxicol. Chem.* 14, pp. 733–742, 1995.

Yamada, H., M. Tateishi, and K. Takayanagi. Bioaccumulation of organotin compounds in the red sea bream (*Pagrus major*) by two uptake pathways: Dietary uptake and direct uptake from water. *Environ. Toxicol. Chem.* 13, pp. 1415–1422, 1994.

Yamaoka, K., T. Nakagawa, and T. Uno. Statistical moments in pharmacokinetics. *J. Pharmacokinet. Biopharm.* 6, pp. 547–558, 1978.

Yang, Y. and C. B. Nelson. *An Estimation of the Daily Average Food Intake by Age and Sex for Use in Assessing the Radionuclide Intake of Individuals in the General Population.* EPA Report No. 520/1–84–021, Office of Radiation Programs, U.S. Environmental Protection Agency, Washington, DC, 1984. p. 30.

Yap, H.H., D. Desaiah, L.K. Cutkomp, and R.B. Koch. Sensitivity of fish ATPases to polychlorinated biphenyls. *Nature (LOND)* 233, pp. 61–62, 1971.

Yarbrough, J.D., R.T. Roush, J.C. Bonner, and D.A. Wise. Monogenic inheritance of cyclodiene insecticide resistance in mosquitofish, *Gambusia affinis. Experientia (BASEL)* 42, pp. 851–853, 1986.

Young, L.B. and H.H. Harvey. Metal concentrations in chironomids in relation to the geochemical characteristics of surficial sediments. *Arch. Environ. Contam. Toxicol.* 21, pp. 202–211, 1991.

Zakharov, V.M. Analysis of Fluctuating Asymmetry as a Method of Biomonitoring at the Population Level, in *Bioindications of Chemical and Radioactive Pollution,* Krivolutsky, D.A., Ed., CRC Press, Boca Raton, FL, 1990.

Zurer, P.S. Chemists solve key puzzle of Antarctic ozone hole. *Chem. Eng. News.* Nov. 30, pp. 25–27, 1987.

Zurer, P.S. Studies on ozone destruction expand beyond Antarctic. *Chem. Eng. News.* May 30, pp. 16–25, 1988.

Zurer, P.S. Arctic ozone loss: Fact–finding mission concludes outlook is bleak. *Chem. Eng. News.* March 6, pp. 29–31, 1989.

Glossary

Abductive Inference. Inference to the best explanation. It uses information gathered about a phenomenon or situation to produce a hypothesis that best explains the data (Ch. 13).

Absolute Bioavailability. The bioavailability of a dose (D) estimated from the AUC for any route or formulation of the compound divided by the AUC after direct injection of the same dose (D) into the bloodstream (Ch. 4).

Absorbed Dose Rate (Radiation). The released energy that is deposited within the tissue (Ch. 14).

Absorption Rate Constant (k_a). A first-order rate constant for absorption calculated as $MAT=k_a^{-1}$ where MAT is the mean absorption time (Ch. 4).

Accelerated Failure Time Model. A survival time model in which the time-to-death ($\ln TTD_i$) of a particular type/class of individual (e.g., smoker) is changed ("accelerated") as some function of a covariate (e.g., classification relative to smoking habits) (Ch. 9).

Acclimatization. Like acclimation, acclimatization is the modification of biological functions, especially those physiological, or structures to maintain or minimize deviations from homeostasis despite change in some environmental quality. However, these shifts are taking place under natural conditions, not controlled laboratory conditions as is often the case with studies of acclimation (Ch. 9).

Acclimation. The modification of biological functions, especially those physiological, or structures to maintain or minimize deviations from homeostasis despite change in some environmental quality such as temperature, salinity, light, radiation, or toxicant concentration. It is an expression of phenotypic plasticity of individuals in response to a sublethal change in some environmental factor (Ch. 9).

Accumulation Factor (AF). The ratio of nonpolar organic compound concentration in the organism to that of sediments ([Organism]/[Sediment]) with the organism's concentration normalized to g of lipid and sediment concentration normalized to g of organic carbon (Ch. 3).

Acetylcholinesterase Inhibitors. Compounds such as many organophosphate and carbamate insecticides that inhibit the normal functioning of the enzyme, acetylcholinesterase which hydrolyzes the neurotransmitter, acetylcholine (Ch. 8).

Acid Precipitation. Precipitation including rain, fog, snow, or other forms of precipitation, with a pH below 5.7 (Ch. 12).

Acid Volatile Sulfides (AVS). Sediment-associated sulfides extracted with cold HCl which are assumed to be primarily iron and manganese sulfides (Ch. 4).

Acidophilic Component. Component of a cell such as the general cytoplasm that is readily stained by an acidic dye (Ch. 7).

Activation. One result of biotransformation in which the effect of an active compound is worsened or an inactive compound is converted to one with an adverse bioactivity (Ch. 3).

Active Transport. Movement of a substance up an electrochemical gradient that requires a carrier molecule and energy (Ch. 3).

Activity (Radio-). A measure of the rate at which a given quantity of radioactive material is emitting radiation (Ch. 14).

Acute Lethality. Death following a short and often intense exposure. The duration of an acute exposure in toxicity testing is generally 96 (4 days) or fewer hours of exposure (Ch. 9).

Additive. The toxicity of two (or more) toxicants in combination is additive if the simple sum of the toxic units of the two (or more) toxicants equals the actual toxicity measured for the mixture, e.g., 0.5 TU of A + 0.5 TU of B should equal 1.0 TU of effect (Ch. 9).

Additive Index. An index for quantifying the joint action of toxicants in mixture (Ch. 9).

Adduct. Modification of the DNA molecule produced when a xenobiotic or a metabolite binds covalently to a base (most often) or another portion of the DNA molecule (Ch. 6).

Adenylate Energy Charge (AEC). An index reflecting the balance of energy transfer between catabolic and anabolic processes. AEC = (ATP + 1/2 ADP)/(ATP + ADP + AMP), where ATP, ADP, and AMP = concentrations of adenosine tri-, di- and monophosphate, respectively (Ch. 8).

Adsorption. The accumulation of a substance at the common boundary of two phases, most often adsorption from a solution onto the solid surface. Adsorption of contaminants can result from processes of ion exchange, weak bonding to surfaces such as that associated with van der Waal forces, or even, molecular orientation of large, dipolar organic compounds relative to solid (less polar) and water (more polar) phases (Ch. 3).

Age Pigment. See Lipofuscin (Ch. 7).

Age-Specific Birth Rate. The mean number of females born to a female of an age class x (Ch. 10).

Age-Specific Death Rate. The probability of dying as tabulated for a life table interval or age class (x) (Ch. 10).

Age-Specific Number of Individuals Dying. The number of individuals dying in a life table interval (x). It is estimated as a simple difference, $d_x = l_x - l_{x+1}$ (Ch. 10).

Aging (of a radionuclide). A decrease in bioavailability of a radioactive contaminant with time. Generally this is due to the radionuclide's increased adsorption to soil particles (Ch. 14).

AHH (aryl hydrocarbon hydroxylase). Activity units of benzo[a]pyrene hydroxylation by aryl hydrocarbon hydroxylase used to reflect cytochrome P-450 monooxygenase activity (Ch. 6).

Alkalinity. The capacity of a natural water to neutralize acid as measured by titration of a water sample with a dilute acid to a specific pH endpoint. Most often, it is a function of carbonate (CO_3^{2-}), bicarbonate (HCO_3^-), and hydroxide (OH^-) concentrations, i.e., the carbonate-bicarbonate buffering of the water. However, dissolved organic compounds, borates, phosphates, and silicates can also contribute to alkalinity (Ch. 12).

Alkoxyradicals. Oxyradicals of organic compounds (R) of the form $RO^\bullet$ (Ch. 6).

Allometry. The study of size and its consequences (Ch. 4).

Allozyme. An allelic variant of an enzyme coded for by a particular locus (Ch. 10).

Alpha (α) Particles. Pieces of the nucleus ejected from a radioactive atom to reduce excess energy and gain stability. They are relatively huge in mass (7,345 times larger than a β particle), consisting of 2 neutrons and 2 protons, and carrying a +2 charge (Ch. 14).

δ-Aminolevulinic Acid Dehydratase (δ-ALAD or ALAD). An enzyme catalyzing the conversion of δ-aminolevulinic acid to porphobilinogen during heme synthesis (Ch. 6).

Analog (Elemental). An element that behaves like, but not necessarily identical to, another element in biological processes, e.g., cesium is an analog of potassium and strontium is an analog of calcium (Ch. 5).

Analysis of Variance (ANOVA). One-way ANOVA breaks down the total variance (total sum of squares) in a data set into the variance among and within treatments, e.g., among the different concentration treatments and within replicates for each concentration treatment. The variance within treatments (mean sum of squares$_{within}$) is assumed to reflect the sampling or error variance, and that among treatments (sum of squares$_{among}$) is thought to estimate the error variance plus any additional variance associated with the treatment. ANOVA is used often to test the null hypothesis of equal means among treatments in sublethal or chronic lethal assays (Ch. 8).

Analysis Plan (in Ecological Risk Assessment). A plan that defines the exact format and design of the assessment, explicitly states the data needed, and describes the methods and design for analyzing these data (Ch. 13).

Androgen Receptor Antagonist. A xenobiotic that acts by blocking androgen receptor-mediated processes (Ch. 8).

Aneuploidy. The deviation by loss or addition of chromosomes from the usual (e.g., 2N) number of chromosomes (Ch. 6).

Annual Limit on Intake (ALI). The activity of a radionuclide which, if inhaled or ingested, would result in a dose equal to the most limiting primary guide (Eckerman et al., 1988). Primary guides are the current recommendations for radiological protection (Ch. 14).

Antagonistic. Here, if the sum of TUs for a mixture of two (or more) toxicants is greater than the actual toxicity of the mixture expressed in TUs, the toxicants are said to be antagonistic (less than additive). The toxic effect of the toxicants is lessened by the presence of the other(s) (Ch. 9).

Antisymmetry. A population quality of bilaterally symmetrical individuals in which the differences (d) between a trait measured from the right and left side of individuals from that population describe a bimodal distribution (Ch. 8). (See Fluctuating Asymmetry for a related concept.)

Apparent Volume of Distribution (V_d). A mathematical volume used in clearance volume-based models. It is expressed in units of volume of a reference compartment, often the blood or plasma compartment (Ch. 3).

Arcsine Square Root Transformation. A common and useful transformation of effects data that often allows one to meet the assumption of homogeneous variances for proportions of exposed individuals responding.

$$Transform = arc \sin \sqrt{P}$$

where P = the measured effect, e.g., proportion of the exposed organisms (Ch. 8).

Assessment Endpoint (in Ecological Risk Assessment). The valued ecological entity that is to be protected and the precise quality to be measured for this entity. Earlier, only the first part of this definition (the valued entity) was considered the assessment endpoint (Ch. 13). (See Measurement Endpoint also.)

Backstripping or Backprojection Procedure. A method used to extract parameter estimates for elimination involving a multiexponential process. The procedure may be implemented graphically or mathematically with a computer program (Ch. 3).

Bartlett's Test. A statistical test used to test the assumption of homogeneity of variances. In this book, it was used to assess this assumption prior to one-way analysis of variance of sublethal and chronic lethal effects data (Ch. 8).

Basophilic Component. Component of a cell such as the nucleus that is readily stained by a basic dye (Ch. 7).

Becquerel (Bq). The official unit of radioactivity now replacing the curie. One curie is 3.7×10^{10} Bq (Ch. 1).

Behavioral Teratology. The study of behavioral abnormalities in otherwise normal-appearing individuals after exposure as an embryo to an agent (Ch. 8).

Behavioral Toxicology. The science of abnormal behaviors produced by exposure to a chemical or physical agent (Ch. 8).

Beta β Particles. Electrons or positrons (positive electron) ejected from the atom during radioactive decay (Ch. 14).

Binomial Method. A method of estimating LC50, EC50, or LD50 if there are no partial kills (Ch. 9).

Bioaccumulation. The net accumulation of a contaminant in and on an organism from all sources including water, air, and solid phases of the environment. Solid phases include food sources (Ch. 3).

Bioaccumulation Factor (BAF). The ratio of the contaminant concentration in the organism to that in the sediment or some other source, i.e., [Organism]/[Sediment] or [Organism]/[Food]. Often, this term is used in a broader sense if the exact source is not known or poorly defined as in a field survey (see Bioaccumulation Factor [BF]). For the purposes of avoiding confusion, BSAF (see Biota-Sediment Accumulation Factor) is used in this textbook to define sediment-associated bioaccumulation and BAF is used in a more general context (Ch. 3).

Bioaccumulation Factor (BF). Like BAF above, the BF is the ratio of contaminant concentration in the organism to that in potential sources. The BF is based on the assumption that water and food (including sediments in some cases) may contribute to differences in concentrations measured for individuals at different trophic levels (Ch. 5).

Bioccumulative Chemicals of Concern (BCCs). See Persistent Organic Pollutants (Ch. 12).

Bioamplification. See Biomagnification (Ch. 5).

Bioavailability. The extent to which a contaminant in a source is free for uptake. In many definitions, especially those associated with pharmacology or mammalian toxicology, bioavailability implies the degree to which the contaminant is free to be taken up by the organism *and to cause an effect at the site of action* (Ch. 4).

Bioconcentration. The net accumulation in and on an organism of a contaminant from water only (Ch. 3).

Bioconcentration Factor (BCF). The ratio of concentrations of contaminant in the organism and dissolved in water (the presumed or explicit source)([Organism]/[Water]) (Ch. 3).

Biological Half-Life ($t_{1/2}$). The time required for the amount or concentration of a contaminant in a biological compartment to decrease by 50% (Ch. 3).

Biologically Determinant. A quality of an element such that its concentration in organisms remains relatively constant over a wide range of environmental concentrations. Many essential elements are biologically determinant due to their metabolic regulation (Ch. 4).

Biologically Indeterminant. A quality of an element such that its concentration in organisms is directly proportional to environmental concentrations (Ch. 4).

Biomagnification. An increase in concentration from one trophic level (e.g., prey) to the next (e.g., predator) due to accumulation of contaminant from food (Ch. 5).

Biomagnification Factor (B). The contaminant concentration at trophic level n (C_n) divided by that at the next lowest trophic level (C_{n-1}) (e.g., Bruggeman et al., 1981; Lasokowski, 1991). This factor may be estimated with individual organisms of known or assumed trophic status (Ch. 5).

Biomagnification Power (b). The exponent of the exponential relationship: concentration in an organism = $ae^{b(\delta\ 15N)}$. It quantifies the proportional increase or decrease in concentration along a trophic food web. The $\delta\ ^{15}N$ reflects the trophic status of the individual (Ch. 5).

Biomarker. A cellular, tissue, body fluid, physiological, or biochemical change in extant individuals that is used quantitatively during biomonitoring to either imply presence of significant pollutant or as an early warning system for imminent effects. A biomarker is also called a bioindicator by some authors (Ch. 2).

Biomineralization. Biologically-mediated deposition of minerals (Preface).

Biominification. See Trophic Dilution (Ch. 5).

Biomonitor. To use organisms to monitor contamination and to imply possible effects to biota or routes of toxicant exposure to humans (Ch. 2).

Biomonitoring. A widely-applied practice of monitoring of a subset of an entire community with the goal of assessing community condition (Ch. 11).

Biomonitoring (Type 1). The monitoring of community changes along a gradient or among sites differing in levels of pollution (Ch. 11).

Biomonitoring (Type 2). The measurement of bioaccumulation in organisms among sites notionally varying in the level of contamination (Ch. 11).

Biomonitoring (Type 3). The measurement of effects on organisms using tools such as biochemical markers in sentinel species or some measure of diminished fitness of individuals (Ch. 11).

Biomonitoring (Type 4). The measurement of genetically-based resistance in populations of contaminanted areas (Ch. 11).

Biota-sediment Accumulation Factor (BSAF). The specific term used for a bioaccumulation factor (BAF) if it is clear that the factor relates accumulated contaminant to that in sediments (Ch. 5). See also Bioaccumulation Factor (BAF).

Biotransformation. The biologically-mediated transformation of one chemical compound to another (Ch. 3).

Biphasic Dose-effect Model. A model of dose-effect that, due to hormesis, is shaped like the threshold model in Figure 7.4, but the curve actually dips down from the control level before increasing with dose. The individuals at these low, subinhibitory concentrations are performing better than individuals exposed to lower or higher concentrations (Ch. 8).

Body Burden. The total mass or amount of contaminant in (and on) an individual (Ch. 3).

Boomerang Paradigm. What you throw away can come back to hurt you (Ch. 1).

Borderline Metal Cations. Metal intermediate between Class A and B metals. These metals have 1 to 9 outer orbital electrons (Ch. 3).

Brillouin Diversity Index. A measure of the species diversity of a sample taken from a community (Ch. 11).

Calcium Sinks. Physical sinks such as arthropod cuticles or bone that render calcium or its analogs less bioavailable during trophic interactions, and consequently, provide a mechanism for trophic dilution (Ch. 5).

Cancer Progression. The change in the biological attributes of neoplastic cells over time that leads to malignancy (Ch. 7).

Carcinogenic. Capable of causing cancer (Ch. 6).

Cardinal Signs of Inflammation. Heat, redness, swelling, and pain although heat is not relevant in the case of poikilotherms (Ch. 7).

Carrier Proteins. Cell membrane-associated proteins that act as carriers to transfer hydrophilic contaminants across the membrane (Ch. 3).

Carrying Capacity (K). The maximum population size expressed as total number of individuals, biomass, or density that a particular environment is capable of sustaining (Ch. 10).

Caseous Necrosis. Necrosis (cell death) in which cells disintegrate and form a mass of fat and protein (Ch. 7).

Catalase (CAT). An enzyme catalyzing the reaction, $2H_2O_2 \rightarrow 2H_2O + O_2$. It is involved in reducing oxidative stress (Ch. 6).

Catalyzed Haber-Weiss Reaction. A greatly accelerated Haber-Weiss reaction ($O_2^{\bullet} \rightarrow H_2O_2 \rightarrow$ $^{\bullet}OH + OH$) catalyzed by metal chelates (Ch. 6).

Cellular Stress Response. An "orchestrated induction of key proteins that form the basis for the cell's protein repair and recycling system" (Sanders and Dyer, 1994) (Ch. 6).

Channel Proteins. Cell membrane-associated transport proteins that form channels to allow solute passage through the membrane (Ch. 3).

Chaperon. Stress proteins that associate with and direct the proper folding and coming together of proteins. They also protect proteins from denaturing and aggregating, and enhance refolding to a functional conformation (Ch. 6).

Chapman Mechanism. The series of reactions by which ozone is formed in the stratosphere (Reactions 12.7 to 12.9) (Ch. 12).

Chelate. A multidentate ligand (Ch. 4).

Chemical Antagonism. Antagonism resulting because two toxicants react with one another to produce a less toxic product (Ch. 9).

Chemicals of Potential Concern. During risk assessments, a list of those chemicals of most concern (chemicals of potential concern) is made from a list of all chemicals present at the site (Ch. 13).

χ^2 **["Chi" square] Ratio**. Here, the ratio of χ^2 values for two candidate models used to select the model that best fits the dose- or concentration-effect data set (Ch. 9).

Chloride Cells. Specialized cells on the gill found predominately on the primary lamellae but also on the secondary lamellae which function in ion regulation (Ch. 7).

Chlorosis. The blanching of green color due to the lack of production or the destruction of chlorophyll (Ch. 8).

Chromatid. Prior to cell division, the DNA in each chromosome is duplicated to produce two chromatids. As the chromosome condenses, the chromosome appears as a pair of chromatids connected to a common centromere (Ch. 7).

Chromatid Aberration. An aberration that occurs if only one strand (chromatid) is broken in the chromosome (Ch. 7).

Chromosomal Aberration. Damage to chromosomes including breakage and loss of segments of DNA, addition of segments of DNA, or chromosomal rearrangements. Chromosomal breaks involve double-strand breaks (Ch. 7).

Chronic Lethality. Death resulting from prolonged exposure. By recent convention, a chronic test should be at least 10% of the duration of the species life span (Ch. 9).

Chronic Reference Dose (Chronic RfD). The RfD for chronic exposure (Ch. 13).

Class A Metal Cation. Metal that has an inert gas electron configuration, high electronegativity, and a hard (difficult to polarize) outer sphere (Ch. 3).

Class B Metal Cation. Metal with filled d orbital of 10 to 12 electrons and low electronegativity. It has a soft (easily polarizable) sphere readily deformed by adjacent ions, i.e., it easily forms covalent bonds with donor atoms such as sulfur (Ch. 3).

Clastogenic. Capable of causing chromosome damage in living cells (Ch. 6).

Clean Air Act (CAA). A U.S. federal act designed to regulate air pollution with the goal of protecting human health and the environment (Appendix 3).

Clearance. As used in modeling, the rate of substance movement among compartments normalized to concentration. Clearance has units of flow, volume time^{-1} (CH. 3).

Clearance Volume-Based Model. A model of substance uptake, elimination, or bioaccumulation based on the distribution of the substance in, and clearance (see Clearance above) from and among compartments of different volumes (Ch. 3).

Coagulation Necrosis. Necrosis (cell death) characterized by extensive cytoplasmic protein coagulation which makes the cell appear opaque. The cell outline and arrangement in the tissue remain for some time after cell death (Ch. 7).

Cold Condensation Theory. Persistent organic pollutants (POPs) in the air will condense onto soil, water, and biota at cool temperatures: consequently, the ratios for POPs concentrations in the air and on condensed phases decreases as one moves from warmer to cooler climates (Ch. 12).

Committed Effective Dose (CED). The dose unit to which risk factors are multiplied in order to estimate the probability of an individual experiencing a deleterious effect from exposure to radiation. CED accounts for the relative biological effectiveness of different types of radiation, characteristics of the radiation such as energy and half-life of the radioactivity, human physiology, and integrates dose from ingested radioactivity over a 50-year period (Ch. 14).

Community. "An assemblage of populations living in a prescribed area or physical habitat: it is an organized unit to the extent that it has characteristics additional to its individual and population components . . . [it is] the living part of the ecosystem" (Odum, 1971). The community is made up of species that interact to form an organized unit (Magurran, 1988) although some species may interact only loosely (Ch. 11).

Community Conditioning Hypothesis. Communities retain information about occurrences in their past and will not return to their original state after perturbation (Ch. 11).

Compensatory Hyperplasia. An excessive amount of hyperplasia occurring in response to injury or irritation (Ch. 7).

Comprehensive Environmental Response, Compensation and Liability Act (CERCLA). A U.S. federal act that gives the EPA the ability to respond to hazardous releases, cleanup or require cleanup of waste sites, and identify liability for released hazardous waste (Appendix 3).

Concentration Factor (CF). The quantitative expression of the change in concentration at different trophic levels relative to the concentration in the ultimate or lowest defined source, e.g., relative to the water concentration, $CF = C_n/C_{water}$. The change in concentration is expressed as a multiple of the source concentration (Ch. 5).

Concept of Strategy. See Principle of Allocation (Ch. 10).

Conceptual Model (in Ecological Risk Assessment). The model that links and interrelates assessment endpoint(s) and stressor. It includes evaluation of potential exposure pathways, effects, and ecological receptors. Conceptual models include hypotheses of risk and a diagram of the conceptual model (Ch. 13).

Conceptual Model Diagram (in Ecological Risk Assessment). Part of the conceptual model. A diagram showing the pathways of exposure and illustrating areas of uncertainty or concern. They are visual aids for communicating the model from which the risk hypotheses emerged to the risk manager (Ch. 13).

Congeners. "[a term used to] point up the relationship among members of a chemical family such as the PCBs" (Bunce, 1991). For example, the individual PCB congeners in a mixture share a common form but vary by the numbers and positions of Cl atoms (Ch. 5).

Contaminant. "a substance released by man's activities" (Moriarty, 1983) (Ch. 2).

Co-Tolerance. See Cross-resistance (Ch. 10).

Council of Environmental Quality. A council established by NEPA to aid and advise the President during preparation of the annual Environmental Quality Report (Appendix 3).

Cough (Gill Purge). An abrupt, periodic reversal of water flow over the gills that dislodges and eliminates excess mucus from the gill's surfaces (Ch. 8).

Cox Proportional Hazard Model. A semiparametric method allowing the examination of proportional hazards without taking on the assumption of any specific model for the underlaying baseline hazard (Ch. 9).

Criteria. Estimated concentrations of toxicants based on current scientific information that, if not exceeded, are believed to protect organisms or a defined use of a water body (Ch. 2).

Critical Life Stage Testing. Toxicity testing focused on the life stage of a species thought to be most sensitive to the toxicant such as newly-hatched individuals (Ch. 9).

Critical Study. During human risk assessment, the human or nonhuman animal study with the lowest-observed-adverse-effect-level (LOAEL) (Ch. 13).

Critical Toxic Effect. In human risk assessment, the effect associated with the critical study (Ch. 13).

Cross-Resistance. The condition in which enhanced tolerance to one toxicant also enhances tolerance to another (Ch. 10).

Curie (Ci). A measure of radioactivity equivalent to 2.2×10^6 dpm (disintegrations per minute) (Ch.1).

Cysteine-Rich Intestinal Protein (CRIP). A protein which serves in the uptake of zinc by cells in the intestine wall (Ch. 4).

Cytochrome P-450 Monooxygenase. A 45 to 60 kDa hemoprotein associated with membranes, especially those of the endoplasmic recticulum, and active in Phase I metabolism of organic compounds, and metabolism of fatty acids, cholesterol, and steroid hormones. (Ch. 6). See also Monooxygenase.

Cytolytic Necrosis. See Liquefactive Necrosis (Ch. 7).

Cytotoxicity. Toxicity causing cell death. (Ch. 7).

Depuration. The loss of contaminant from an organism that is measured after the organism has been placed into a clean environment and allowed to eliminate contaminant (Ch. 3).

Developmental Reference Dose (RfD$_{dt}$). A RfD determined for developmental consequences of a single, maternal exposure during development (Ch. 13).

Developmental Stability. The capacity of an organism to develop into a consistent phenotype in an environment (Ch. 8).

Developmental Toxicity. A broad area of toxic effect that considers altered growth and functional deficiencies in addition to classic teratogenic effects (Ch. 8).

Diffusion. The movement of a contaminant down an electrochemical gradient that requires no energy (Ch. 3).

Dilution Paradigm. The solution to pollution is dilution (Ch. 1).

Directional Asymmetry. The deviation for a population from a mean of 0 for the difference between a trait measured from the right and left sides of bilaterally symmetrical individuals from that population. For example, measurement of the difference in weights of left and right arms of humans would display directional asymmetry because most humans are right-handed and have larger right arms. (Ch. 8). (See also Fluctuating Asymmetry as a related concept.)

Discrimination Ratio (or Factor). A ratio measuring the degree of isotopic discrimination (see definition below) with a ratio of 1 indicating no discrimination. In the context of discrimination between elemental analogs such as cesium and potassium in a trophic exchange, a discrimination factor or ratio is expressed as $[Cs]_{food}/[K]_{food}$ divided by $[Cs]_{body}/[K]_{body}$ (Ch. 5).

Disintegration. The event in which a radioactive element releases photons or particles to gain stability, i.e., radioactive decay (Ch. 14).

Disposable Soma Theory of Aging. Aging is a consequence of the gradual accumulation of cellular damage via random molecular defects (Ch. 10).

Dispositional Antagonism. Antagonism involving toxicant mixture effects on the uptake, movement within the organism, deposition at specific sites, and elimination of the toxicants. The presence of the two toxicants together shifts one or more of these processes to lower the impact of the toxicants at the site(s) of action or target organ(s) (Ch. 9).

Dobson Units (DU). A measure of atmospheric ozone levels that is the equivalent of 0.001 mm thickness of pure ozone at 1 atmosphere (Ch. 12).

Doubling Time. The doubling time of a population is the estimated time required for the population to double its present size. It (t_d) is estimated from the intrinsic rate of increase (r) as (ln 2)/r (Ch. 10).

Dry Deposition. The flux of particles and gases such as SO_2, HNO_3 and NH_3 to surfaces (Ch. 12).

Dunnett's Test. A parametric, post-ANOVA test used often in the analysis of sublethal and chronic lethal effects data (Ch. 8).

Dynamic Energy Budget (DEB) Approach. A theory-rich approach utilizing energy budgeting for individuals as the central theme around which survival, growth and reproduction under the influence of toxicants are modeled. Standard toxicity test data may be incorporated directly into the DEB approach (Ch. 10).

Early Life Stage (ELS) Test. A critical life stage test using early life stages such as embryos or larvae based on the observation or assumption that the early life stage is the most sensitive stage in the species' life cycle (Ch. 9).

Ecological Epidemiology. The name given to epidemiological methods applied to determining the cause, incidence, prevalence, and distribution of adverse effects to nonhuman species inhabiting contaminated sites. It is frequently associated with retrospective ecological risk assessment (Ch. 10).

Ecological Mortality (or Death). The toxicant-related diminution of fitness of an individual functioning within an ecosystem context that is of a magnitude sufficient to be equivalent to somatic death (Ch. 8).

Ecoregion. A relatively homogeneous region in an ecosystem or association between organisms and their environment. Ecoregions are usually defined on maps (Ch. 2, 12).

Ecosystem. The functional unit of ecology including the biotic community and its abiotic environment functioning together as a unit to direct the flow of energy and cycling of materials (Ch. 11).

Ecosystem Incongruity. The traditional, but now too confining, bias toward the ecosystem or lower levels in ecotoxicology (Ch. 12).

Ecotone. Area of transition between two or more community types (Ch. 12).

Ecotoxicology. The science of contaminants in the biosphere and their effects on constituents of the biosphere, including humans (Ch. 1).

Edge Effect. Ecotones often have species assemblages with high species richness and high abundance of individuals relative to those of the adjacent communities (Ch. 12).

Effective Dose. A term used in pharmacology to define the amount of drug entering the blood and available to have a pharmacological effect. It is used in the context of drug bioavailability from different routes of administration (Ch. 4).

Effective Dose Equivalent (Radiation). Recognizing that biological effects from a uniform irradiation of the whole body are different than effects from a similar dose concentrated in specific tissues, effective dose equivalent weights the radiation dose to different organs or tissues. Thus, the fractional contribution of organs and tissues to the total risk is normalized to when the entire body is uniformly irradiated (Ch. 14).

Effective Half-life (k_{eff}). An estimated half-life in a compartment model that has numerous elimination mechanisms, each with an associated k_i. It is equal to $(\ln 2)/\Sigma k_i$ (Ch. 3).

Elasticity (Community). The ability of a community to return to its prestressed condition (Ch. 11).

Elimination. The loss or metabolism of a contaminant resulting in a decrease in the amount of contaminant within an organism (Ch. 3).

Elutriate Test. A test in which a nonbenthic species such as *Daphnia magna* is exposed to an elutriate produced by mixing the test sediment with water and then centrifugating the mixture (Ch. 9).

Emergent Properties. Properties emerging in hierarchical systems such as ecological communities or ecosystems that cannot be predicted solely from our limited understanding of the system's parts or components (Ch. 11).

Endocytosis. Uptake of solids (phagocytosis) or liquids (pinocytosis) by cells through a process of engulfing the material and enclosure in a cellular vacuole (Ch. 3).

Enrichment Factor (EF_{crust}). A measure of anthropogenic enrichment of an element above natural levels. It is an element's concentration (X) measured in air samples divided by that expected in the earth's crust: $EF_{crust} = [X/Al]_{air}/[X/Al]_{crust}$. Both air and crustal concentrations are normalized to Al concentrations (Ch. 12).

Enterohepatic Circulation. Recirculation of toxicant back to the liver after passage into the intestine via the bile and reabsorption in the intestine (Ch. 3).

Environmental Assessment (EA). A short, preliminary assessment of potential environmental damage used to determine if a full environmental impact statement (EIS) is required (Appendix 3).

Environmental Epidemiology. A subdiscipline of human epidemiology concerned with diseases caused by chemical or physical agents (Ch. 10).

Environmental Impact Statement (EIS). A document required by NEPA that outlines possible impacts, describes impacts to resources, and details alternatives for any major federal action (Appendix 3).

Environmental Law. "a body of federal, state, and local legislation, in the form of statutes, bylaws, ordinances, and regulations, plus court-made principles known as common law." (McGregor, 1994) (Appendix 3).

Environmental Quality Report. An annual report from the U.S. President's office discussing current environmental conditions and trends. Under NEPA, the Council on Environmental Quality aids and advises the President in the preparation of this report (Appendix 3).

EPA Weight of Evidence Classification. A classification used in human risk assessment involving carcinogenic effects that suggests to the risk manager the strength of the evidence supporting the risk calculation (Ch. 13).

Epidemiology. The science concerned with the cause, incidence, prevalence, and distribution of infectious and noninfectious diseases in populations (Ch. 10).

Epizootic. Outbreak of disease in a population or in a large number of individuals of a species (Ch. 10).

EROD (ethoxyresorufin O-deethylase). Units of activity for O-deethylation of ethoxyresorufin by ethoxyresorufin O-deethylase. Used to reflect cytochrome P-450 monooxygenase activity (Ch. 6).

Essential Element. An element essential for the normal functioning of a living organism. Mertz (1981) lists the following as the essential elements: H, Na, K, Mg, Ca, V, Cr, Mo, Mn, Fe, Co, Ni, Cu, Zn, Cd(?), C, B, Si, Sn(?), N, P, As, O, S, Se, F, Cl, and I (Ch. 4).

Estrogenic Chemicals. Contaminants possessing biological activities like estrogen that cause changes in the sexual characteristics of individuals (Ch. 8).

Etiological Agent. An agent responsible for causing, initiating, or promoting a disease (Ch. 10).

Euler-Lotka Equation. An equation used to estimate the intrinsic rate of increase from life table data (Ch. 10).

European Community (EC). An organization formed in 1957 to foster a stronger union among European countries and, presently, is composed of Austria, Belgium, Denmark, France, Finland, Germany, Greece, Ireland, Italy, Luxembourg, Netherlands, Portugal, Spain, Sweden, and the United Kingdom. The EC has its own budget, and the power to make and enforce laws including environmental laws. With the Maastricht agreement (December 11, 1991), the EC became the European Union (EU). (Appendix 3).

Exchange Diffusion. Diffusion across a membrane by means of a carrier molecule which requires no energy, and involves the exchange of two ions across the membrane (Ch. 3).

Exocytosis. Fusion of intracellular vesicles with the cell membrane followed by the emptying of the vesicle contents to the cell exterior (Ch. 3).

Expected Life Span. The calculated life span for individuals of age class x of a life table. It can be estimated as T_x/l_x (Ch. 10).

Exploitation Competition. Interspecies competition in which species compete for some limiting resource such as food (Ch. 11).

Exponential Relationship. A mathematical relationship in which the Y variable is related to some base raised to the X variable, i.e., $Y = a10^{bX}$ (Ch. 4). (Compare to Power Relationship.)

Exposure. In risk assessment, contact with the contaminant or stressor (Ch. 13).

Exposure Characterization (in Ecological Risk Assessment). A description of the presence and characteristics of contact between the contaminant and the ecological entity of concern, and a summary of this information in an exposure profile (Ch. 13).

Exposure Pathways. The avenues by which an individual is exposed to a contaminant including the source and route to contact (Ch. 13).

Exposure Profile (in Ecological Risk Assessment). A profile that "quantifies the magnitude and spatial and temporal pattern of exposure for the scenarios developed during problem formulation" (Norton et al., 1992) (Ch. 13).

Facilitated Diffusion. Diffusion down a gradient not requiring energy, but occurring at a rate faster than expected by simple diffusion alone (Ch. 3).

Fat Necrosis. Necrosis (cell death) which involves deposits of saponified fats in dead fat cells (Ch. 7).

Feasibility Study (FS). Part of a Remedial Investigation and Feasibility Study that explores the various options for remediation (Ch. 13).

Fecundity Selection. A component of the life cycle of an individual in which natural selection can occur, involving the production of more offspring by matings of certain genotype pairs than produced by other genotype pairs (Ch. 10).

Federal Insecticides, Fungicide and Rodenticide Act (FIFRA). A U.S. federal law designed to control the production, distribution, labeling, sale, and use of pesticides (Appendix 3).

Fertilization Effect. At low levels, the nitrogen and sulfur added at low levels to a forest in acid precipitation can enhance growth (Ch. 12).

FETAX (Frog Embryo Teratogenesis Assay - Xenopus). A teratogenesis assay using embryos of the frog, *Xenopus laevis* (Ch. 8).

Filaments. See Primary Lamellae (Ch. 7).

Finding of No Significant Impact (FONSI). A statement of no significant impact of a major federal action concluded after an environmental assessment (EA). (Appendix 3).

Finite Rate of Increase (λ). The rate of increase of population size measured over set intervals such as between age classes of a life table or generations of a population with nonoverlapping generations, e.g., an annual plant (Ch. 10).

Fission Products. Radioactive fragments produced by nuclear fission (Ch. 14).

Flory-Huggins Theory. A quantitative theory relating solubility (partition coefficient) of compounds in dilute solutions to solvent molecular size (volume). In environmental toxicology, it has been used to explain the nonideal behavior of the K_{OW} in reflecting partitioning between water and lipids for very lipophilic compounds (Ch. 4).

Flow-through Test. An aquatic toxicity test that has a constant (continuous flow-through test) or nearly constant (intermittent flow-through test) flow of the toxicant solutions through the exposure tanks (Ch. 9).

Fluctuating Asymmetry. Deviation from perfect bilateral symmetry for a population that is thought to reflect developmental instability. A quality is measured from the right and left sides of a bilaterally symmetrical species and the difference (d = Right - Left) calculated. The variance in d for the population is a measure of fluctuating asymmetry (Ch. 8).

Free Ion Activity Model (FIAM). "The universal importance of free metal ion activities in determining the uptake, nutrition and toxicity of all cationic trace metals" (Campbell and Tessier, 1996) (Ch. 4, 14).

Free Radical. A molecule having an unshared electron. (The electron is usually designated by a dot, •.) Free radicals are extremely reactive (Ch. 6).

Freundlich Isotherm Equation. An empirical relationship quantifying adsorption (Ch. 3).

Functional Antagonism. Antagonism resulting from two chemicals eliciting opposite physiological effects and, as a consequence, counterbalancing each other (Ch. 9).

Functional Redundancy. An apparently unaltered maintenance of community functioning despite changes in structure (Ch. 11).

Functional Response. A change in some predator function, such as prey consumption rate, in response to changes in prey density (Ch. 11).

Fundamental Niche. A species has a certain (Hutchinsonian) niche in which it could exist and function based on its physiological and other limits. In contrast to the realized niche, this fundamental niche includes all of the possible niche volume (Ch. 11).

Gaia Hypothesis. An hypothesis forwarded by James Lovelock that the earth's temperature, albedo, and surface chemistry are homeostatically regulated by the sum of all the biota of the earth (Preface).

Gametic Selection. A component of the life cycle of an individual in which natural selection can occur, involving differential success of gametes produced by heterozygotes (Ch. 10).

Gamma (γ) Rays. Electromagnetic photons emitted from the nucleus (Ch. 14).

Gangrenous Necrosis. Combination of coagulation and liquefactive necrosis often resulting from puncture and subsequent infection (Ch. 7).

Gastric Emptying Rate. The rate at which the contents of the stomach are emptied into the small intestine (Ch. 4).

Gastrointestinal Excretion. Excretion through the intestinal mucosa by active or passive processes. This may involve loss by normal cell sloughing of the intestine wall. Metals such as cadmium and mercury can experience significant levels of gastrointestinal excretion (Ch. 3).

General Adaptation Syndrome (GAS). The specific syndrome associated with Selyean stress composed of three phases: the alarm reaction, adaptation or resistance, and exhaustion phases. The goal in all phases of the GAS is to regain or resist deviation from homeostasis (Ch. 8).

Genetic Hitchhiking. Used in this book to describe the situation in which a scored locus is acting only as a marker for a closely-linked gene that is actually responsible for the difference in tolerance among genotypes. More generally, it is the condition "in which a given allele changes in frequency as a result of linkage or gametic phase disequilibrium with another selected locus" (Endler, 1986) (Ch. 10).

Genetic Risk. The risk to the progeny of the exposed individual of an adverse effect associated with heritable genetic damage, e.g., damage to germ cells leading to a nonviable fetus or an offspring with a birth defect (Ch. 7).

Genotoxicity. Damage by a physical or chemical agent to genetic materials, e.g., chromosomes or DNA (Ch. 6).

Geographic Information Systems (GIS). Computerized systems to handle spatial data at a reasonable cost. Most allow one to archive, organize, integrate, statistically analyze, and display many kinds of spatial information using a common coordinate system (Ch. 12).

Global Distillation. A process by which persistent and relatively volatile organochlorine compounds are distilled from warmer regions of use to cooler regions of the globe (Preface, Ch. 12).

Global Fractionation. Because POPs differ in their individual rates of degradation, vapor pressures, and lipophilities, a fractionation occurs in which some POPs move more rapidly than others toward the polar regions. The net result is a redistribution of the different POPs from the Equator or site of origin toward the cold polar regions of the Earth (Ch. 12).

Global Warming. A general warming of the Earth thought to result from the increased atmospheric carbon dioxide (CO_2) concentrations from fossil fuel burning, release of other greenhouse gases, and the worldwide destruction of forests (Ch. 12).

Glucose Regulated Proteins (grps). Proteins that, under low glucose or oxygen conditions, are part of the cellular stress response. The grps are structurally similar to heat shock proteins, are present at basal levels in unstressed cells, and are induced in glucose- or oxygen-deficient cells exposed to toxicants which modify calcium metabolism, e.g., lead (Sanders, 1990) (Ch. 6).

Glucuronic Acid. A carbohydrate which can be conjugated to xenobiotics by UDP- glucuronosyl-transferase (Ch. 6).

Glutathione (GSH). A tripeptide composed of cysteine, glutamate and glycine; specifically, γ-glutamyl-L-cysteinyl-glycine (Ch. 6).

Glutathione Peroxidase. An enzyme catalyzing the reaction, 2 Reduced Glutathione (GSH) + $H_2O_2 \rightarrow$ Oxidized Glutathione (GSSG) + H_2O. It is involved in reducing oxidative stress (Ch. 6).

Glutathione S-transferase (GST). A Phase II enzyme which conjugates glutathione with a xenobiotic or its metabolite (Ch. 6).

Granulation Tissue. During the repair stage of the inflammation process, small blood vessels begin to form and connective tissue begins to grow in a mass called the granulation tissue (Ch. 7).

Grasshopper Effect. Global distillation of persistent organic pollutants can involve seasonal cycling of temperatures such that movement toward the higher latitudes occurs in annual pulses. This is called the grasshopper effect (Ch. 12).

Greenhouse Effect. Greenhouse gases are relatively transparent to light but absorb long-wave, infrared radiation radiating back from the earth's surface. The net balance for sunlight influx, infrared radiation absorption by greenhouse gases, and infrared efflux from the earth's surface determines the steady state temperature of the earth. The net warming of the earth is called the greenhouse effect (Ch. 12).

Greenhouse Gases. Atmospheric gases that are relatively transparent to sunlight entering the atmosphere but absorb infrared radiation being generated at the earth's surface. They include water vapor, carbon dioxide, methane, nitrous oxide, CFCs, methylchloroform, carbon tetrachloride and the fire retardant, halon. Ozone *in the troposphere* may also act as a greenhouse gas (Ch. 12).

Growth Dilution. The decrease in contaminant concentration in a growing organism because the amount of tissue in which the contaminant is distributed is increasing (Ch. 3).

Guild (Ecological). A "group of functionally similar species whose members interact strongly with one another but weakly with the remainder of the community" (Smith, 1986) (Ch. 11).

Haber-Weiss Reaction. The reaction, $O_2^{\bullet} \rightarrow H_2O_2 \rightarrow {}^{\bullet}OH + OH^-$ (Ch. 6).

Half-life (Radionuclide). The amount of time required for one-half of the number of radioactive atoms to decay (Ch. 14).

Hardness (of Water). The sum of the concentrations of dissolved calcium and magnesium (Ch. 9).

Hardy-Weinberg Equilibrium. The frequency of genotypes will remain constant through time if the following conditions are met: (1) the population is large ("infinite") and composed of randomly mating, diploid organisms with overlapping generations, (2) no selection is occurring, and (3) mutation and migration are neglible (Ch. 10).

Hazard Assessment. An assessment that compares the expected environmental concentration (EEC) to some estimated threshold effect (ETT) with the intent of deciding if (1) a situation is safe, (2) a situation is not safe, or (3) there isn't enough information to decide (Ch. 13).

Hazard Quotient. A crude indicator of hazard calculated as the expected environmental concentration (EEC) divided by some estimated threshold effect concentration (ETT): HQ = EEC/ETT (Ch. 13).

Health Advisory Concentrations. Concentrations below which no impact on human health is expected for the specified duration of exposure. They often are applied by public officials in dealing with spills, short-term exposures, or similar situations although longer term advisory concentrations are also available (Ch. 13).

Heat Shock Proteins (hsp). Stress proteins induced by an abrupt shift in temperature that function to reduce associated protein damage in cells (Ch. 6).

Henderson-Hasselbalch Equation. The relationship between pH and the ratio of conjugate base (B^-) to acid (BH), $pH = pK_a + \log ([B^-]/[BH])$ (Piszkiewicz, 1977) (Ch. 4).

Heterosis. The superior performance of heterozygotes (Ch. 10).

Histopathology. The change in cells and tissues associated with a communicable or noncommunicable disease (Ch. 7).

Homeopathic Medicine. A branch of medicine, founded by the Samuel Hahnemann, that is based on the law of similars, a drug that induces symptoms similar to those of the disease will aid the body in defending itself by stimulating its natural responses (Ch. 8).

Homolog. A pair of homologous chromosomes (Ch. 7).

Hormesis. A stimulatory effect exhibited with exposure to low, subinhibitory levels of some toxicants or physical agents. Hormesis is not normally a toxicant-specific response (Ch. 8).

Hormonal Oncogenesis. Tumor production resulting from high levels of hormones (a promoter) with associated hyperplasia (Ch. 7).

Hutchinsonian Niche. A niche is ". . . the certain biological activity space in which an organism exists in a particular habitat. This space is influenced by the physiological and behavioral limits of a species and by effects of environmental parameters (physical and biotic, such as temperature and predation) acting on it." (Wetzel, 1982) (Ch. 11).

Hyperplasia. The capacity of cells to multiply and increase in tissues and organs (Ch. 7).

Hypertrophy. An increase in cell size (and function) resulting from an increase in the mass of cellular structural components often as a compensatory response (Ch. 7).

Imposex. The development (imposition) of male characteristics such as a penis or vas deferens in females (Ch. 8).

Incidence. Number of new individuals scored as having the disease in a time interval (Ch.10).

Incidence Rate. Incidence rate of a disease for a nonfatal condition is calculated as the number of individuals with the disease divided by the total time that the population had been exposed. It is expressed in units of individuals or cases per unit of exposure time, e.g., 10 new cases per year (Ch. 10).

Incipient Median Lethal Concentration. The concentration below which 50% of individuals will live indefinitely relative to the lethal effects of the toxicant (Ch. 9).

Index of Biological Integrity (IBI). A composite index combining 12 qualities of fish communities of warm-water, low-gradient streams to determine the level of stream degradation. This index has been modified and widely used in the U.S. (Ch. 11).

Individual Effective Dose (IED) Concept. A concept forming the basis for most dose-response models which holds that there exists a smallest dose needed to kill any particular individual. The IED is a characteristic of an individual (Ch. 9).

Individual Tolerance Concept. See Individual Effective Dose Concept (Ch. 9).

Industrial Melanism. The gradual increase to predominance of melanic forms in industrialized regions (Ch. 10).

Inertia (Community). A community's ability to resist change (Ch. 11).

Inflammation. A response to cell injury or death that attempts to isolate and destroy the offending agent and any damaged cells (Ch. 7).

Initiator. An agent producing cancer by converting normal cells to latent tumor cells (Ch. 7).

Innovative Science. An activity within any scientific discipline that questions existing paradigms and formulates new paradigms (Ch. 2).

Integrated Risk Information System (IRIS). A large data base containing reference doses, slope factors, drinking water health advisories (one-day, ten-day, longer term, and lifetime advisories), and associated information compiled by the EPA (Ch. 13).

Interference Competition. Interspecies competition in which one species interferes with another as might occur with territoriality or aggressive behavior (Ch. 11).

Interspecies Competition. The interference with or inhibition of one species by another (Ch. 11).

Intrinsic (or Malthusian) Rate of Increase. The rate of increase in size of a population growing under no constraints (Ch. 10).

Ionization (Radiation-Induced). Ion formation in materials such as tissue caused by energetic rays and particles released during radioactive decay. Ionization occurs in biological material exposed to radiation, resulting in some probability of molecular or genetic damage. Part of the process often involves the formation of free radicals (Ch. 14).

Ischemia. Localized inadequacy of blood supply to or anemia of tissue resulting from an obstruction of blood flow such as that associated with a wound (Ch. 7).

Isotopes. Nuclides with the same number of protons but different numbers of neutrons are called isotopes. The number of protons determines the chemical identity of an atom. For example all atoms with 82 protons are lead; however, lead can have 122, 124, 125, or 126 neutrons. Thus ^{204}Pb (122 neutrons + 82 protons), ^{206}Pb, ^{207}Pb and ^{208}Pb are all isotopes of lead (Ch. 14).

Isotopic Discrimination. The differential behavior of isotopes occurring if the rate or extent of participation in some biological or chemical process depends significantly on the mass of the isotope. Also called the isotope effect (Ch. 5).

Itai-itai Disease. An epidemic of cadmium poisoning (1940-1960) linked to water contaminated with mine wastes used to irrigate rice fields. Itai-itai literally means "ouch-ouch" and reflects the extreme joint pain of victims (Ch. 1).

Kaplan-Meier Method. See Product-Limit Method (Ch. 9).

Karnofsky's Law. Any agent will be teratogenic if it is present at concentrations or intensities producing cell toxicity (Ch. 8).

Karyolysis. The disintegration of the cell nucleus with necrosis (cell death) (Ch. 7).

Keystone Species. A species that influences the ecological community by its activity or role, not its numerical dominance (Ch. 11).

K_{OA}. The partition coefficient for a compound between n-octanol and air. Like the K_{OW}, it is a measure of lipophilicity (Ch. 12).

K_{ow}. The partition coefficient for a compound between n-octanol and water, i.e., concentration in octanol/concentration in water at equilibrium. It or its log-transformed value is used to reflect lipophilicity of compounds (Ch. 3).

k-Strategy. An equilibrium strategy for species involving effective interactions with each other in the community, allowing coexistence of many species. Equilibrium species are more effective competitors than opportunistic species (Ch. 11).

Landscape. The sum total aspect of any geographical area.

Langmuir Isotherm Equation. Theoretically-derived relationship quantifying adsorption (Ch. 3).

Latent (Latency) Period. The time or lag between exposure to the carcinogenic agent and the appearance of cancer (Ch. 7).

Law (U.S. Environmental). "a body of federal, state, and local legislation, in the form of statutes, bylaws, ordinances, and regulations, plus court-made principles known as common law" (McGregor, 1994) (Ch. 2).

Law of Frequencies. There exists a relationship between the number of species and the number of individuals in a community (Ch. 11).

Law of Similars. The foundation premise of homeopathic medicine that a drug which induces symptoms similar to those of the disease will aid the body in defending itself by stimulating the body's natural responses (Ch. 8).

Lesion. Alterations in cells, tissues, or organs indicating exposure or damage (Ch. 7).

Liebig's Law of the Minimum. A population's size (number of individuals or biomass) is limited by some essential factor in the environment that is scarce relative to the amounts of other essential factors, e.g., phosphorus limited algal growth in a lake (Ch. 10).

Life Cycle Studies. Comprehensive studies to determine the impact of a substance or mixture on the survival, growth, reproduction, development, or other important qualities at all stages of a species life cycle (Ch. 9).

Ligand. An anion or molecule that forms a coordination compound or complex with metals (Ch. 4).

Limited Lifespan Paradigm. An inherent quality of an individual is a genetically-defined maximum lifespan (Ch. 10).

Linear Energy Transfer. The average energy released by ionizing radiation per unit path length through a medium, usually expressed in thousands of electron volts per micron of path length. For example, the LET (keV m^{-1}) in water for a 1.2 MeV gamma ray emitted from ^{60}Co, a 0.6 keV beta particle from tritium, and a 5.3 MeV alpha particle from polonium is 0.3, 5.5, and 110.0, respectively. The more energy released, the greater the probability of damage (Ch. 14).

Linear-No Threshold Theory. This theory, relating incidence or risk of cancer to dose, is based on several radiation-induced cancer studies suggesting no threshold dose. It assumes that any lack of cancers below a certain dose reflects our inability to measure low incidences at these exposure levels, not a threshold of effect. The dose-response curve is a straight line (Ch. 7).

Linear Solvation Energy Relationship (LSER). A class of quantitative structure-activity relationships based on molecular volume, ability to form hydrogen bonds, and polarity or ability to become polarized (Ch. 4).

Lipid Peroxidation. The oxidation of polyunsaturated lipids in membranes resulting in cell damage during xenobiotic exposures (Ch. 6).

Lipofuscin (Age Pigment). A degradation product of lipid oxidation that accumulates in cell vacuoles with age or exposure to some toxicants such as copper (Ch. 7).

Liquefactive (Cytolytic) Necrosis. Necrosis characterized by a rapid breakdown of the cell as a consequence of the release of cellular enzymes (Ch. 7).

Litchfield Method. A simple, semigraphical method for analyzing survival time data and estimating LT50 values (Ch. 9).

Litchfield-Wilcoxon Method. A semigraphical method for estimating an LC50, EC50, or LD50. Although very easy to perform, it is the most subjective method for such estimations because it involves fitting a line to data by eye (Ch. 9).

Logit. A metameter used for dose- or concentration-response data under the assumption of a log logistic model. Although it has the form logit(P) = ln [P/(1-P)], its transform ([logit as just calculated]/2 + 5) is often used because the associated values are very close to those of the probit metameter (Ch. 9).

Log Normal Model. A model fit to species abundance curves that is thought to reflect a community structure in which several factors influence species interactions and subsequent allocation of resources (Ch. 11).

Lorax Incongruity. The delusion of selfless motivation in environmental stewardship or advocacy. The Lorax is a character in a popular children's book by Dr. Suess who "speaks for the trees, for the trees have no tongues." (Preface).

Lordosis. The extreme and abnormal forward curvature of the spine (Ch. 8).

Loss of Life Expectancy (LLE). A calculated estimate of loss in life time associated with a risk factor. It is estimated as the simple difference between life expectancy without the risk factor and life expectancy with the risk factor (Ch. 13).

Lowest Observed Effect Concentration (or Level) (LOEC or LOEL). The lowest concentration in a test with a statistically significant difference in response from the control response (Ch. 8).

l_x. From a life table, the number of individuals in a cohort alive at age or stage class, x (Ch. 10).

l_x **Life Table or Schedule.** A life table that summarizes mortality data for populations (Ch. 10).

$l_x m_x$ **Life Table.** A life table that summarizes both mortality and natality data for populations (Ch. 10).

MacArthur-Wilson Model of Island Colonization. The model, $S_t = S_{EQ} (1 - e^{-Gt})$ where $S_t =$ the number of species present at time t, $S_{EQ} =$ the equilibrium number of species for the island, and $G =$ the rate constant for colonization of the island (Ch. 11).

Male-mediated Toxicity. Disease or birth defects produced by a father's exposure to a physical or chemical agent (Ch. 8).

Malondialdehyde. A breakdown product of lipid peroxidation used as an indicator of oxidative damage (Ch. 6).

Malthusian Theory. A series of assumptions and observations regarding limitations on human populations developed by Thomas R. Malthus (1766-1834) (Ch. 10).

Marine Protection, Research and Santuaries Act (MPRSA). A U.S. federal act that recognizes the unsoundness of ocean dumping, states the U.S. policy regarding dumping, and establishes limitations and prohibition on dumping (Appendix 3).

Maturity Index. An index for pollution based on the proportions of species in a soil nematode community that fell into various categories ranging from colonizers (r-strategists) to persisters (K-strategists) (Ch. 11).

Maulstick Incongruity. The incongruous assignment of ecological or biological significance of a contaminant's effect based primarily on statements of statistical significance (Ch. 8).

Maximum Acceptable Toxicant Concentration (MATC). ". . . an undetermined concentration within the interval bounded by the NOEC and LOEC that is presumed safe by virtue of the fact that no statistically significant adverse effect was observed" (Weber et al., 1989) (Ch. 8).

Maximum Contaminant Levels (MCLs). Primary drinking water standards established under the SDWA for a specified list of contaminants that can adversely affect health (Appendix 3).

Maximum Likelihood Estimation (MLE). Here, a parametric method used to fit dose- or concentration-effect data to the log normal, log logistic, or other models. Probit and logit approaches are most often applied with MLE methods (Ch. 9).

Mean Absorption Time (MAT). The mean time required for absorption of a drug or contaminant calculated as the difference in mean residence time (MRT) of the material introduced by the (noninstantaneous) route of interest and the MRT for the same material injected intravenously (Ch. 4).

Mean Generation Time (T_c). The predicted generation time for a population estimated from life tables as the sum of the $xl_x m_x$ column divided by R_0 (Ch. 10).

Mean Residence Time (τ or MRT). An estimated mean time that a particle (molecule or atom) remains in a compartment (Ch. 3).

Measurement Endpoint (in Ecological Risk Assessment). That measurable response to the stressor (e.g., fledglings produced per nest each year) that is related to the valued qualities of the assessment endpoint (e.g., reproductive success of bald eagles) (Ch. 13).

Median Effective Concentration (EC50). For sublethal or ambiguously lethal effects, the concentration affecting 50% of exposed individuals by a predetermined time, e.g., 96 h (Ch. 9).

Median Effective Time (ET50). For sublethal or ambiguously lethal effects, the time until 50% of the exposed individuals respond (Ch. 9).

Median Lethal Concentration (LC50). The concentration resulting in death for 50% of exposed individuals by a predetermined time, e.g., 96 h (Ch. 9).

Median Lethal Dose (LD50). The dose resulting in death for 50% of the exposed individuals by a predetermined time, e.g., 96 h (Ch. 9).

Median Lethal Time (LT50). The time resulting in death for 50% of the exposed individuals (Ch. 9).

Median Time-to-Death (MTTD). Like the LT50, the time resulting in death for 50% of the exposed individuals (Ch. 9).

Meiotic Drive. A component of the life cycle of an individual during which natural selection can occur, involving the differential production of gametes by different heterozygous genotypes (Ch. 10).

Membrane Transport Proteins. Cell membrane-associated proteins involved in transport of solutes (Ch. 3).

Mesocosms. Relatively large experimental systems designed to simulate some component of an ecosystem. Mesocosms are delimited and enclosed to a lesser extent than are microcosms. They are normally used outdoors, or in some manner, incorporated intimately with the ecosystem that they are designed to reflect (Ch. 11).

Metallothionein. A relatively small (circa 7,000 Da) protein with approximately 25 to 30% of its amino acids being cysteine, having no aromatic amino acids or histidine, and having the capacity to bind six to seven metal atoms per molecule (Ch. 3).

Metallothionein-like Proteins. Poorly characterized, metal-binding proteins or proteins not conforming precisely to the classic properties of metallothioneins (Ch. 6).

Metameter. A measurement or a transformation of a measurement used in the analysis of biological tests, e.g., the probit metameter (Ch. 9).

Metastasis. The process in which pieces of a cancerous growth dislodge, move to other tissues via the circulatory or lymphatic system to establish other loci of cancerous growth. This process leads to the spread of a cancer from the site of origin to other sites in the body (Ch. 7).

Methemoglobinemia. Referred to as the blue-baby syndrome because of the initial skin color of afflicted babies, it is caused by the reaction of nitrite (and some drugs) to hemoglobin (oxidation of ferrous iron to ferric iron) to produce methemoglobin that is incapable of the normal transport of molecular oxygen in the blood of the newborn. It can be caused directly by nitrite in drinking water or by the conversion of nitrate to nitrite in the baby's anaerobic stomach. It can also be caused in ruminants by the consumption of plants with high nitrate content (Ch. 1).

Method of Multiple Working Hypotheses. A method proposed by Chamberlin (1897) to reduce precipitate explanation by considering all plausible hypotheses simultaneously in testing so that equal amounts of effort and attention are provided to each (Ch. 2).

Microcosms. Laboratory systems designed to simulate some component of an ecosystem such as multiple species assemblages (Ch. 11).

Micronuclei. Membrane-bound masses of chromatin separate from the nucleus proper (Ch. 6).

Microtox® Assay. A rapid, bacterial assay in which a decrease in bioluminescence is thought to reflect toxic action (Ch. 9).

Minamata Disease. An epidemic in Minamata and then Niigata, Japan resulting from organic mercury release from industrial sources (acetaldehyde and vinyl chloride production) and consequent contamination of seafood. The first case was reported in 1953 and almost 1,000 victims were identified by 1975 (Ch. 1).

Minimal Time to Response. For any toxicant effect, there can be a minimum time required to get an effect. Regardless of the toxicant concentration, the effect cannot occur any faster than this minimum time (Ch. 9).

Mixed Function Oxidase (MFO). The P-450 complex composed of cytochrome P-450, NADPH-cytochrome P-450 reductase, NADPH, and O_2 (Ch. 6). See Monooxygenase.

Modifying Factor. A factor based on expert opinion used to decrease the NOAEL (or LOAEL) during risk assessment (Ch. 13).

Monodentate Ligand. A ligand sharing one pair of electrons with a cation (Ch. 4).

Monogenic Control. Control of some quality by a single gene (Ch. 10).

Monooxygenase. One of a general class of enzymes involved in Phase I reactions with xenobiotics. Their action involves the addition of an oxygen atom (from O_2) to the xenobiotic and reduction of the remaining O atom to produce water. Also called mixed function oxidases and abbreviated MFO (Ch. 3, 6).

Montreal Protocol. An international treaty to limit and eventually eliminate the use of CFCs that was signed into law in 1988 (Ch. 12).

Most Sensitive Species Approach. An ecotoxicological approach in which results for the most sensitive of all tested species are used as an indicator of that concentration most likely to protect the entire community (Ch. 11).

Moving Average Method. Here, a method of estimating LC50, EC50, or LD50. It may be implemented with straightforward equations if the toxicant concentrations are set in a geometric series, and there are equal numbers of individuals exposed in each treatment (Ch. 9).

Multidentate Ligand. A ligand sharing more than one pair of electrons with the cation. They are also called chelates (Ch. 4).

Multiple Heterosis. A generally higher fitness of an individual as a composite or summed effect of heterozygote superiority (heterosis) at each of a series of loci (Ch. 10).

Multiplicative Growth Factor per Generation. See Finite Rate of Increase.

Mutagen. A physical or chemical entity capable of producing mutations (Ch. 7).

Mutagenic. Capable of causing mutations (Ch. 6).

NAS Paradigm. A paradigm used for both human and ecological risk assessments. There are four components to this paradigm: hazard identification, exposure assessment, dose-response assessment, and risk characterization (Ch. 13).

National Ambient Air Quality Standards (NAAQS). Standards of air quality that Section 109 of the Clean Air Act requires the EPA to establish and periodically revise (Appendix 3).

National Contingency Plan. A plan to identify sites for cleanup, and to identify the required cleanup activities as mandated by CERCLA (Appendix 3).

National Environmental Policy Act (NEPA). A U.S. federal law that, combined with the Environmental Quality Improvement Act, forms the cornerstone for U.S. environmental improvement policy. NEPA established the federal government's commitment to judge their actions relative to environmental impacts and to action that fostered harmony between human activities and nature. It established the Council on Environmental Quality and the requirement of an environmental impact statement (Appendix 3).

National Pollutant Discharge Elimination System (NPDES). A state or EPA permitting system mandated by the Clean Water Act that imposes limits on discharges (Appendix 3).

Natural Radiation Background (Radiation). Cosmic radiation emitted from stars and radiation from long-lived terrestrial radionuclides that are ubiquitously present in the earth's soils (Ch. 14).

Necrosis. Cell death resulting from disease or injury (Ch. 7).

Neoplasia. "Hyperplasia which is caused, at least in part, by an intrinsic heritable abnormality in the involved cells" (La Via and Hill, 1971) (Ch. 7).

Neoplastic Hyperplasia. Hyperplasia resulting from a hereditary change in the cell such that it no longer responds properly to chemical signals that normally control cell growth. Such cells can result in cancerous growth (Ch. 7).

Net Reproductive Rate (R_0). The expected number of females to be produced during the lifetime of a newborn female as estimated with a life table (Ch. 10).

Niche Preemption. A rapid use and preemption of resources by a species that exploits them to the exclusion or severe disadvantage of another species (Ch. 11).

Nine Aspects of Disease Association. Hill (1965) defined nine aspects of evidence fostering the accuracy of linkage between a risk factor and disease: strength of association, consistency of association, specificity of association, temporal association, biological gradient (dose-response) in the association, biological plausibility, coherence of the association, experimental support of association, and analogy (Ch. 10).

No Action Alternative (to remediation of the site). A scenario in which one assesses if the contaminants at the waste site presents, or will present in the future, a risk if left alone (Ch. 13).

Non-Stochastic Health Effects. In contrast to stochastic health effects of radiation, non-stochastic health effects are those dependent on the magnitude of the dose in excess of a threshold. Some non-stochastic health effects of radiation include acute radiation syndrome, opacification of the eye lens, erythema of the skin, and temporary impairment of fertility (Ch. 14).

No Observed Effect Concentration (or Level) (NOEC or NOEL). The highest concentration in a test for which there was no statistically significant difference in response from that of the control (Ch. 8).

Normal Equivalent Deviation (NED). As used in this book, the proportion dying in a toxicity test expressed in terms of standard deviations from the mean of a normal curve (Ch. 9).

Normal Science. A major activity of any scientific discipline that works within the framework of established paradigms, increasing the amount and accuracy of knowledge within that framework (Ch. 2).

Normit. The metameter equal to the normal equivalent deviation (NED). The resulting analysis of dose- or concentration-effect data with the normit metameter is often called normit analysis and is essentially equivalent to probit analysis (Ch. 9).

Nuclear Fission. The splitting of atomic nuclei with neutrons, resulting in the release of energy (Ch. 14).

Numerical Response. A change in predator or grazer number through increased reproductive output, decreased mortality, or increased immigration in response to changes in prey or food densities (Ch. 11).

Octaves. Log_2 classes (e.g., 1-2, 2-3, 4-7, 8-15, 16-31, . . . individuals) used in species abundance curves and representing doublings of the numbers of individuals in a species (Ch. 11).

Odds Ratio. A measure of relative risk in case-control studies in epidemiology. The number of disease cases that (a) were or (b) were not exposed, and the number of controls that (c) were and (d) were not exposed to the risk factor such as an etiological agent are used to estimate the odds ratio: odds ratio = (a/b)/(c/d) or (ad)/(bc) (Ch. 10).

Oklo Natural Reactors. Naturally-occurring nuclear reactors arising through biogeochemical processes approximately 1.8 billion years ago in Oklo (Gabon, Africa) (Preface).

Oncogene. A gene involved in cancer. Cancer results from the mutation of this gene that was involved in the normal growth and differentiation of cells (Ch. 7).

One-Hit Risk Model (for carcinogenic effect). In risk assessment, this model (Equation 13.11) is used to predict risk if the estimated risk is 0.01 or higher (Ch. 13).

Optimal Foraging Theory. The theory that the ideal forager will obtain a maximum net rate of energy gain by optimally allocating its time and energy to the various components of foraging (Ch. 11).

Optimal Stress Response. The optimal stress response involves a shift in the balance in energy allocation between somatic growth rate and longevity (survival) to optimize Darwinian fitness under stressful conditions (Ch. 10).

Organization for Economic Cooperation and Development (OECD). An organization founded in 1960 with the purpose of enhancing economic growth, living standards and financial stability, and development of and holding forums on associated policies. Member countries are Australia, Austria, Belgium, Canada, Denmark, Finland, France, Germany, Greece, Iceland, Ireland, Italy,

Japan, Luxembourg, Mexico, Netherlands, New Zealand, Norway, Portugal, Spain, Sweden, Switzerland, Turkey, United Kingdom, and United States (Appendix 3).

Oxidative Stress. The damage to biomolecules from free oxyradicals (Ch. 6).

Oxyradical. A free radical involving an unshared electron of oxygen, e.g., RO$^\bullet$ (Ch. 6).

Ozone Hole. A hole or extreme thinning of ozone above the Antarctic due to the combined effects of circulation patterns above the Antarctic and ozone destruction as a consequence of CFC accumulation in the stratosphere (Ch. 12).

Paradigms. Generally accepted concepts that, in a healthy science, have withstood rigorous testing and are given enhanced status as explanations of fact and observation (Ch. 2).

Partial Kill. A treatment in a toxicity test in which some, but not all, exposed individuals are killed (Ch. 9).

Pedosphere. That part of the earth made up of soils and where important soil processes are occurring (Ch. 12).

Peroxyradicals. Oxyradicals of organic compounds (R) of the form, ROO$^\bullet$ (Ch. 6).

Persistent Organic Pollutants (POPs). Those organic pollutants that are long-lived in the environment and tend to increase in concentration as they move through foodchains (Ch. 12).

Persistent Toxicants that Bioaccumulate (PTBs). See Persistent Organic Pollutants (Ch. 12).

Pesticide. A substance used to prevent, destroy, repel, or mitigate any pest. Under FIFRA, insecticides, fungicides, rodenticides, dessicants, and defoliants are included (Appendix 3).

Pharmacokinetics. The study and predictive modeling of the internal kinetics of drugs (Ch. 3).

Phase I Reactions. Reactions in the metabolism of organic contaminants in which reactive groups are added or made available. Although oxidation reactions are the most important Phase I reactions, hydrolysis and reduction reactions are also significant (Ch. 3).

Phase II Reactions. Reactions in the metabolism of organic contaminants in which conjugates are formed that inactivate the compound and foster elimination (Ch. 3).

Photo-Induced Toxicity. Toxicity of a chemical in the presence of light due to the production of toxic, photolysis products (Ch. 9).

Photosensitivity. Sensitivity of cutaneous tissues to the effects of light evoked by a chemical (Ch. 9).

pH-Partition Hypothesis. Bioavailability is determined by the diffusion of the unionized form through the gastrointestinal lumen as determined by pK_a and pH (Ch. 4).

Physiologic Hyperplasia. Nonpathological hyperplasia in response to a variety of usual stimuli such as that involved in the tissue repair process (Ch. 7).

Physiologically-Based Pharmacokinetics (PBPK) Model. A pharmacokinetics model that includes physiological and anatomical features in describing internal kinetics (Ch. 3).

Phytochelatin. A class of peptides in plants that are induced by and bind to metals. They may function in the regulation and detoxification of metals by plants (Ch. 3).

Pielou's J'. A measure of species evenness for a sample from a community (Ch. 11).

Pielou's J'. A measure of species evenness for a community (Ch. 11).

Pollutant. "A substance that occurs in the environment at least in part as a result of man's activities, and which has a deleterious effect on living organisms" (Moriarty, 1983) (Ch. 2).

Pollution-induced Community Tolerance (PICT). An increase in tolerance to pollution resulting from species composition shifts in the community, acclimation of individuals, and genetic changes in populations in the community (Ch. 11).

Polygenic Control. Control of some quality by several genes (Ch. 10).

Population. A group of individuals of a species occupying a defined space at a particular time (Ch. 10).

Porins. Pores in the cell membrane that are nonspecific among ions (Ch. 3).

Porphyrins. Molecules having a tetraphyrrole ring (four simple, heterocyclic nitrogen ring compounds bound together to form a ring) that are produced as intermediates during heme synthesis in animals (Ch. 6).

Positron. A particle with the same mass of an electron but a positive charge. Also called a positive electron (Ch. 14).

Potentiation. Enhanced toxicity of a chemical in the presence of a second chemical that is not itself toxic at its concentration in the mixture (Ch. 9).

Power Relationship. A mathematical relationship in which the Y variable is related to the X variable raised to some power. For example, $Y = aX^b$ (Ch. 4). (Compare to Exponential Relationship.)

Precipitate Explanation. The obsolete and unreliable scientific practice of uncritical or untested acceptance of an explanation based on some ruling theory (Ch. 2).

Predictive Risk Assessment. A risk assessment dealing with a planned or proposed condition (Ch. 13).

Prevalence. The incidence rate of a disease multiplied by the amount of time that individuals were at risk (Ch. 10).

Prevention of Significant Deterioration (PSD). A program in the Clean Air Act with the goal of no significant deterioration of air quality in areas attaining NAAQS (Appendix 3).

Primary Lamellae (Filaments). Gill structures extending outward at right angles from the branchial arches (Ch. 7).

Principle of Allocation. There exists a cost or trade-off to every allocation of energy resources. Energy spent by an individual organism on one function, process, or structure cannot be spent on another. Optimal allocation of resources enhances Darwinian fitness (Ch. 10).

Probabilistic (Bayesian) Induction. A type of induction that uses probabilities associated with competing theories or explanations to decide which is the most probable. Certainties or credibilities are then assigned to competing explanations based on their associated probabilities. Instead of the quantal (accept or reject) falsification of a working hypothesis, Bayesian induction considers a hypothesis falsified if it were sufficiently improbable (Ch. 13).

Probit. A metameter produced by adding 5 to the normal equivalent deviation (NED). Forming the basis of probit analysis, it was first proposed in the 1930s to avoid negative numbers (Ch. 9).

Problem Formulation (in Ecological Risk Assessment). The planning and scoping phase that establishes the framework around which the risk assessment is done (Ch. 13).

Procarcinogen. A compound that is converted to a carcinogen (Ch. 6).

Product-Limit (Kaplan-Meier) Method. A nonparametric method for analyzing time-to-death or survival time data that does not require a specific model for the survival curve (Ch. 9).

Promoter. An agent producing cancer by enhancing the growth of mutated cells (Ch. 7).

Proportional Diluter. A special apparatus used in flow-through toxicity tests to mix and deliver a series of dilutions of a toxic solution to exposure tanks (Ch. 9).

Proportional Hazard Model. A survival or time-to-death model that relates the hazard (proneness to die or risk of dying at any time, t of one group (e.g., smokers) quantitatively to that of a reference group (e.g., nonsmokers) (Ch. 9).

Proteinuria. The presence of protein in the urine (Ch. 13).

Proteotoxicity. A toxic or adverse effect of a chemical or physical agent with an underlying mechanism of protein damage (Ch. 6).

Proto-Oncogene. A gene involved in some way with the normal growth (enhancement) and differentiation of cells which, upon mutation, becomes an oncogene (Ch. 7).

Ptolemaic Incongruity. The false paradigm that any particular level of biological organization holds a more central or important role than another in the science of ecotoxicology (Ch. 7).

Quantitative Structure-Activity Relationship (QSAR). A quantitative, often statistical, relationship between a molecular quality or molecular qualities and some activity, i.e., bioavailability or toxicity (Ch. 4).

Quotient Method. The use of the hazard quotient as a crude indicator of hazard (Ch. 13).

Radiosensitizers. Chemicals, such as derivatives of nitro imidazoles, used during radiation treatment to enhance the production of free radicals which then kill cancer cells (Ch. 6).

Rarefaction Estimate of Richness. An estimate of species richness expressed relative to that of a sample with a standard number of individuals in it (Ch. 11).

Rate Constants. Constants used in mathematical models that have units of $time^{-1}$. They describe the rate at which a contaminant transfers from one compartment to another. Use of a rate constant implies a first-order kinetic process where the rate of transport is assumed proportional to the amount of contaminant in the compartment (Ch. 14).

Rate Constant-based Model. A compartment model that employs rate constants to quantify the rate of change in concentration or amount of toxicant (Ch. 3).

Rate of Living Theory of Aging. The total metabolic expenditure of a genotype is generally fixed and longevity depends on the rate of energy expenditure (Ch. 10).

Rate Ratio. The ratio of disease incidence rates for two populations. Rate Ratio = I_A/I_0 where I_A = incidence rate in population A, and I_0 = incidence rate in the reference or control population (Ch. 10).

Realized Niche. That portion of a species' fundamental niche that it actually occupies (Ch. 11).

Reasonable Maximum Exposure (RME). Exposure calculated for a chemical of potential concern during a risk assessment. It is a conservative estimate of exposure that is computed differently, depending on the route of exposure (Ch. 13).

Receptor Antagonism. Antagonism that involves the binding of the toxicants to the same receptor and one toxicant blocking the other from fully expressing its toxicity (Ch. 9).

Redox Cycling. In the context of contaminant (quinones, aromatic nitro compounds, aromatic hydroxylamines, bipyridyls, and some chelated metals) involvement in the generation of oxyradicals, redox cycling occurs if contaminants are reduced to radicals and then participate in redox reactions to produce the superoxide radical from molecular oxygen. The contaminant exists in its original form at the end of the redox reactions and is available to recycle many times through this process and produce more oxyradicals (Ch. 6).

Redundancy Hypothesis. Many species are redundant and their loss will not influence the community function as long as crucial (e.g., keystone and dominant) species populations are maintained (Ch. 11).

Reference Dose (RfD) (for noncarcinogenic effects). The best estimate of the daily exposure for humans, including the most sensitive subpopulation, that will result in no significant risk of an adverse health effect if it is not exceeded (Ch. 13).

Region (Geographical). An "area of the earth's surface differentiated by its specific characteristics" (Monkhouse, 1965) (Ch. 12).

Regulation (Legal). Specific regulations derived by U.S. federal agencies for ensuring that the intent of a law is met. Environmental regulations "specify conditions and requirements to be met, . . . provide a schedule for compliance, and . . . record any exceptions to the regulated community" (Mackenthun and Bregman, 1992) (Appendix 3).

Relative Bioavailability. The bioavailability estimated for a dose administered by any route or formulation relative to a dose administered in a reference (or alternate) route or formulation (Ch. 4).

Relative Risk. In survival time analysis, the risk of one group expressed as a multiple of that of another. Relative risk is usually estimated with a hazard model. In epidemiology, it is the ratio of occurrences of the disease in two populations (Ch. 9, 10).

Remedial Investigation (RI). Part of a Remedial Investigation and Feasibility Study that has three parts: characterization of the type and degree of the contamination, human risk assessment, and ecological risk assessment (EPA, 1994) (Ch. 13).

Remedial Investigation and Feasibility Study (RI/FS). For a Superfund site, a study that has as its goal the implementation of "remedies that reduce, control, or eliminate risks to human health and the environment" or, more specifically, the accumulation of "information sufficient to support an informed risk management decision regarding which remedy appears to be most appropriate for a given site" (EPA, 1989d) (Ch. 13).

Remote Sensing. Technologies that allow the acquisition and analysis of data without requiring physical contact with the land or water surface being studied. Most determine qualities or characteristics of areas of interest based on measurements of visible light, infrared radiation, or radio energy coming from them (Ch. 12).

Repair Fidelity (of DNA). The accuracy in repairing and returning the DNA to its original state after damage (Ch. 7).

Reproductive Value (V_A). The expected contribution of offspring during the life of an individual of an age class x in a life table (Ch. 10).

Residual Body. A cell vacuole containing lipofuscin, a degradation product of lipid oxidation (Ch. 7).

Resilience (Community). The number of times a community can return to its normal state after perturbation (Ch. 11).

Resistance. The term "resistance" is often reserved for the enhanced ability to cope with a factor due to genetic adaptation. The term "tolerance" is often reserved for enhanced abilities associated with physiological acclimation. Tolerance is used in this book for both acclimation and genetic adaptation (Ch. 10).

Resource Conservation and Recovery Act (RCRA). A U.S. federal law that regulates production, treatment, storage, and disposal of hazardous wastes (Appendix 3).

Respiratory Lamellae. See Secondary Lamellae (Ch. 7).

Retention Effect. The K_{OW}s of persistent organic pollutants (POPs) influence their global movement toward higher latitudes. Those POPs with high lipophility tend to be held more firmly in solid phases such as soil and vegetation than less lipophilic POPs. Consequently, they spend less time in the atmosphere and are less available for transport in that medium (Ch. 12).

Retroactive Risk Assessment. A risk assessment dealing with an existing condition (Ch. 13).

Ricker Model. A difference equation model (Eq. 10.10) for growth of populations with nonoverlapping generations or experimental designs with discrete intervals of population growth (Ch. 10).

Risk. As used in risk assessments, the probability (or likelihood) of some adverse consequence occurring to an exposed human or to an exposed ecological entity (Ch. 13).

Risk Assessment. The process by which one estimates the probability or likelihood of some adverse effect(s) of a present or planned release to either human or ecological entities (Ch. 13).

Risk Assessor. A person or group of people "who actually organizes and analyses site data, develops exposure and risk calculations, and prepares a risk assessment report" (EPA, 1989d) (Ch. 13).

Risk Characterization (in Ecological Risk Assessment). The last step of the ecological risk assessment that draws together the information generated from previous steps to produce a statement of the likelihood of an adverse effect to the assessment endpoint (Ch. 13).

Risk Factor. Any quality of an individual (e.g., age) or an etiological factor (e.g., chronic exposure to high levels of the toxicant) that modifies an individual's risk of developing the disease in question (Ch. 10). Specific to chemical risk assessment (Ch. 13), a chemical concentration multiplied by a toxicity value estimates the risk factor. Relative to radiation's effects (Ch. 14), a risk factor gives the probability of a deleterious effect for each mSv of dose received.

Risk Hypotheses (in Ecological Risk Assessment). As part of the conceptual model, they are clear statements of postulated or predicted effects of the contaminant on the assessment endpoint (Ch. 13).

Risk Manager. "The individual or group who serves as primary decision-maker for a [waste] site" (EPA, 1989d) (Ch. 13).

Rivet Popper Hypothesis. Species in a community are like rivets that hold an airplane together and contribute to its proper functioning. The loss of each rivet weakens the structure (Ch. 11).

Roentgen (R). A measure of the amount of energy deposited in some material by a certain amount of radiation. It is expressed relative to energy dissipation in 1 cc of dry air. Use of R to express dose allows one to normalize for the different amounts of energy that are deposited in materials such as tissue by different types of radiation (Ch. 10).

Roentgen Equivalent Man (Rem). A measure of radiation that takes into account the differences in potential biological effects of different types of radiation. It relates the dose received to potential damage (Ch. 1).

r-Strategy. An opportunistic strategy favoring species that establish themselves quickly, grow quickly to exploit as many resources as possible, and produce many offspring (Ch. 11).

Safe Concentration. "The highest concentration of toxicant that will permit normal propagation of fish and other aquatic life in receiving waters. The concept of a 'safe concentration' is a biological concept, whereas the 'no observed effect concentration' is a statistically defined concentration." (Ch. 8).

Safe Drinking Water Act (SDWA). A U.S. federal law that protects the water supply of the public (systems supplying more than 25 people or with more than 15 service outlets) by setting water quality standards, and protects source aquifers from contamination by establishing permitting requirements for underground disposal of waste (Appendix 3).

Saprobien Spectrum. The characteristic change in community composition at different distances below the discharge of putrescible organic waste to a river or stream (Ch. 11).

Scaling. The handling or transformation of allometric data to produce a quantitative relationship between organism (or species) size and some characteristic such as metabolic rate, gill surface area, lung ventilation rate, or biochemical activity (Ch. 4).

Scoliosis. Lateral curvature of the spine (Ch. 8).

Scope of Activity. The difference between the rates of oxygen consumption under maximum and minimal activity levels. It reflects the respiratory capacity available for the diverse demands on and activities of an organism (Ch. 8).

Scope of Growth. An index (P = production) calculated as the amount of energy taken into the organism in its food (A) minus the energy used for respiration (R) and excretion (U): P = A − R − U. It is the amount of energy available for growth or production of young (Ch. 8).

Secondary Lamellae (Respiratory Lamellae). These gill structures are parallel rows of projections on the dorsal and ventral sides of each primary lamellae. They are the primary sites of gas exchange of the gills (Ch. 7).

Selection Components. Components of the life cycle of an individual upon which natural selection can act. They are viability selection, sexual selection, meiotic drive, gametic selection, and fecundity selection (Ch. 10).

Selyean Stress. A nonspecific response of the body when extraordinary demands are made of it. ". . . the state manifested by a specific syndrome which consists of all the nonspecifically induced changes within a biological system" (Selye, 1956) (Ch. 8).

Sentinel Species. A feral, caged, or endemic species used in measuring and indicating the level of contaminant or effect during a biomonitoring exercise. The proverbial canary in the coal mine is an example of a sentinel species (Ch. 6).

Sexual Selection. A component of the life cycle of an individual in which natural selection can occur, involving differential mating success of individuals (Ch. 10).

Shannon Diversity Index. A measure of species diversity of an entire community (Ch. 11).

Shapiro-Wilk's Test. A statistical test of the null hypothesis that data are normally distributed. Used in this book to test this assumption prior to performing one-way analysis of variance on sublethal or chronic lethal effects data (Ch. 8).

Shelford's Law of Tolerance. A species' tolerance(s) along an environmental gradient (or series of environmental gradients) will determine its population distribution and size in the environment (Ch. 10).

Sister Chromatid. At the metaphase plate, chromosomes are composed of two chromatids called sister chromatids (Ch. 7).

Sister Chromatid Exchange (SCE). The exchange of DNA between sister chromatids as a consequence of DNA breakage followed by reunion and crossing over of DNA segments of the chromatids (Ch. 7).

Slope Factor (SF). In human risk estimation, it is the slope (risk or probability of occurrence per unit of dose or intake) for the risk-dose model used to estimate the probability of a cancer at a specified exposure (Ch. 13).

Solvent Drag. The movement of a solute (contaminant) along with the bulk movement of the solution (Ch. 3).

Somatic Death. Death of an individual organism (Ch. 7).

Somatic Risk. The risk of an adverse effect to the exposed individual associated with genetic damage to somatic cells, e.g., damage leading to cancer (Ch. 7).

Sorption. A term used instead of adsorption if the specific mechanism by which a compound in solution becomes associated with a solid surface is unknown or undefined (Ch. 3).

Spearman-Karber Method. A nonparametric method to estimate the LC50, EC50, or LD50 when it is difficult or unnecessary to assume a specific model for the dose- or concentration-effect data (Ch. 9).

Species Assemblage. See Taxocene (Ch. 11).

Species Diversity (= Heterogeneity). The heterogeneity or diversity of the community considering both species richness and evenness (Ch. 11).

Species Evenness. The degree to which the individuals in the community are evenly or uniformly distributed among species (Ch. 11).

Species Richness. The number of species present in the community (Ch. 11).

Specific Activity Concept (for radiotracer use). The radionuclide used to trace or quantify the movement of a stable nuclide (e.g., ^{14}C for stable C) is assumed to behave identically in chemical and biological processes as its nonradioactive analog (e.g., stable C) (Ch. 5).

Spiked Bioassay Approach (SB). A sediment toxicity test method to generate a concentration-response model for or test hypotheses regarding effects to individuals placed in sediments spiked with different amounts of toxicant (Ch. 9).

Spillover Hypothesis. Based on the assumption that binding by metallothionein sequesters toxic metals away from sites of action, this hypothesis states that toxic effects will begin to be seen after exceeding the capacity of the metallothionein present at any time to bind metals. The unbound metals then "spill over" to interact at sites of adverse action (Ch. 6).

Stable Population. If conditions do not change with time, a population with a particular r will eventually establish a stable distribution of individuals among the various age classes. Such a population is called a stable population (Ch. 10).

Standard. Legal limits (concentration or intensity) permitted for a specific water body, based on criteria and the specified use of a water body (Ch. 2).

State Implementation Plan (SIP). State plans required by the Clean Air Act that outline steps toward meeting and maintaining national air quality standards (Appendix 3).

Static-renewal Test. A modified static toxicity test in which solutions are completely or partially replaced with new solutions at set periods during exposures, or organisms are periodically transferred to new solutions (Ch. 9).

Static Toxicity Test. A type of aquatic toxicity test in which the exposure water is not changed during the test (Ch. 9).

Steel's Many-One Rank Test. A nonparametric, post-ANOVA test often employed in the analysis of sublethal and chronic lethal effects data (Ch. 8).

Stochastic Health Effects. Cancer and genetic disorders for which initiation of effects by radiation is probabilistic, and that the risk of incurring cancer or genetic effects is proportional, without a threshold, to the dose in the relevant tissue (Ch. 14).

Stress. "At any level of ecological organization, a response to or effect of a recent, disorganizing or detrimental factor" (Newman, 1995). See also Seylean Stress for the definition of stress in an individual organism (Ch. 2).

Stressor. That which produces stress (Ch. 2). In ecological risk assessment, "any chemical, physical, or biological entity that can induce adverse effects on *ecological components*, that is, individuals, populations, communities, or ecosystems" (Norton et al., 1992) (Ch. 13).

Stress Proteins. A class of proteins involved in lessening the damage to proteins (see Cellular Stress Response) associated with a variety of stressors including heat, anoxia, UV radiation, arsenate, metals, and some xenobiotics. This term is also used by several ecotoxicologists (e.g., Sanders and Dyer, 1994) in a more generic context to mean a protein induced in response to a stressor. Such a definition would include proteins such as metallothioneins. (See Heat Shock Proteins also.) (Ch. 6).

Stress Protein Fingerprinting. The proposed use of the patterns of stress protein induction seen in the field to suggest the particular toxicant inducing the response. Patterns from organisms sampled from the field can be compared to those obtained with single candidate toxicants in the laboratory (Ch. 6).

Stress Theory of Aging. Stress shortens longevity by accelerating energy expenditure (see Rate of Living Theory of Aging). Selection takes place for resistance to stress, and as an epiphenomenon, individuals resistant to stress will predominate in extreme ages classes of a population. The diminution of homeostasis under stress with age should be slowest in individuals with highest longevity (Ch. 10).

Structure-Activity Relationship (SAR). A relationship between molecular qualities and some activity such as bioavailability or toxicity (Ch. 4).

Subcooled Liquid Vapor Pressure (P_L). The liquid vapor pressure corrected or adjusted for the heat of fusion, the energy needed to convert a mole of a compound from a solid to a liquid phase. Its use allows the expression of *liquid* vapor pressures at a specific temperature for organic compounds with widely varying melting temperatures (Ch. 12).

Subchronic Reference Dose (RfD$_s$). A RfD derived from short-term exposure data (Ch. 13).

Sublethal Effects. Effects seen at concentrations below those producing direct somatic death, e.g., slowed growth of an individual or diminished reproduction (Ch. 8).

Sulfotransferase. A Phase II enzyme which conjugates sulfates to xenobiotics or their metabolites (Ch. 6).

Superoxide Dismutase (SOD). An enzyme catalyzing the reaction, $2 O_2^{\bullet-} + 2H^+ \rightarrow H_2O_2 + O_2$. It functions to reduce oxidative stress (Ch. 6).

Suppressor Gene. A gene that functions normally to suppress cell growth and may inhibit abnormal growth (Ch. 7).

Synergistic. Here, if the sum of TUs for a mixture of two (or more) toxicants is less than the actual toxicity of the mixture expressed in TUs, the toxicants are said to be greater than additive or synergistic. The toxic effect of the toxicants is enhanced by the presence of the other(s) (Ch. 9).

Target Organ. The specific or characteristic organ in which lesions from a toxicant occur or are expected to occur based on toxicant transport to, accumulation in, or activation by that organ (Ch. 7).

Taxocene. A taxonomically-defined subset of the entire ecological community (Ch. 11).

Temperature of Condensation (T_C). The temperature (°C) at which the compound condenses or partitions from the gaseous to the nongaseous phase (Ch. 12).

Teratogen. A chemical or physical agent capable of causing a developmental malformation (Ch. 8).

Teratogenic. Capable of causing developmental malformations (Ch. 6).

Teratogenic Index (TI). The mortality of eggs expressed as an LC50 divided by the TC50 (EC50 for production of abnormal embryos). The TI is thought to reflect the developmental hazard of a contaminant (Ch. 8).

Teratology. The science of fetal and embryonic abnormal development (Ch. 8).

Tetrad. Two homologous chromosomes composed of two chromatids each come together at the metaphase plate to form a tetrad (Ch. 7).

Threshold Theory. This theory assumes no response (dose-related incidence or risk of cancer) below a certain low dose. Above the threshold, the slope of the response versus dose curve increases rapidly. The dose-response curve takes on the appearance of a hockey stick (Ch. 7).

Tolerance. This term is often reserved for enhanced ability to cope with a factor due to physiological acclimation. Resistance is used if the enhanced abilities are associated with genetic adaptation. Tolerance is used in this book for both acclimation and genetic adaptation (Ch. 10).

Toxicity Value. In risk assessment, it is a factor used to estimate a risk factor. Its estimation can involve one of two methods. One can use the slope of a published effect-dose relationship to estimate the risk factor: R = Slope * C where C = toxicant concentration. The other uses a reference dose (RfD) value in a similar manner (Ch. 13).

Toxicokinetics. The study and predictive modeling of the internal kinetics of poisons (Ch. 3).

Toxic Substances Control Act (TSCA). A U.S. law that regulates, through the EPA, the manufacture, processing, transport, use, import, and disposal of chemicals or chemical mixtures that may pose an unreasonable risk to health or the environment (Appendix 3).

Toxic Unit (TU). Amount or concentration of a toxicant expressed in units of lethality such as LD50 or LC50. For example, if toxic units are based on the LC50, a chemical with an LC50 of 20 mg L^{-1} would be present at 0.5 TU in a 10 mg L^{-1} solution (Ch. 9).

Tracer. Radioactive contaminants are often present in such low concentrations, compared to the concentrations of similar nonradioactive elements, that the behavior of the contaminant is governed by that of the similar element rather than its own mass characteristics. Thus the radionuclide traces or mimics the normal behavior of the similar element (Ch. 14).

Trophic Dilution. The decrease in contaminant concentration as trophic level increases. Trophic dilution results from a net balance of ingestion rate, uptake from food, internal transformation, and elimination processes favoring loss of contaminant that enters the organism via food (Ch. 5).

Trophic Enrichment. See Biomagnification (Ch. 5).

t-Test with a Bonferroni Adjustment. A parametric, post-ANOVA test used often in the analysis of sublethal and chronic lethal effects data (Ch. 8).

t-Test with a Dunn-Šidák Adjustment. A parametric, post-ANOVA test used rarely in the analysis of sublethal and chronic lethal effects data that has slightly better statistical power than the t-test with a Bonferroni adjustment (Ch. 8).

Twin Tracer Technique. A technique that introduces together a radiotracer of the substance being assimilated and an inert tracer which will not be assimilated and to which assimilation is compared (Ch. 5).

Type A Organism. According to the scheme of Campbell et al. (1988), an organism in contact with the sediments but unable to ingest particulates. The implication is bioavailability from interstitial water but not sediment-associated particulates. Some examples include rooted macrophytes and benthic algae (Ch. 4).

Type B Organism. According to the scheme of Campbell et al. (1988), an organism in contact with the sediments capable of ingesting particulates. The implication is bioavailability from interstitial water and sediment-associated particulates. Some examples include detritivores and suspension feeders (Ch. 4).

Uncertainty Factors. In risk assessment, factors to decrease the NOAEL (or LOAEL) in order to compensate in a conservative direction for uncertainty (Ch. 13).

Uridinediphospho glucuronosyltransferase (UDP-glucouronosyltransferase, UDP-GT). A Phase II enzyme that transfers glucuronic acid from uridine diphosphate glucuronic acid to electrophilic xenobiotics or their metabolites. It also binds covalently with electrophilic compounds such as PAHs (Ch. 6).

Uptake. The movement of a contaminant into or onto an organism (Ch. 3).

Viability Selection. A component of the life cycle of an individual in which natural selection can occur through the differential survival of individuals. It begins at the formation of the zygote and continues throughout the life of the individual (Ch. 10).

Vital Rates. Rates at which important processes such as birth, migration, and death occur in populations (Ch. 10).

Vulnerability (Community). Susceptibility to irreversible damage by toxicants (Ch. 11).

Wahlund Effect. There will be a net deficit of heterozygotes when two populations, each in Hardy-Weinberg equilibrium but with different allele frequencies, are mixed and the genotype frequencies quantified in a combined population sample (Ch. 10).

Waldsterben. "The widespread and substantial decline in growth and the change in behavior of many softwood and hardwood forest ecosystems in central Europe" (Schütt and Cowling, 1985) (Ch. 12).

Weakest Link Incongruity. An incongruous extension of critical life stage concept that protection of the most sensitive stage will ensure protection of all life stages. The dubious extension is made in which one assumes that exposure of field populations to concentrations identified in testing as causing significant mortality at a critical life stage will result in significant impact on the field population. This may or may not be true (Ch. 9).

Weibull Metameter. A metameter used occasionally in dose- or concentration-effect data analysis. It has the form, $U = \ln(-\ln(1-P))$ where P is the proportion dead (Ch. 9).

Weight of Evidence. In risk assessment this phrase appears to refer to whether a *reasonable person* reviewing the available information *could* agree that the conclusion was plausible. The more the evidence supports the conclusion, the stronger the "weight of evidence." It could mean a quantitative, semiquantitative, or qualitative estimate of the degree to which the evidence supports or undermines the conclusion (Ch. 13).

Wet Deposition. Deposition in precipitation of pollutants that were formed in the liquid media of the precipitation and that were incorporated into the precipitation during rain out (Ch. 12).

Wilcoxon Rank Sum Test with Bonferroni's Adjustment. A nonparametric, post-ANOVA test often employed in the analysis of sublethal and chronic lethal effects data (Ch. 8).

Williams' Test. A parametric test that is more powerful than other post-ANOVA tests used to analyze sublethal and chronic lethal data. It assumes a monotonic trend (increase or decrease) may occur with increasing concentration (Ch. 8).

Working Hypothesis. A hypothesis that is used to determine fact during scientific inquiry. It is not assumed to be true and only serves to test facts (Ch. 2).

Xenobiotic. A "foreign chemical or material not produced in nature and not normally considered a constitutive component of a specified biological system. [It is] usually applied to manufactured chemicals" (Rand and Petrocelli, 1985) (Ch. 2).

X-rays. Electromagnetic photons emitted from the shells of electrons that surround the nucleus (Ch. 14).

Zenker's Necrosis. Necrosis (cell death) which occurs in skeletal muscle and is similar to coagulation necrosis (Ch. 7).

Appendix 1
International System (SI) of Units Prefixes

TABLE AI. SI Unit Prefix Convention

Factor	Unit Prefix (Symbol)	Factor	Unit Prefix (Symbol)
10^{12}	tera- (T)	10^{-1}	deci- (d)
10^{9}	giga- (G)	10^{-2}	centi- (c)
10^{6}	mega- (M)	10^{-3}	milli- (m)
10^{3}	kilo- (k)	10^{-6}	micro- (μ)
10^{2}	hecto- (h)	10^{-9}	nano- (n)
10^{1}	deka- (da)	10^{-12}	pico- (p)
		10^{-15}	femto- (f)

Appendix 2
Miscellaneous Conversion Factors

(Extracted from CRC Handbook (Lide, 1992) and miscellaneous sources.)

TABLE A2. Conversion Factors for Miscellaneous Units

1 acre (a) $= 4.046873 \times 10^3$ meter^2 (m^2)

1 atmosphere (atm) $= 1.013250 \times 10^5$ pascal (Pa)

1 bar (bar) $= 1.000000 \times 10^5$ pascal (Pa)

1 barrel of oil (bbl) $= 4.2 \times 10^1$ gallons (gal) $= 1.589873 \times 10^{-1}$ meter^3 (m^3)

1 Becquerel (Bq) = 1 disintegration per minute (dpm) $= 27.027 \times 10^{-12}$ curies (Ci)

1 British thermal unit (Btu) $= 1.055 \times 10^3$ joule (J)

1 calorie (Cal) = 4.186800 joule (J)

1 cubic foot (ft^3) $= 2.831685 \times 10^{-2}$ cubic meter (m^3)

1 curie (Ci) $= 2.2 \times 10^6$ disintegrations per minute (dpm) $= 3.700 \times 10^{10}$ becquerel (Bq)

Dalton (Da) = an arbitrary atomic mass unit that is 1/12 the mass of a carbon atom with mass number of 12.

Degrees Celcius (C) = degrees Kelvin (K) - 273.16

Degrees Celcius (C) = (5/9)(degrees Fahrenheit –32) (F)

1 Electron volt (eV) $= 1.602 \times 10^{-12}$ erg $= 1.602 \times 10^{-19}$ Joule

1 foot (ft) $= 3.048000 \times 10^{-1}$ meter (m)

1 gallon (gal) $= 3.785412 \times 10^{-3}$ cubic meter (m^3)

1 Gray (Gy) = 1 Joule of energy deposited per kg of tissue

1 Liter (L) of crude oil ≈ 868.63 gram (g) based on the density of Kuwait crude oil at 20 C

1 micron (µ) $= 1.000000 \times 10^{-6}$ meter (m)

1 mile (mi) $= 1.6093 \times 10^3$ meter (m)

1 millibar (mbar) $= 1.000000 \times 10^2$ Pascal (Pa)

1 millimeter of mercury $= 1.33322 \times 10^2$ Pascal (Pa)

1 pound (lb) $= 4.535924 \times 10^{-1}$ kilogram (kg)

1 rad (absorbed radiation dose) $= 1.000000 \times 10^{-2}$ gray (Gy)

1 rem (Roentegen equivalent man) $= 1.000000 \times 10^{-2}$ sievert (Sv)

1 Roentgen (R) $= 2.58 \times 10^{-4}$ coulomb per kg of air

1 seivert (Sv) = 100 mrem

1 ton (tn) $= 2.000 \times 10^3$ pound (lb) $= 9.071847 \times 10^2$ kilogram (kg)

1 torr (mm Hg) $= 1.33322 \times 10^2$ Pascal (Pa)

1 yard (yd) $= 9.144000 \times 10^{-1}$ meter (m)

Appendix 3
Summary of U.S. Laws and Regulations

Laws and regulations are important aspects of practical ecotoxicology that deserve discussion. More detail may be acquired using the suggested readings at the end of Chapter 2 and the references provided for the specific laws. By necessity, discussion is limited to U.S. federal law. Laws pertinent to Canada and the **European Community (EC)** are outlined in Foster (1985) for water, Elsom (1987) for air, and Frankel (1995) for marine resources. European community activities and the **Organization for Economic Cooperation and Development (OECD)** policies are described in Blok and Balk (1995) and Grandy (1995), respectively.

Once the law is established, a federal agency such as the EPA is charged with developing the associated regulations that are intended to enforce the particulars of the law. When the EPA is so charged, its proposed rules are published initially in the *Federal Register* (*FR*) and remain open for public comment for a specified period of time. After that time has elapsed, the EPA repeats the internal process of rule development with consideration of any public comment and other input. Final rules are then established and published in the *FR*. These regulations may be open to interpretation by affected parties but redress occurs through the courts after this point in the process. During the next revision of the *Code of Federal Regulations* (*CFR*), the regulations are incorporated into this compilation of federal code.

The acts and laws established by the U.S. Congress are identified using references such as 42 USC §§6901 to 6992k or PL 91-190 for the National Environmental Policy Act (NEPA). USC and PL are abbreviations for U.S. Code and Public Law, respectively. As mentioned, a law[1] requires a federal agency such as the EPA to develop **regulations** that provide specific details ensuring that the intent of the law is met. According to Mackenthun and Bregman (1992), regulations "specify conditions and requirements to be met, . . . provide a schedule for compliance, and . . . record any exceptions to the regulated community." Within the text of these materials, legal definitions are also provided for terms such as contaminant (40 CFR Sections 230.3, 300.6, 310.11, etc.), point source (40 CFR Sections 233.3 and 260.10), or pollutant (40 CFR Section 122.2) that may vary among laws and deviate from scientific definitions. Therefore, it is important to understand the legal context of such terms. Legal dictionaries of environment-related terms (e.g., King, 1993) should be consulted to avoid confusion when working under a specific set of

---

[1]Although we are discussing only federal law here, **environmental law** is much more inclusive. McGregor (1994) defines environmental law as "a body of federal, state, and local legislation, in the form of statutes, bylaws, ordinances, and regulations, plus court-made principles known as common law."

regulations. Regulations from the EPA are published in the *CFR* under Title 40 (Protection of Environment) and updated as needed in the *FR*. The daily issuances of the *FR* augment the *CFR*, and consequently, these sources together define current federal regulations. These specific codifications of environmental regulation are referenced with numbers such as 40 CFR Parts 1500 to 1517. The numbers 1500 to 1517 identify specific parts of Title 40 of the *CFR*; here, a section providing details on National Environmental Policy Act.

With this background established, specific laws and associated regulations can now be outlined. Each is presented with relevant references noted from the *CFR* and the *FR*.

National Environmental Policy Act or NEPA (1969, PL 91-190; 42 USC §§6901-6992k; see 40 CFR 1500-1517) was signed into law on January 1, 1970 by President Richard Nixon. Combined with the Environmental Quality Improvement Act (EQIA, 1970), the NEPA formed the cornerstone for U.S. environmental policy (Foster, 1985). The NEPA established a federal commitment to judge the government's actions relative to environmental impacts and to take action fostering "harmony" between human activities and nature. It established the **Council on Environmental Quality** to aid and advise the President during preparation of the annual **Environmental Quality Report**. This council assesses federal programs relative to the NEPA and researches various topics on environmental quality.

The NEPA requires federal agencies to prepare an **environmental impact statement** (EIS) for any major federal action that could have an adverse environmental impact. In addition to direct actions, federal actions include issuance of federal permits, leases, or licenses, and granting, contracting, or loaning of federal monies (McGregor, 1994). The NEPA does not cover nonfederal actions and there are no defined rights of individuals to live without environmental damage included in the NEPA (Freedman, 1987). The EIS outlines possible adverse impacts, details alternatives to the action including the "no-action" alternative, and describes irreversible impacts to resources (McGregor, 1994). Prior to an EIS, a shorter and preliminary **environmental assessment** (EA) may be used to determine if a full EIS is necessary. A statement of no significant effect ("**finding of no significant impact**" or FONSI) is made if no EIS is thought to be needed. Otherwise, a full EIS is developed. This process of incorporating environmental impact into the decision making of the federal government is open to public comment, and involves relevant state, local, and federal agencies with the EPA overseeing much of the process.

The **Clean Air Act** or CAA (1977, PL 9595, 42 USC §§7401-7671q; see 40 CFR50-99; also 1990 Clean Air Act Amendments (CAAA), PL 101-549, 104 Stat. 2399) is designed to regulate air pollution so as to protect human health and the environment. It is overseen and administered by the EPA but responsibility for control and prevention of air pollution is passed to state and local governments. This act sets maximum allowable levels of pollution and dictates emission levels designed to achieve these limits. It establishes national emission standards for pollutants. Stationary sources are controlled by a state permit system. As detailed in the CAA, mobile sources are regulated through processes such as motor vehicle emissions standards or vehicle inspection programs. States are required to develop **State Implementation Plans** (SIPs) outlining steps toward meeting and maintaining **national ambient air quality standards** (NAAQS). Section 109 of this act requires the EPA to establish and periodically revise NAAQS. The act also includes details for prevention of significant deterioration (PSD) in geographical areas that have attained the NAAQS.

The **Federal Insecticide, Fungicide and Rodenticide Act** or FIFRA (1972, 7 USC 136, PL 95396; see also 40 CFR 150-189) controls the production, labeling, shipping, sale, and use of **pesticides** ("substances used to prevent, destroy, repel, or mitigate any pest") (Mackenthun and Bregman, 1992). Pesticides may be individual substances or mixtures including plant regulators, insecticides, fungicides, rodenticides, desiccants, and defoliants (Freedman, 1987). The FIFRA defines certification of restricted-use pesticide applicators, defines the conditions for canceling the registration for a pesticide, and establishes pesticide disposal requirements (Mackenthun and Bregman, 1992). The FIFRA requires submittal of materials for pesticide registration which includes the specific labeling and directions for use, and results of product tests. Specifics of ecological effects testing and environmental fate are given in 40 CFR 158.145 and 40 CFR 158.130, respectively (Touart, 1995). Submitted information must demonstrate that the product will work as intended without unacceptable human or ecological risk. Registration is renewed periodically and, if a pesticide is found after registration to cause unreasonable risk, the EPA can cancel its registration under Section 6 of FIFRA (see 40 CFR 154).

The **Marine Protection, Research and Sanctuaries Act** or MPRSA(1972, 33 USC 1401; see 40 CFR 220-233) recognizes that the practice of ocean dumping is unsound, states the U.S. policy regarding dumping, and establishes limitations and prohibitions on dumping. It prohibits or limits dumping of various materials from the U.S., or by U.S. vehicles (Freedman, 1987). The EPA limits dumping of wastes except dredge spoils through a permit system. The U.S. Army Corp of Engineers uses EPA criteria and regulates dredge spoils dumping (Mackenthun and Bregman, 1992). The Ocean Dumping Ban Act (1988) amended the MPRSA and put a time limit on dumping of materials such as sewage sludge, industrial waste. and medical waste (Mackenthun and Bregman, 1992). Also prohibited is the dumping of radioactive waste, and radioactive, chemical, or biological weapons material (Freedman, 1987).

The **Clean Water Act** or CWA (1977, PL 95-217, 33 USC §§1251 to 1387; see 40 CFR Sections 100 to 140, 400 to 699; see also PL 100-4, Feb. 4, 1987 amendments) limits the discharge of pollutants (primarily industrial and municipal discharges) into navigable waters. This is accomplished by imposing limits on discharges using either a state or EPA permitting system (**National Pollutant Discharge Elimination System** or NPDES). State limits are developed under the guidance of the EPA. The immediate goal of this legislation was to improve water quality so as to ensure that waters are swimmable and fishable (McGregor, 1994). The overall goal is to "restore and maintain the chemical, physical, and biological integrity" of U.S. water bodies.

The **Safe Drinking Water Act** or SDWA (1974, ; 42 USC §§ 300f to 300j-26; see 40 CFR 141-149; amended in 1986) protects the public water supply (systems supplying more than 25 people or possessing more than 15 service outlets) by setting quality standards, and it protects source aquifers from contamination by establishing permitting requirements for underground disposal of waste. The EPA has oversight, but states are assigned responsibility for enforcement. The EPA establishes primary and secondary, "at the tap" standards for public water which the states can either use without modification or use to establish even more stringent standards (Mackenthun and Bregman, 1992). Primary standards ("maximum contaminant levels" or MCLs) are those for a list of specific contaminants that adversely affect health; secondary standards are EPA recommendations for qualities that protect the public welfare (e.g., water taste, color, smell, etc.) (McGregor, 1994). The present MCL list includes qualities under the categories of volatile organic compounds, organic compounds, inorganic chemicals including radionuclides, microbiology, and turbidity (Mackenthun and Bregman, 1992). This list of substances of concern is reviewed and revised every three years by the EPA. In the 1986 amendment, a ban which specified limits was added to restrict lead use in public water system components such as solder and fixtures.

Disposal of wastes into potential underground drinking water sources is also closely controlled under the SDWA through the use of a permit system or designation of aquifers as "sole source aquifers" to be protected. Federal activities are restricted in areas with sole source aquifers. The EPA also sets strict regulations for underground waste injection activities under the SDWA.

The **Toxic Substances Control Act** or TSCA (1976, 15 USC 2601; see 40 CFR 700-799) regulates, through the EPA, the manufacture, processing, transport, use, import, and disposal of chemicals or chemical mixtures that may pose an unreasonable risk to human health or the environment. Consideration is given to the quality and quantity of substances being produced. It does not cover items already regulated by other laws such as cosmetics, drugs, firearms and ammunition, food additives, pesticides, and nuclear materials (Mackenthun and Bregman, 1992). However, it does include products of biotechnology such as genetically-altered microbes intended for release into the environment. It specifically regulates some existing chemicals not already regulated by other laws, e.g., PCBs. The EPA requires producers to give 90-day notification of intent to manufacture or import, or to put a chemical product to new use. It requires testing to clarify the potential for unreasonable risk and a statement of the intended use of the material. The EPA may then permit, deny permission within 90 days of notice, or delay permit issuance until further study has been completed. The EPA may inspect facilities or investigate any actions that may violate this law. The law provides for criminal liability under specific conditions (knowing and willful violation) (McGregor, 1994). There are also record keeping requirements and an obligation to notify the EPA if new information arises relative to the risk associated with the chemical or mixture (Freedman, 1987).

The **Resource Conservation and Recovery Act** or RCRA (1976, 42 USC 6901-6987; see 40 CFR 240-299 and 1984 Hazardous and Solid Waste Amendments, PL 98-616) regulates production, storage, treatment, and final disposal of hazardous wastes (McGregor, 1994), i.e., "cradle-to-grave" regulation. The EPA is required to develop standards for generation, transport, storage, and treatment of hazardous waste and to establish a permit system for storage, treatment, and disposal (Mackenthun and Bregman, 1992). Disposal requirements for small-quantity generators, such as research laboratories are also defined by this law. The RCRA requires accurate record keeping. With the 1984 amendment, clear legal penalties were specified for knowing violation of a permit, illegal transport or treatment of waste, or falsifying or altering of records. Disposal of waste into underground formations was also restricted or prohibited by the 1984 Amendment.

The **Comprehensive Environmental Response, Compensation and Liability Act** or CERCLA or Superfund Act (1980, 42 USC 9601-9615, 9631-9633, 9651-9657, 26 USC 4611-4612, 4661-4662, 4681-4682; see 40 CFR 300-374, amended in 1986 with the Superfund Amendments and Reauthorization Act or SARA) gives the EPA the authority to respond to hazardous releases, require cleanup of hazardous waste sites, and assign liability for released hazardous waste (Mackenthun and Bregman, 1992). Although costs are first sought from responsible parties, the CERCLA also established two funds (Hazardous Substance Response Trust Fund and Post-Closure Liability Trust Fund) with various tax sources (oil, chemical, and hazardous waste activities) to cover some cleanup costs (Freedman, 1987). The CERCLA also establishes a **National Contingency Plan** to identify sites for cleanup, and define the required cleanup activities (Freedman, 1987). At this time, the EPA has compiled a listing (National Priority List or NPL) of sites for cleanup under the CERCLA.

Appendix 4
Derivation of Units for Simple Bioaccumulation Models

(Provided by P. Landrum and discussed in Chapter 3)

Assume a closed system containing water and a fish in which both the amounts in the fish and water are changing over time. The total amount in the system (A) is $Q_w + Q_f$ where the subscripts w and f refer to water and fish, respectively and Q denotes the amount (mass) in the compartment. There is no direct reference to the size of the fish or water compartments. The amount in the fish can be described by the following equation,

$$\frac{dQ_f}{dt} = k_1 \ Q_w - k_2 \ Q_f \tag{A4.1}$$

where k_1 = rate constant for the fractional reduction in the amount in the water (fraction of mass t^{-1}) and k_2 is the rate constant for the fractional reduction in the amount in the fish (fraction of mass t^{-1}). These are conditional rate constants because they depend on the specific experimental conditions, i.e., the sizes of the compartments. For example, assume two systems differing only in the relative sizes of the associated fish and water compartments. If one system had fish and water compartments of identical size, the rate constants would be different from another system in which the fish compartment was much smaller than the water compartment. Although the flux of material into the fish is the same for the two systems, the fractional reduction in the amount in the waters for the two systems (i.e., the rate constants) would be different. A larger fraction of the amount in the water of the system with equal compartment sizes for the fish and water would be removed per unit time than from the system with the much larger water compartment. The integrated form of Eq. A4.1 is Eq. A4.2.

$$Q_f = \frac{k_1 \ A(1 - e^{-(k_1 + k_2) \ t})}{k_1 + k_2} \tag{A4.2}$$

The units analysis for this is the following,

$$Q_f(ng) = \frac{k_1(h^{-1})A(ng) \ (1 - e^{-(k_1(h^{-1}) + k_2(h^{-1})) \ t(h)})}{k_1(h^{-1}) + k_2(h^{-1})} \tag{A4.3}$$

or

$$ng = ng$$

Now, the information is referenced to the size of the fish and the concentration in the water with the formulation of the equation for the change of concentrations.

$$C_f = \frac{k_u \; C_w}{k_e} (1 - e^{-(k_e t)})$$

(A4.4)

Solving the units,

$$C_f\left(\frac{ng}{g}\right) = \frac{k_u\left(\dfrac{mL}{g\ h}\right) C_w\left(\dfrac{ng}{mL}\right)}{k_e\ (h^{-1})} (1 - e^{-(k_e(h^{-1})\ t(h))})$$

(A4.5)

or

$$\frac{ng}{g} = \frac{ng}{g}$$

It is clear from this analysis of units that the units for k_u are mL (g^{-1} h^{-1}) and those of k_e are h^{-1}. The k_u is the clearance rate of the water by the fish of a specific size.

Demonstration of the associated units for k_u can also be done with the relationships assuming steady state.

$$\frac{dQ_f}{dt} = k_1 \; Q_w - k_2 \; Q_f$$

(A4.6)

At steady state, $k_1 Q_w = k_2 Q_f$. Or $Q_f/Q_w = k_1/k_2$ for this relationship describing changes in mass. To get to a bioconcentration factor (BCF = concentration in fish expressed as mass per unit of mass of fish divided by concentration in water expressed as mass per unit of volume of water), the relationship just derived between masses and constants is multiplied by volume/mass.

$$\left(\frac{volume}{mass}\right)\left(\frac{Q_f}{Q_w}\right) = \left(\frac{volume}{mass}\right)\left(\frac{k_1}{k_2}\right) = \frac{\dfrac{Q_f}{mass}}{\dfrac{Q_w}{volume}} = \frac{C_f}{C_w}$$

(A4.7)

Rearranging Eq. A4.7 and linking the rearranged equation to the relationship between constants for the concentration-based formulation (i.e., k_u/k_e = BCF; see Eq. 3.23 for more detail),

$$\frac{k_1\left(\dfrac{volume}{mass}\right)}{k_2} = \frac{C_f}{C_w} = \frac{k_u}{k_e} \tag{A4.8}$$

Because both k_2 and k_e describe the fractional change in chemical in the fish per unit time, they are equivalent. But the relationship between k_1 and k_u is k_1(volume/mass) = k_u. The units of k_u for the model of change in concentration over time are $mL(g^{-1}h^{-1})$, not h^{-1}.

Appendix 5
Equations for the Estimation of Contaminant Exposure

Equations and example values for variables were extracted directly from EPA guidelines (EPA, 1989d). Each is a derivation of the general equation provided in Chapter 13 (Equation 13.3). Note that units may change for variables among the equations.

1. Imbibed via drinking water or beverages made with drinking water

$$Intake(mg \; per \; kg - day) = \frac{(CW)(IR)(EF)(ED)}{(BW)(AT)} \quad \text{(A5.1)}$$

where   CW = concentration in the drinking water (mg L^{-1}),
   IR = imbibing or ingestion rate (L day^{-1}), e.g., 2 L day^{-1} for an adult,
   EF = exposure frequency (days year^{-1}), e.g., 365 days year^{-1} for a resident,
   ED = exposure duration (years), e.g., 70 years for a lifetime,
   BW = body weight or mass (kg), e.g., 70 kg for an adult, and
   AT = averaging time (days), e.g., ED $\times$ 365 days for a noncarcinogen or 70 years lifetime^{-1} $\times$ 365 days year^{-1} for a carcinogen.

2. Imbibed from surface water while swimming

$$Intake(mg \; per \; kg - day) = \frac{(CW)(CR)(ET)(ER)(ED)}{(BW)(AT)} \quad \text{(A5.2)}$$

where   CW = concentration in the drinking water (mg L^{-1}),
   CR = contact rate (L day^{-1}), e.g., 0.050 L day^{-1},
   ET = exposure time (hours event^{-1}),
   EF = exposure frequency (days year^{-1}), e.g., 7 days year^{-1} as an average,
   ED = exposure duration (years), e.g., 70 years for a lifetime,
   BW = body weight or mass (kg), e.g., 70 kg for an adult, and
   AT = averaging time (days), e.g., ED $\times$ 365 days for a noncarcinogen or 70 years lifetime^{-1} $\times$ 365 days year^{-1} for a carcinogen.

3. Absorbed during dermal contact with water

$$Absorbed\ Dose(mg\ per\ kg - day) = \frac{(CW)(SA)(PC)(ET)(EF)(ED)(CF)}{(BW)(AT)} \qquad \text{(A5.3)}$$

where   CW = concentration in the drinking water (mg L^{-1}),
         SA = surface area of skin in contact (cm^{-1}) which is dependent on sex, age, and contact type,
         PC = dermal permeability constant (cm hour^{-1}) which is highly dependent on the specific chemical,
         ET = exposure time (hours day^{-1}),
         EF = exposure frequency (days year^{-1}), e.g., 7 days year^{-1} as an average,
         ED = exposure duration (years), e.g., 70 years for a lifetime,
         CF = volumetric conversion factor for water (1 L (1000 cm^3)$^{-1}$),
         BW = body weight or mass (kg), e.g., 70 kg for an adult, and
         AT = averaging time (days), e.g., ED × 365 days for a noncarcinogen or 70 years lifetime^{-1} × 365 days year^{-1} for a carcinogen.

4. Inhaled as a vapor in the air

$$Intake(mg\ per\ kg - day) = \frac{(CA)(IR)(ET)(EF)(ED)}{(BW)(AT)} \qquad \text{(A5.4)}$$

where   CA = concentration in the air (mg m^{-3}),
         IR = inhalation rate (m^3 hour^{-1}), e.g., 30 m^3 hour^{-1} for an adult,
         ET = exposure time (hours day^{-1}),
         EF = exposure frequency (days year^{-1}),
         ED = exposure duration (years), e.g., 70 years for a lifetime,
         BW = body weight or mass (kg), e.g., 70 kg for an adult, and
         AT = averaging time (days), e.g., ED x 365 days for a noncarcinogen or 70 years lifetime^{-1} × 365 days year^{-1} for a carcinogen.

5. Ingested in fish or shellfish from the contaminated area

$$Intake(mg\ per\ kg - day) = \frac{(CF)(IR)(FI)(EF)(ED)}{(BW)(AT)} \qquad \text{(A5.5)}$$

where   CF = concentration in the food (mg kg^{-1}),
         IR = ingestion rate (kg meal^{-1}), e.g., 0.284 kg meal^{-1} (upper 95% confidence level for fish) consumption,
         FI = fraction ingested from the contaminated source (no units to this fraction),
         EF = exposure frequency (meals year^{-1}),
         ED = exposure duration (years), e.g., 70 years for a lifetime,
         BW = body weight or mass (kg), e.g., 70 kg for an adult, and
         AT = averaging time (days), e.g., ED × 365 days for a noncarcinogen or 70 years lifetime^{-1} × 365 days year^{-1} for a carcinogen.

5. Ingested in vegetable matter from the contaminated area

$$Intake(mg\ per\ kg - day) = \frac{(CF)(IR)(FI)(EF)(ED)}{(BW)(AT)} \tag{A5.6}$$

where   CF = concentration in the food (mg kg^{-1}),
   IR = ingestion rate (kg meal^{-1}),
   FI = fraction ingested from the contaminated source (no units to this fraction),
   EF = exposure frequency (meals year^{-1}),
   ED = exposure duration (years), e.g., 70 years for a lifetime,
   BW = body weight or mass (kg), e.g., 70 kg for an adult, and
   AT = averaging time (days), e.g., ED $\times$ 365 days for a noncarcinogen or 70 years lifetime$^{-1} \times$ 365 days year^{-1} for a carcinogen.

Study Questions

Chapter 1

1. Define the terms highlighted in the text and check your definitions against those provided in the glossary.
2. Briefly describe the transition and events leading to the shift from the Dilution Paradigm to the Boomerang Paradigm.
3. How does one maintain conceptual coherency in dealing with a hierarchical field of study such as ecotoxicology?

Chapter 2

1. Define the terms highlighted in the chapter and check your definitions against those provided in the glossary.
2. Define the scientific, technical, and practical goals of ecotoxicologists.
3. Explain the evolution of scientific inquiry from the ruling theory to the working hypothesis to the multiple working hypotheses stages. Is the multiple working hypothesis prevalent in ecotoxicology today?
4. Compare and contrast normal and innovative science. What are their relative values in young versus mature sciences?
5. What are the abnormal practices called precipitate explanation, the tyranny of the particular and *idola quantitatus*? Are they a problem in ecotoxicology today?
6. What qualities are valued in scientific, technological, and practical ecotoxicology?

Chapter 3

1. Define the terms highlighted in the chapter and check your definitions against those provided in the glossary.
2. Distinguish between steady state and chemical equilibrium. Which term is most accurate in describing bioaccumulation? Why?
3. What contaminants are most likely to pass into an organism via the lipid route? Which are least likely?
4. What is the reaction order of the following equation? $dC/dt = kC^1$? Is this reaction order common or uncommon for elimination kinetics?
5. Describe the role of biomineralization in the bioaccumulation of metals and metalloids.
6. Describe Phase I and II reactions in the metabolism of organic contaminants.
7. Distinguish among elimination, depuration, and clearance.
8. What are the three phases of renal elimination?
9. Describe the differences and similarities of rate constant-based, clearance volume-based, and fugacity models.

10. The rate constant for a single component, rate constant-based elimination model is 0.1 h⁻¹. What are the biological half-life and mean residence time predicted for this model?

11. Describe the backstripping method. Does it have weaknesses?

12. Explain the clearance volume concept.

13. Describe the models given in Equations 3.20 to 3.22. What are the meanings of the associated variables and constants?

Chapter 4

1. Define the terms highlighted in the chapter and check your definitions against those provided in the glossary.

2. Discuss the definition of bioavailability relative to ecotoxicology and pharmacology.

3. Describe the various approaches to estimating bioavailability.

4. Discuss the FIAM and exceptions to this model.

5. What factors influence bioavailability of contaminants in food?

6. What measures can be used to reflect the bioavailability of metals from oxic and anoxic sediments?

7. Describe the QSAR approach. What is the basis for the extensive use of K_{ow} in these models?

8. Explain the pH-partition hypothesis and relate the Henderson-Hasselbalch relationship to the bioavailability of a monobasic acid using this hypothesis.

9. How would you normalize for animal size effects on bioaccumulation?

10. A series of clam populations are sampled and analyzed for tissue cadmium concentrations. Cadmium concentrations are found to be directly dependent on ambient cadmium concentrations. In this case, is cadmium acting as a biologically determinant or indeterminant element? Explain your answer.

Chapter 5

1. Define the terms highlighted in the chapter and check your definitions against those provided in the glossary.

2. Describe how isotopic ratios of nitrogen, sulfur, and carbon might be used to define trophic structure of a lake community. What are the advantages and disadvantages of each of these elements for such use?

3. Describe how you would estimate biomagnification using body mass-weighted mean concentrations.

4. What elements might biomagnify? Why would these specific elements biomagnify?

5. What is "liquid" digestion by zooplankton and how does it relate to metal bioavailability from algae?

6. How do calcium sinks influence transfer of calcium analogs?

7. What qualities of organic compounds would foster biomagnification?

Chapter 6

1. Define the terms highlighted in the chapter and check your definitions against those provided in the glossary.

2. What are the qualities of the ideal molecular biomarker?

3. What biomarkers are associated with phase II reactions?
4. Give the qualities of metallothioneins.
5. What is the spillover hypothesis? Are there any situations that you can identify where this concept would not be valid, e.g., mixtures of metals, etc? Why?
6. Describe proteotoxicity and the roles that stress proteins play in lessening the effects of proteotoxic agents.
7. Describe the types of molecular damage that can be produced by oxyradicals. How does the cell lessen the damage of oxyradicals?
8. Hydrogen peroxide is not an oxyradical yet it contributes to oxidative stress. How?
9. How might you measure lipid peroxidation in a field population of a species thought to be experiencing oxidative stress?
10. How would you measure DNA damage?
11. A group of subsistence fishermen are concerned that they are subject to high mercury exposure from their diet. They hire you to determine if this is a legitimate concern. Design a biomarker study to assess the potential exposure.

Chapter 7
1. Define the terms highlighted in the chapter and check your definitions against those provided in the glossary.
2. Define and contrast the holistic and reductionist approaches to hierarchical subjects such as ecotoxicology. Which is the correct or best approach to ecotoxicology?
3. Why are cellular, tissue, and organ biomarkers valuable in ecotoxicology?
4. Describe some of the characteristics of necrosis.
5. Describe the process of inflammation including its four cardinal signs.
6. If you suspected that a genotoxic agent was affecting mice inhabiting an area, what biomarkers would you use to test your suspicion? In your answer, explain why you would use each biomarker.
7. An individual was exposed to high concentrations of a chemical with the properties of an initiator. Will that individual develop cancer? Explain your answer.
8. You are asked to determine if trout in a lake modified by acid precipitation (low pH and high aluminum concentrations) have been affected at the cellular, tissue, or organ level. What would you examine to provide the answer to this question? What changes might you expect? Link these biomarkers to consequences to individuals and to the trout population.

Chapter 8
1. Define the terms highlighted in the chapter and check your definitions against those provided in the glossary.
2. What is Selyean stress? Which of the following responses is not associated with Selyean stress: increased adrenal size, induction of EROD activity, kidney damage by Cd, elevated blood pressure due to an emotional shock, or depletion of secretory granules in cells of the adrenal cortex?
3. A data set for growth retardation of an endemic, endangered fish suggests that concentrations of zinc in the range measured in your discharge consis-

tently stimulates growth but local officials responsible for regulating your discharge insist on a linear-no threshold extrapolation of effect down to no more than a 5% reduction in growth (i.e., extend the dose-effect line downward from high effect concentrations to a concentration that is predicted to have only a 5% reduction in growth based on a linear-no threshold model). Extrapolation predicts 5% inhibition but your data set suggests 20% enhancement of growth. Give your reasoning in deciding to conform to or contest the regulators' decision.

4. You are told that voles near a contaminated site have necrotic lesions in their livers. Discuss one reason why this observation *may* suggest that abnormal development will also be occurring in the young of this rodent.

5. Hatchlings of an endangered and long-lived sea turtle are closely monitored by state fish and wildlife technicians. Data from twenty years ago describe a sex ratio of 50:50 male:female of hatchlings for many populations of these turtles. Now, almost all turtles are females. Describe what you would do to determine the cause of this apparent shift and what you might do to counterbalance the shift. Should you intercede?

6. Periodic episodes of meadow lark mortality are noted near an agricultural field. The field was used for many years to grow cotton and, consequently, has a history of DDT and lead arsenate application. At present, carbamate pesticides are periodically applied to the fields. How would you determine which, if any, of these potential toxicants is producing the mortality?

7. Why are behavioral abnormalities used less often than other sublethal effects in the assessment of contaminant effects to individuals?

8. Tabulate the advantages and disadvantages of the post-ANOVA methods described in this chapter.

9. Compare and contrast the hypothesis testing and regression methods for analyzing sublethal effects data.

10. Given the following table of NOEC and LOEC values for toxicant X, estimate the concentration ("safe concentration") that will protect an endangered siren (amphibian) living below a continuous discharge containing toxicant X.

| Species | Effect | Concentration ($\mu g\ L^{-1}$) | | Exposure Duration |
		NOEC	LOEC	
Fathead minnow	Growth	1	10	14 d
Bluegill sunfish	Egg hatch success	5	46	7 d
Leopard frog	Tadpole growth	0.09	0.60	7 d
FETAX assay	Heart development	0.17	1.7	4 d
Daphnia magna	Fecundity	0.001	0.008	7 d

Chapter 9

1. Define the terms highlighted in the chapter and check your definitions against those provided in the glossary.

2. The most critical (sensitive) life stage of a particular invertebrate is its first instar which has an incipient LC50 of 10 $\mu g\ L^{-1}$ and a NOEC of 5 $\mu g\ L^{-1}$ for growth for toxicant X. Can a viable population of this invertebrate be maintained below an effluent with a maximum concentration of 5 $\mu g\ L^{-1}$ for this toxicant?

3. Compare the advantages and disadvantages of static, static-renewal, and flow-through (continuous and intermittent) tests.

4. What are the relative advantages of the dose- or concentration-response versus the survival time approach to measuring toxicity?

5. A channel in a heavily-industrialized estuary must be deepened. How would you assess the potential for acute and chronic toxicity to endemic species resulting from the dredging activities?

6. Defend the concept of individual effective dose.

7. You are required to accurately estimate the LC50, LC10, and LC5 for a toxicant. Describe the methods you would use to calculate these numbers. Give the reasons for selecting one approach over another.

8. Would normit and probit analyses (MLE method) of a data set produce the same LC50 estimate? Would they produce the same LC5 estimate? Would the probit and logit analyses produce the same LC50 and LC5 estimates? Give reasons for each of your responses.

9. Describe how you would estimate toxic incipiency.

10. Could a pair of chemicals that are functional antagonists also be receptor antagonists? Could they also be dispositional antagonists?

11. What is the difference between proportional hazard and accelerated failure time models? Are there any differences in the data collected to fit these two models?

12. Describe the Bliss method of accounting for the effect of individual size on toxic effect.

13. What is the difference between photo-induced toxicity and photosensitivity?

14. What is a split probit? What factors would result in a split probit plot?

Chapter 10

1. Define the terms highlighted in the chapter and check your definitions against those provided in the glossary.

2. Review the previous chapters and identify ten specific risk factors. Five should be qualities of individuals and five should be etiological factors.

3. Liver tumors in flounder are surveyed in a contaminated bay. Calculate the incidence rate as cases per 1000 flounder years of exposure. Assume the tumors kill each diseased individual within the sampling year. What was the mean prevalence rate over the five years of sampling?

Year	Number of Flounder Sampled	Number of Diseased Flounder
1	1,583	112
2	2,032	100
3	1,911	216
4	2,118	96
5	1,625	97

4. The above population is compared to a reference population ("control") that has the qualities tabulated below. Use the mean incidence rate for both populations to calculate the rate ratio. Assuming that the contamination in the

bay was the "treatment", calculate the odds ratio for each of the five years of the survey.

Year	Number of Flounder Sampled	Number of Diseased Flounder
1	3,008	3
2	5,025	11
3	4,091	7
4	3,557	6
5	4,253	17

5. Describe the nine aspects of disease association. Include an example of each.
6. Explain the predisposition of ecotoxicologists to using laboratory sublethal and lethal test results to predict the viability or distribution of populations about a contaminated field site.
7. In one year, a population increased from 100 to 150 individuals. Calculate λ and r for this population for that time. Assuming no change in growth dynamics, calculate the population size in 20 years of growth beginning with 100 individuals.
8. How could knowledge of V_a for the age classes of a population aid in estimating the viability of a population exposed to a toxicant over long periods of time?
9. Using the principle of allocation and concept of strategy, relate how life histories of individuals in a population may shift in response to a toxicant.
10. Contrast how a stressor might influence longevity of an individual under the disposable soma and stress theories of aging.
11. How might a toxicant influence the genetics of a population?
12. What factors influence the rate of tolerance acquisition in a population?
13. Describe the concept of selection components.

Chapter 11
1. Define the terms highlighted in the chapter and check your definitions against those provided in the glossary.
2. What are the differences among the terms, community, species assemblage, and guild?
3. Describe the community-level NOEC concept, including its advantages and disadvantages.
4. How might optimal foraging theory help us to understand the effects of toxicants on populations in a community?
5. Describe the factors contributing to the ability of a community to successfully avoid irreversible damage by a pollutant.
6. What is the major difference between the Shannon and Brillouin indices of species diversity? Which one will have the largest value for any sample?
7. Describe the rivet popper hypothesis versus redundancy hypothesis debate. What is the significance of this debate relative to assessing an adverse effect of pollutants to ecological communities?
8. In general, how would the log normal, species abundance curve change with

the introduction of a toxicant to an ecosystem? What is (are) the possible mechanism(s) for this shift in the curve?

9. What are the differences between microcosms and mesocosms? What are the relative advantages and disadvantages of these tools?

10. Assuming that the MacArthur-Wilson model accurately reflects colonization, predict the number of species inhabiting a beach devastated by an oil spill if the number of species originally on the beach was 23 (S_{EQ} = 23?), G = 0.10 (units of years), and 10 years have passed since the devastation occurred. How long would it take for 95% of the species to reestablish themselves? Would the community (species assemblage) be back to normal at that point?

Chapter 12

1. Define the terms highlighted in the chapter and check your definitions against those provided in the glossary.

2. Give three reasons why the traditional ecosystem context for ecotoxicology should be expanded to a larger scale in order to address ecotoxicological problems facing us today.

3. What factors would influence the sensitivity of a watershed to acid precipitation? What effects might you expect to aquatic and terrestrial biota in a watershed with a granite bedrock geology?

4. How does the increased use of CFCs result in decreased ozone levels in the stratosphere above the Antarctic? Why is the problem less noticeable above the Arctic?

5. Describe the processes by which POPs are transported and differentially distributed from the equator to the earth's poles.

Chapter 13

1. Define the terms highlighted in the chapter and check your definitions against those provided in the glossary.

2. What are the four components of the NAS paradigm for risk assessment? What are their specific goals?

3. What is the major distinction between a hazard assessment and a risk assessment? As described in this chapter, is the assessment done for noncarcinogenic effects hazard or risk assessment? Explain your answer.

4. Outline the steps in a human risk assessment. Check your outline against that shown in Figure 13.2.

5. Groundwater below a closed landfill contains high concentrations of trichloroethylene (TCE) and the contaminated water is predicted to outcrop to a stream within the next two years. You are asked to do an assessment of the situation. Are you doing a predictive or retroactive risk assessment? Give the reasons for your answer.

6. What is the function of uncertainty factors? Is it the same as the slope factor? Of the modifying factor?

7. Get onto IRIS to find the RfD and associated information for hexavalent chromium, i.e., chromium (IV). If the concentration in drinking water taken from a representative well near a hazardous waste site was 6 mg L^{-1}, what

would the hazard quotient be? Provide the qualifiers needed by the risk manager in order to use this calculated value intelligently.

8. Contrast and give examples of assessment and measurement endpoints. What are the qualities of good assessment and measurement endpoints? Give some good and bad examples of both endpoints.

9. Describe the steps and products of an ecological risk assessment.

Chapter 14

1. Define the terms highlighted in the chapter and check your definitions against those provided in the glossary.

2. Why are alpha particles of little concern from an external irradiation standpoint, but of major concern if inhaled?

3. Assume that one batch of milk is contaminated with 1000 Bq of ^{131}I and another batch with 1000 Bq of ^{134}Cs. The half lives of ^{131}I and ^{134}Cs are 8 d and 2.4 y, respectively. Recognizing that the milk is contaminated we opt to make cheese from it, and store the product until it is safe for human consumption (which we'll arbitrarily set at a level of 100 Bq). How long would you have to store each batch of cheese before it has a safe activity? Hint: Convert Equation 14.3 into natural logarithms and solve for "t".

4. Explain the concept of a screening level model.

5. A hypothetical individual is exposed to a ^{137}Cs plume of 900 kBq m^{-3}. The individual is outside, directly in the plume for 30 h, and indoors for an additional 100 h. The residence structure provides a shielding factor of 0.1. The air dose coefficient for ^{137}Cs is 7.74 E -18 Sv-m^3 /Bq-s. Calculate the probability that this individual will acquire a fatal cancer from the external exposure.

6. How do risk analyses for human exposures to radioactive contaminants differ from exposure to nonradioactive contaminants? Consider the pathways of contaminant transport and derivation of the risk factors.

7. Which provides the more rigorous quantitative answer—human or ecological risk analyses? Explain why.

8. Why do some radiation protection agencies think that if levels of radiation are kept low enough to protect humans then the natural biota living in the same contaminated environment should be adequately protected as well?

Index